Catastrophic Incidents

This interesting book offers an analysis of man-made catastrophes and asks why they continue to occur. 87 catastrophes or near-catastrophes, including high profile cases such as the Bhopal gas disaster, Grenfell Tower, Shoreham Air Show crash, Brumadinho dam collapse and Fukushima Daiichi, are described together with the reasons why they occurred and why over 50 different safety management approaches and techniques failed to prevent them.

Featuring 63 eye opening stories from the author's own personal experience and over 200 pitfalls in safety management approaches, this title is illustrated by 24 hypothetical cases in which the reader is asked to consider the approach they would take. Safety management techniques discussed include operating practices, personnel selection and emergency response. Safety management approaches including safety governance in organisations, along with the role of government and local authorities using the instruments of the law are extensively discussed. The work concludes with imaginative and creative ways forward with the aim to make considerable progress and to potentially eliminate man-made catastrophes for good.

This title will be an ideal read for safety managers and engineers, students of chemical engineering and engineering in general, community leaders in civic duties or labour union roles and professionals tasked with stopping and mitigating the impacts of man-made catastrophes, along with non-technical readers who are curious and concerned.

Catastrophic Incidents
Prevention and Failure

Trevor J. Hughes

CRC Press
Taylor & Francis Group
Boca Raton London New York

CRC Press is an imprint of the
Taylor & Francis Group, an **informa** business

Designed cover image: © Shutterstock

First edition published 2023
by CRC Press
4 Park Square, Milton Park, Abingdon, Oxon, OX14 4RN

and by CRC Press
6000 Broken Sound Parkway NW, Suite 300, Boca Raton, FL 33487-2742

CRC Press is an imprint of Informa UK Limited

British Library Cataloguing-in-Publication Data
A catalogue record for this book is available from the British Library

ISBN: 978-1-032-21029-2 (hbk)
ISBN: 978-1-032-41967-1 (pbk)
ISBN: 978-1-003-36075-9 (ebk)

DOI: 10.1201/9781003360759

Typeset in Sabon
by MPS Limited, Dehradun

Contents

SECTION IV
Vulnerability to Natural Catastrophes, Sabotage, Terrorism, and War

SECTION V
Preventing Catastrophes – Potential Solutions

Foreword

"The appearance of *Catastrophic Incidents—Prevention and Failure* is timely, arriving at an important inflection point for the safety profession. Large catastrophic accidents may be less numerous—thank goodness—but are much more devastating. We don't need to enumerate the spate of large incidents that we see around the world and across every major industry almost daily. Just open a media source and it hits you in the face. Massive oil spill in the Gulf of Mexico with many deaths, billions in losses and entire livelihoods ruined along the coastline. Though air travel is one of the safest, we still see crashes due to poor training and quality control. Some of the most famous global mining companies have seen large accidents where hundreds have been killed in a single catastrophic event. The chemical industry certainly is not immune to these large incidents with explosions, fires, and toxic clouds released forcing nearby communities to evacuate in every country around the world. The commercial nuclear power industry, though with only a few large accidents, nonetheless dramatic ones, have left regions uninhabitable with controlled exclusion zones, causing some countries to eliminate the entire industry from its power grid.

Global climate change has only accelerated these disasters—turbocharged their potency. Severe weather from massive heat waves to flooding is now all too common in all countries around the world. Hurricanes have impacted the price of gasoline at the pump and broken power grids when most needed. Why are these incidents still occurring and why are they even more potent than one hundred years ago?

We have progressed tremendously over the last hundred years in how we protect people, the environment, and assets, but clearly not enough. Accidents, incidents, and natural disasters are now transnational—coronavirus is one such example. Covid-19 not only killed millions around the world but also broke fragile global supply chains, severely impacting each one of us. Technological advances are moving so fast that society struggles to absorb its benefits and mitigate its vices. We must ask ourselves: are we trying to solve 21st century problems using 20th century thinking?

Trevor Hughes has written an important book to help us understand why incidents still occur and how to prevent them—prevent the precursors that lead to these catastrophic events. He helps us understand what can cause a simple event or action and how it can quickly grow into an uncontrolled catastrophic event. *Catastrophic Incidents—Prevention and Failure* delves into why current approaches are failing and what we should do about it. Most importantly, his book focuses on prevention. By going into the anatomy of incidents in various industries he helps us understand how they evolve and thus where and how to prevent incidents from the very beginning. Analyzing prevention failures in myriad industries, he teaches us lessons from those industries, not just showing no one is immune, but also how we can take the best from other industries and apply to our own.

Catastrophic Incidents—Prevention and Failure is important to read now because our industries are confronting increasing operational complexity. Technology is advancing at increasing speed. Regulations are continuously changing, not just at home but around the world. Many industries are facing, or about to face, massive-scale retirements of experienced industry experts, creating a large knowledge gap in operational safety understanding, as baby-boomers leave the market. And lest we forget, global climate change means that natural disasters are getting bigger, more severe, and much more frequent with records constantly beaten. *Catastrophic Incidents—Prevention and Failure* helps us to understand how and why this is occurring and what to do about it. This is the thinking we need for the 21st century."

Nicholas Bahr
Global Managing Director for Operational Risk Management
dss+
Author of *System Safety Engineering: A Practical Approach*

"Society continues to be burdened by the high cost of losses in the hydrocarbon industry. The latest 2022 (27th) edition of the Marsh publication '100 Largest Losses in the Hydrocarbon Industry' demonstrates the roller coaster ride of financial losses in the various industry sectors. The operating company will initially bear the cost, which may be partially offset by claims recoveries from insurers. Still, ultimately the cost is borne by society – manifesting in the cost of our goods, services and indeed in employment.

In this book, Trevor draws on his experiences, many of which were gained whilst working as a risk engineer with the Marsh Global Energy Risk Engineering ('GERE') team, to highlight how catastrophes happen and to demonstrate how organisations can prevent such losses occurring when they do things right.

Marsh's GERE team works with energy companies with operations in the oil, gas, petrochemical, and other energy fields to reduce the likelihood and consequence of incidents and losses and, when losses do occur, that financial exposure to these losses is mitigated to within tolerable limits. Marsh GERE has developed unique tools, services and innovative approaches to traditional loss prevention that support this work. Trevor's contribution through his book should help towards the ultimate goal that everyone goes home safely from their place of work every day."

Andrew Herring, CEO
Marsh Energy and Power (London)

Preface

How do you feel when you have done something 'good'? This might be something which others have appreciated, or it might be something you simply recognise within yourself that you did well. Most people would use words like 'elated', 'pleased', 'on an emotional high'…

How do you feel when you have not done something well? This might be something you omitted to do, a missed opportunity, a mistake. Most people would use words like 'annoyed with myself', 'sad', even 'depressed'.

If I think about my career in the management and control of high hazard facilities, such as chemical plants, I feel a mixture of emotions. In truth, if I am absolutely open, my emotions mixed, and can be summarised as "could have done better". As my working life comes to a close, I wanted to write a book in the hope of leaving a legacy of my varied experiences, which might help those who want to make a difference in the future.

Readers who want to make a difference could be anyone in society willing to ask the right questions. Industrial leaders are influenced by people like you who ask the right questions, but to do so you need to be well informed to ask questions with depth. Otherwise, your question will be easily be dismissed – of course we do everything safety around here! Other readers who want to make a difference will be already managing or otherwise involved in the control of high hazard facilities. This book is for you also. I dwell a lot on how management and control of high hazard facilities can be done badly – so that you can ask the well-informed questions and do things better.

Acknowledgements

To my wife who endured six years of me talking about writing this book, and who toiled through reading every chapter, yellow highlighter in hand.

To our friend Betty Field who proofread the manuscript and gave her valuable feedback as a non-technical reader.

My thanks to DSS$^+$ (formerly Du Pont Sustainable Solutions) for the opportunity to carry out consulting in so many varied locations, providing so much valuable material for "Notes from Trevor's Files".

And finally, to Marsh, Global Energy Risk Engineering, who took me on late in my career, and provided a unique window into the world of risk, and methods which should be working to reduce that risk.

Author Biography

Trevor John Hughes has 50 years of experience in the chemical and oil industries, working in various roles including process engineer, plant manager and site director. For 28 years he had direct supervisory and management responsibility for several manufacturing facilities in the U.K. This was followed by 14 years as a consultant, primarily in process safety, working in offshore in the North Sea, and then on assignment in Russia, Kazakhstan, Ukraine and Saudi Arabia. For the last eight years he has worked primarily as an insurance risk engineer in Azerbaijan, Bahrain, Uzbekistan, Romania, most Western European countries and the U.S.A.

He has graduate and post-graduate qualifications in both Chemical Engineering, and Health and Safety Management. Trevor has had a lifelong interest in the concept of risk. Currently he is an active Professional Process Safety Engineer.

Chapter 1

Introduction

This book is about preventing disasters – major man-made incidents. It also is about the power of people like you to help drive improvements in all types of industrial and non-industrial operations to reduce the risk of a significant incident.

There are a lot of books around posing the question 'Why have we not learned'. These are mostly written by 'spectators'. People who have not had responsibility for hazardous operations, criticising those who have had such responsibility. These books provide many valuable insights, but they often don't sit well with those who have had such responsibilities. What makes this book different it is indeed that I have had, in the past, such responsibilities. I have made some mistakes. If you like I am one of the 'guilty' which outsiders understandably criticise. Given this background, I believe this book provides a deeper opportunity for learning from the past.

With all the modern engineering design and safety management techniques, one might think that modern civilisation has at least started to get this under control. For most of my working life, I believed this. Now I question it. I have some challenging solutions to this problem which I will explain at the end of this book in Section 5. I ask you to play a part in moving towards this solution.

Who can play this part in finding a solution? All of us, as I will try to explain.

The status quo is unacceptable. Tragic catastrophes will continue to occur.

Why conclude that the current situation is unacceptable? Humankind has an inbuilt mechanism to forget painful and difficult times. It can seem that because there was no major catastrophe in the last month or two, we are moving in the right direction. Recall some recent disasters:

1. On 4 August 2020, a large amount of ammonium nitrate stored at the port of Beirut exploded, causing at least 204 deaths, 6,500 injuries, and $ 15 billion in property damage. An estimated 300,000 people were left homeless. There is nothing mysterious about ammonium nitrate, a common fertiliser component, and the hazards of the material and storage precautions are well documented.

DOI: 10.1201/9781003360759-1

2. Two commercial airline crashes in 2018-19 resulted in 346 deaths. Both crashes involved the Boeing 737 Max. It is understood that the plane was produced with previously unidentified software design faults which conflicted with the pilot's intentions. Highly skilled engineers design aircraft and subject them to rigorous testing and certification.
3. On 25 January 2019, a tailings dam at an iron ore mine close to Brumadinho, Brazil, suffered a catastrophic failure. Over 300 people died.
4. On 21 March 2019, a chemical plant in Jiangsu, China, exploded, causing 78 fatalities.

    Lest the reader jump to the conclusion that such catastrophes are only to be found in 'third world' countries, consider also:
5. A train derailment in Quebec, Canada, on 6 July 2013, killed 42 people.
6. 17 April 2013, 15 people were killed when a fertiliser (ammonium nitrate) plant caught fire and exploded in West, Texas.

I started my career in 1972 after three years at one of the best Universities for Chemical Engineering studies in the U.K. I worked in manufacturing in the chemical industry for most of my career. Looking back, I had undue faith in the competence of myself and those around me. I found out quickly that there were some ethical barriers to my success.

Notes from Trevor's Files:

One of my first ethical challenges concerns a plastics mouldings shop I was assigned to supervise. The factory operated a piecework incentive scheme. The ladies who did the plastic mouldings finishing complained about the difficulty of achieving a decent bonus. A colleague, a supervisor who was well respected by management for 'getting things done', suggested that I fabricate some work-study results in order to achieve a satisfactory bonus. This falsification was an ethical line I was unwilling to cross and did not do so. This particular event did not have any safety consequences, but it was an early learning which did not bode well for my career to come.

Subsequently, there have been more ethical challenges in my career than I would like to admit. I have not always been 'Mr Clean', although some people recognise me by that description. Over the years, my faith in the competence of industrial management has unfortunately declined progressively. That's why I decided to write this book. The sins of the past need to be corrected in many ways. Later I will describe some parallels between catastrophe prevention and other man-made catastrophes such as global warming.

The research undertaken in the preparation of this book further questioned, in my mind, the actual competence of people in charge. I take one example here of many examples discussed in subsequent chapters. This particular example involved a nuclear plant. The fact that it is nuclear is not relevant to the story. Fortunately, investigations into issues at nuclear plants are often available in public domain documents which are often not available for other facilities of lesser hazard. The case concerned the practice at a nuclear power plant in the U.S.A. of entirely unloading all of the spent fuel rods from the reactor core at once, during refuelling operations rather than one-third at a time as regulations required. The Nuclear Regulatory Commission (N.R.C.) apparently knew about the practice but ignored it. They considered the method safe, as the plant's managers clearly did. The regulator was practising 'enforcement discretion'.

However, further consideration exposed issues with the spent fuel pool into which the spent fuel rods were placed. It was not designed to take a full load of fuel rods at once, and this practice was well outside the facilities licence. More importantly, failure of the spent fuel pool cooling could be catastrophic and cause a widespread discharge of radiation. This discovery brings me to an astounding revelation when the N.R.C. Inspector General wrote the following in 1996:

> "We shouldn't have regulations on the books and then ignore or wink at them".

Shocking. And this is a problem for the whole of industry, not just this example from the nuclear industry.

As an illustration of the transient nature of human memory of catastrophes, let's take the example of earthquakes and the appetite for earthquake insurance. Earthquakes occur due to the movement of tectonic plates beneath us. The moving tectonic plates create pressures resulting in faults in the earth's crust. The movement of the earth's crust along a fault line produces a seismic event, and if it is large enough, an earthquake. The nature of earthquake science is that the risk is lowest immediately after an earthquake and increases as time passes. The earthquake relieves pressure along the fault line, following which the pressure typically progressively increases in the years following the earthquake. However, I understand that actual sales of earthquake insurance are highest immediately following the earthquake event and declines as the memory of the earthquake fades.

The human mind, it seems, concludes that if it is easy to recall examples of 'something', then the more common that 'something' must be. Recollection fades with time. This tendency has a profound influence on the interpretation of risk.

There is also a human tendency to attribute undesirable events to some feature that distances the risk of that event from impacting oneself or one's family. For example, with the Beirut warehouse explosion in 2020, people tend to consider this as a consequence of the social and political situation in Lebanon. People fail to consider the history of ammonium nitrate explosions throughout the world, including in the U.S.A. and France. The Brumadinho tailings dam failure is considered something to do with a faraway place in Brazil, even though catastrophic tailings dam failures have occurred in countries such as Australia, Canada, the U.S.A., and the U.K. A chemical plant disaster in China is considered due to poor legislative control or lack of education in China. In contrast, chemical plant explosions are happening with alarming frequency worldwide. Check the Chemical Safety Board website (www.csb.gov) right now to fact-check my claim.

As our primary source of information, the media plays a significant role in determining our assessment of risk. As summarised by Dan Gardner (2008),

> One of the most consistent findings of risk perception research is that we overestimate the likelihood of being killed by the things that make the evening news and underestimate those that don't. What makes the evening news? The rare, vivid, and catastrophic killers. Murder, terrorism, fire, and flood. What doesn't make the news are the common causes of death that kill one person and don't lend themselves to strong emotions and pictures. Diabetes, asthma, heart disease.

It is now 36 years since Bhopal, where the immediate death toll exceeded 2000 due to the accidental release of a highly toxic chemical. Longer-term injuries and fatalities resulting from the exposure may have exceeded 500,000. It is considered the world's worst industrial disaster. Are we so much smarter now to be reassured that such a disaster will not happen again? The following pages will cause you to doubt.

Just eight months after Bhopal, an incident occurred at a U.S. plant similar to that at Bhopal, albeit involving a somewhat less hazardous material than that involved at Bhopal. The Occupational Safety and Health Administration (OSHA) conducted a careful examination. It concluded that this was an 'accident waiting to happen' and cited hundreds of 'constant, wilful violations' at the plant. But this plant had been inspected by OSHA before this incident and given a clean bill of health (Perrow p. 359). I doubt that this is a rare occurrence where a government authority, insurance surveyor, third party auditor, or other entity entitled to examine an organisations facility has found them suitable, only to judge them as inadequate after a tragic event.

This book is intended to be relevant to a broad spectrum of industries and technologies. The author's background is in chemical engineering, involving work in many process industries globally. This book includes many personal experiences, primarily from process industries, including oil, gas, and chemical. The author recognises that this book is richer in the field of process industries but nevertheless is relevant to all high-hazard facilities. In some cases, the author has done additional research to examine technologies beyond the process industries, such as air and ship transportation.

It is also recognised that some of the legislative approaches given are related to U.K. legislation since the author is more familiar with U.K. safety law than with the law of other countries. Alternative legal frameworks in other countries are mentioned by comparing their approaches to that of the U.K.

There are five sections to this book:

Section 1 attempts to help the reader answer the question 'what makes risks tolerable or intolerable'.

Section 2 of this book is on 'High-Risk Technologies and Their Catastrophes'. It reviews the hazards of a broad range of industries and the catastrophic incidents which characterise those risks.

Section 3 addresses the issues of 'Catastrophe Prevention and Why It Is Failing'. We look at some of the many tools and approaches to catastrophe prevention. Whilst they are excellent tools, if they were working the incidents of catastrophe in our world would not be this high. This section addresses why these techniques and approaches are failing.

Section 4 considers the consequences of natural catastrophes, sabotage, and terrorism to hazardous facilities. Natural disasters will continue to occur and may even be exacerbated by global warming. This section examines how high-hazard facilities are protected or should be protected against natural disasters. Pandemics also find a place here, as a naturally occurring catastrophe aided by the activities of mankind and impacting considerably our ability to staff hazardous facilities with people of the appropriate training and competence.

Section 5 makes some positive proposals as to how to improve our ability to prevent catastrophes. Some of the recommendations are challenging, and some readers may consider them impractical. To make progress, we need to be imaginative. We are already being creative over global warming and sustainable development, even if these initiatives are late in receiving the attention they deserve. This section discusses the need for change and how we might achieve that in a robust way.

Throughout this book, particular points are illustrated as follows:
    Double Border to text:

---

1. Notes from Trevor's Files:

These are actual experiences from my working life either as a plant manager over various facilities or as a consultant visiting and advising facilities across the globe.

Dotted Border to text:

> ## 2. Incidental Case Studies
>
> Incidents are briefly described to illustrate a particular point (additional to those given in more depth in Section 2).

Dashed Border to text:

> ## 3. Illustrative Hypothetical Case
>
> These are short mental exercises for the reader to think through particular examples of risk reduction at either:
>
> a. An existing small whisky distillery
> b. An existing small fuels terminal (gasoline/petrol, diesel, L.P.G., fuel oil)
>
> The reader should put him or herself in the place of someone trying to evaluate the potential for a severe incident at the plant. The reader represents a stakeholder – a resident in the adjacent community, an owner of adjacent property, an elected councillor or local politician, or even a potential enterprise buyer. In these paragraphs, the kinds of questions that the reader should ask are considered. Suggestions are also made regarding the sorts of answers that the reader should expect.
> For these Illustrative Hypothetic Cases, refer to the plant layouts and diagrams in Appendix A.

In the case studies on actual incidents, I have used many references from information freely available on the internet. My initial thoughts were to make these anonymous since it is what happened and why it happened which is important. However, many of the incidents are easily recognisable and already named in the public domain by the name of the owner of the asset where the incident occurred. For these and lesser incidents, detail is readily available from publicly available databases such as the invaluable U.S. Chemical Safety and Hazard Investigation Board but can only be easily extracted if the name of the owning company is known.

Some organisations may react to my opinions expressed here in this book by asking why I did not ask them directly. Firstly, I have only used sources which describe the causes of incidents where they can be traced to respected investigations, typically the CSB or government-related accident investigation board reports. Secondly, it is, in my opinion, the owner/operator of the hazardous facility who needs to manage its public relations, and its profile on the internet. Where balanced arguments are available, I have used them.

One of the inspirations for writing this book was the reading of books by Charles Perrow, in particular, 'Normal Accidents: Living in High-Risk Technologies', and Anthony Hopkins, including 'Lessons from Longford', 'Failure to Learn: The BP Texas City Disaster', and 'Disastrous Decisions: The Human and Organisational Causes of the Gulf of Mexico Blowout'. These are compelling books that appeal to both the technical and non-technical readers. Both Perrow and Hopkins are sociologists by training and profession. In this book, I intend to appeal to the same audience of both technical and non-technical readers with enquiring minds. This requires the non-technical reader to absorb some basics of the technologies which are provided in the various chapters. The more technical reader is advised to skip past technology descriptions with which they are already familiar.

The author of this book has spent over 48 years working in high-hazard industries as an engineer, manager, business director, management consultant, and insurance risk engineer. I should have all the tools to prevent a catastrophe. After reading this book, you will find that this is not the case.

First, in Section 2, let us examine how catastrophes happen ....

How Do Catastrophes Happen and What Can Prevent Them?

Chapter 2

Man-Made Catastrophes and Safety Risk

WHAT IS A MAN-MADE CATASTROPHE?

Dictionary definitions of catastrophe include 'an event causing great and usually sudden damage or suffering' and 'an unexpected event that causes great suffering or damage'.

For the purposes of this book, we are interested in catastrophes that are caused by man, as opposed to naturally occurring disasters. We are interested in 'big' catastrophes such as events immediately causing one or more of the following consequences:

- Multiple fatalities to the employees of the organisation within which the incident occurred.
- Fatalities to the public.
- Widespread illness or injury to members of the public.
- Extensive damage both financially and in loss of facilities for an extended period.

We will also discuss incidents that were not catastrophic. These incidents are selected to demonstrate the failure of safety management systems. In different circumstances, a tragedy would have occurred. We also need to look for potential future catastrophes. History can only look back. When pressure vessels were invented as part of the steam engine, there had been no associated explosions. Future explosions had not been foreseen until a series of vessel disintegrations occurred and pressure vessel regulations were introduced. A catastrophe might not have happened yet, but the potential needs to be identified.

Catastrophes considered in this book include:

- Explosions, major fires, and toxic gas clouds from industrial facilities, and utilities such as power stations.
- Explosions and radioactive releases from nuclear power stations and other nuclear facilities.

DOI: 10.1201/9781003360759-3

- Mining incidents where collapse or explosions have caused multiple deaths.
- Dam collapses, causing flooding, which threatens life downstream.
- Aircraft crashes.
- Ship collisions.

This book does not intend to cover workplace safety and occupational health where the maximum potential is one or two fatalities to employees or contractors.

CATASTROPHIC RISK

Hazards surround us. A hazard is anything that can harm or damage someone or something. A risk occurs when someone is exposed to the hazard.

For example, dams present a hazard because they create a barrier and hold up water either as a reservoir for irrigation or water supply. The dam's collapse could lead to widespread flooding, damage, and loss of life. I live about 15 km from a large river, the River Severn. It has several large dams for water capture upstream. However, there is no risk to me where I live close to the top of the Ridgeway, at least 50 metres above the broad river valleys.

However, some 100 metres from my house is a small industrial estate. A little research shows that among the 25 small employers on the site, activities include oil supplies, cleaning supplies, and metal coatings. Further investigation shows that a cleaning product supplier mixes and blends materials that are strong oxidising agents capable of explosive reaction if mixed with the wrong materials. Some other materials in use on the industrial estate are potentially damaging to the environment if spilled. Others are lung irritants. At other units, oils are blended, and metalwork is coated using flammable paints. There is potential for the wrong materials to be mixed accidentally and for a significant air or water pollution event. In my view, this industrial estate does present a risk to me and those who live close by.

In Chapter 15, we will examine the risks created by industrial estates. Whilst there have not been any significant incidents in the industrial estate close to my home, there have been notable incidents at industrial estates elsewhere.

Illustrative Hypothetical Case: Hazards

Suppose the reader was evaluating either:

 a. An existing small whisky distillery
 b. An existing small fuels terminal (gasoline, diesel, LPG, fuel oil)

The reader is trying to assess the potential for a severe incident at the plant.

We will return to these cases throughout this book. Information on the hypothetical distillery and fuels terminal is shown in Appendix A.

Here let us start by identifying the kinds of hazards we will encounter in each facility.

Both facilities:

- Moving equipment is a crush hazard
- Flammable liquids are a fire hazard
- Flammable vapours are a fire and explosion hazard
- Trucks are a collision hazard

Whisky distillery hazards additionally include:

- Wastewater, including potentially toxic components
- Solid materials handling (wheat and barley)
- Dust (from wheat and barley crushing)
- Boiling liquids
- Hot surfaces
- Natural gas (fuel)
- Flammable gas (alcohol vapour)

Fuels Terminals hazards are additionally:

- Railway operations (from trains bringing in the fuels)
- Rail Tank Car unloading of the fuels
- Road Tanker Filling with the fuels
- Road Tanker Vehicle Operation with many vehicle movements
- Pipelines of flammable materials within the terminal and from the refinery.
- Pressurised and liquified flammables (LPG)

Yes, fires and explosions have occurred in Whisky distilleries! In the U.K., most Whisky distilleries are registered under the Control of Major Accident Hazards Legislation (COMAH) see Appendix 1. Many fuels terminals are also registered under COMAH. The Health and Safety Laboratory in the U.K. (HSL, 2003) points out that explosions are infrequent in whisky distilleries but then lists seven blasts, including some incidents with multiple fatalities. We are using the whisky distillery as a hypothetical case here because the process is relatively simple to describe and be understood by the non-technical reader.

So far, we have only looked at the hazards. Some of these hazards do not have the potential to result in a catastrophe as defined above. Potential consequences would be limited to a single injury or fatality. Let's consider this in the coming paragraphs.

Risk involves exposure to the hazard. A lion is hazardous – it can cause grievous harm – but if I don't go to any places where lions live, I am taking no risk of being harmed by a lion! Furthermore, if I drive through lion country in a safari vehicle for an hour, my risk is low because the exposure time is low, at least by comparison with the risk to those who live in lion country. The magnitude of the consequence of an incident is also significant. If instead of two people in a safari vehicle, 50 people travel in a coach which breaks down at the lions watering hole, then the consequences could be extremely high!

Risk is the chance of something happening that will have a negative effect. The level of risk reflects:

- the likelihood of the unwanted event
- the potential consequences of the unwanted event.

Illustrative Hypothetical Case – Whisky Distillery

We have evaluated some hazards at our hypothetical whisky distillery and fuels terminal. It is now time to go through those hazards and consider the extent of exposure and the potential consequences so that we can understand the risks from an incident. Remember that we have defined catastrophe as one or more of the following:

- Multiple fatalities to the employees of the organisation within which the incident occurred.
- Fatalities to the public.
- Widespread illness and/or injury to members of the public.
- Extensive damage both financially and in loss of facilities for an extended period.

Listing our hazards identified previously, examine each one by one for both exposure and consequences.

The maps in Appendix A help with this exercise.

Table 2.1 Illustrative Hypothetical Case – Hazards in a Whisky Distillery

Hazard	Risk with Catastrophic Potential	Explanatory Note
Moving equipment is a crush hazard	No	Limited to individual fatality or injury
Flammable liquids are a fire hazard	Yes	Fire could destroy much of the distillery and potentially adjacent property
Flammable vapours are a fire and explosion hazard	Yes	Explosion could destroy much of the distillery, and potentially adjacent property
Trucks are a collision hazard	No	Limited to individual fatality or injury
Wastewater, including potentially toxic components	Yes	Accidental release of contaminated waste could cause fatalities and damage to aquatic and livestock downstream
Solid materials handling (wheat and barley)	No	Limited to individual fatality or injury
Dusts (from wheat and barley crushing)	Yes	Dust explosion could destroy much of the distillery and potentially adjacent property
Boiling liquids	No	Limited to individual fatality or injury
Hot surfaces	No	Limited to individual fatality or injury
Natural gas (fuel)	Yes	Fire could destroy much of the distillery and potentially adjacent property.

So far, we have only looked at the hazards. Some of these hazards present risks at the distillery, which would not result in a catastrophe rather, they would be limited to a single injury or fatality. Let us consider this in the coming paragraphs.

Here is a similar analysis for the Fuels Terminal

Illustrative Hypothetical Case – Fuels Terminal

Table 2.2 Fuels Terminal Hazards

Hazard	Risk with Catastrophic Potential	Explanatory Note
Moving equipment is a crush hazard	No	Limited to individual fatality or injury
Flammable liquids are a fire hazard	Yes	Fire could destroy much of the terminal and potentially adjacent property
Flammable vapours are a fire and explosion hazard	Yes	An explosion could destroy much of the terminal and potentially adjacent property.
Trucks are a collision hazard	No	Limited to individual fatality or injury
LPG	Yes	Fire/explosion could destroy much of the terminal and potentially adjacent property

Incidental Case Study: Gasoline Station

Gasoline (Petrol) gives off highly flammable vapour even at very low temperatures. Because of the flammability of the vapours, service stations carry a risk of fire or explosion. Ignition of petrol vapours can happen if vapour comes into contact with a heat source capable of igniting it. An ignition spark might come from an electrical switch, a cigarette, or a static electrical discharge. Fires have occurred the driver picked up a static charge simply by shuffling out of his seat. Gasoline vapour is heavier than air and tends to sink to the lowest possible level of its surroundings. It can gather in tanks, cavities, drains, pits, or other low points and will travel across the ground due to gravity (downhill) or be carried in the direction of the wind. The author has not identified such a fire that resulted in a fatality to date. In order to generate one of the catastrophic consequences above, one or more of the following is needed:

- more fuel, for example, from a road tanker being unloaded
- more people, for example, a coach or minibus full of people being refuelled
- congestion limiting opportunity to escape
- failure to activate the emergency switch. All gasoline stations should have such an emergency switch. Check it out next time you are filling up with gasoline/petrol.

In Section 2 on High-Risk Technologies and Their Catastrophes, many examples of incidents associated with many industries are examined. Hazards and the way these hazards become risks are identified.

Chapter 3

What Makes Risk Tolerable, or Intolerable?

During the coronavirus pandemic of 2020, it was common to hear government officials and politicians referring to absolute safety. The health minister at the time appeared on television and reassured us that vaccinations would not be carried out unless the government was convinced that the available vaccines were absolutely safe. I recall hearing a member of the teachers' union saying that their members would not return to the classroom until it was absolutely safe to do so.

There is no such thing as absolute safety. Everything we do has some risk. Even sitting watching the television leads to risks from obesity and lack of exercise. Therefore, it seems strange that the concept of absolute safety is often spoken about in the workplace by people who should understand well the concept of risk. For example, captions such as 'Safety First' and 'We won't carry out any operation unless it is safe to do so'. Whilst such statements are well-meaning, they imply an unachievable state. It is good to see that in many organisations, such statements have been replaced by more achievable words such as 'Pursue the goal of no harm to people' (Health and Safety Policy, Shell). Nevertheless, there are still safety policies that say, for example, only safe working methods will be employed.

Notes from Trevor's Files

While engaged in some consultancy work at a remote oil field in Kazakhstan, I was invited to conduct a safety tour. A Drilling Superintendent informed us, with a broad Texas drawl, that "here in this organisation we have a 'Stop the Job Policy'. If anyone sees anything unsafe, they should stop the job – it is both their right to do so and their responsibility".

During the site tour we found several activities that we thought put the operator at risk and which could be improved. Reporting this back to the Drilling Superintendent, he asked in a challenging manner if we had 'Stopped the Job' and was quite dismissive of our comments. To him, it seemed that if the safety improvement opportunities we had seen did not justify us

---

'Stopping the Job', then they were not worthy of further consideration. Had we 'stopped the job', I doubt that we would have been popular!

Some 'Stop the Job' policies are a little more cautiously worded. For example, personnel may be asked to 'Stop The Job' where they observe work activities that pose an **imminent** danger to life, health, and the environment. This statement is, in my opinion, a considerable improvement.

If we discard the concept of absolute safety, how do we determine an acceptable level of safety? The determination of an acceptable level of safety has been the subject of considerable study in the fields of sociology and psychology.

As an example of how human psychology responds to events and influences assessing risk, take the example of the fear that understandably arose after the tragic terrorist act known as 9/11 or the 'twin towers' in the U.S.A. A total of four aeroplanes were highjacked, two were deliberately crashed into skyscrapers in New York, one was intentionally crashed into the Pentagon in Washington, D.C. A fourth crashed into a field after passengers attempted to regain control of the aircraft. For a prolonged period of time, the response of many American people was to abandon flying and use their automobiles in its place. Yet any understanding of the risk of driving on the road shows that this decision is misguided.

As Dan Gardner (2008) comments, this irrationality is partly caused by those who promote fear for their own gain. Who might these be? Politicians, activists, and the media are prone to using biased presentation of the facts to pursue their own ends. As for the media, it is said that 'sex sells'. It is also the case that, when it comes to the media, 'fear sells'. A quick review of some of the more populist newspapers will confirm this.

It does seem that society is getting more and more concerned about risk, even though life expectancy has dramatically increased in recent years. This improved life expectancy is due, at least in part, to improved environmental regulations, worker safety, and advances in public health. Acceptability of risk is a societal one. It is created in the human mind from various societal influences. (Ropeik and Gray, 2002, p. 423) list the 'one-year odds' of dying of a long list of potential causes. For the U.S.A., motor vehicle accidents are by far the highest at 1 of chance in 6700. Only certain medical conditions such as cancer and heart disease come higher. We choose to travel on the roads, so does this provide a benchmark for acceptable risk? Would you accept the same risk of harm from a nearby nuclear power station as you would from driving a car? In a city of 1 million people, 1 chance in 6700 equates to 150 fatalities per year. We accept this for motor vehicles but not for the nuclear power station. Some of the key differences are presented in Table 3.1.

Table 3.1 Key Differences in Risk between Road Transport and the Risk from a Nuclear Power Station

Risk from Road Accident	Risk from Nuclear Power Station
• Voluntary – choose to drive	• Involuntary – no choice but to live at the location
• Control of risk – driver has at least some control over how risky he is driving	• Control of risk is entirely in the hands of the Nuclear Power Station management.
• Beneficial to person at risk	• Not generally beneficial to person at risk.*

Note

* Some hazardous facilities have extensive social programmes in the local community, which helps the perception that the risk is beneficial to the inhabitants of the community.

Most people are more afraid of new risks, compared with those they have lived with for a while. Familiarity with a risk seems to progressively reduce the fear of that risk. This is a considerable challenge in risk management. We can become blind to risks which, when we were first introduced to them, we well recognised.

- People are generally less afraid of natural risks compared to those which are man-made. Many people are more afraid of the radiation from nuclear waste than they are from radiation from the sun. However, radiation from the sun poses a far greater risk, as can be verified from the measurement of the radiation dose. Sunbathers beware!
- Voluntary risks are more acceptable than risks which are imposed on us. Smokers are less afraid of smoking than they are afraid of asbestos and other indoor air pollution in their workplace. The latter being something over which they have little choice.
- Risks are more palatable if they bring benefits to those exposed to the risk. People risk injury or death in an earthquake by living in San Francisco or Los Angeles. Those people find benefits in living in those locations, such as finding their preferred line of work, and desirable climate.
- We are particularly averse to risks which cause harm or death in particularly awful ways, like being eaten by a shark.
- We are typically less afraid of risks over which we have some control, like driving. Where we are put in situations where we have no control, we are more fearful. Examples might be flying, or living close to a chemical plant.
- Trust is an important indicator of fear. We are less afraid of risks from people, places, and organisations we trust. More fear comes if the threat is from a source we do not trust. A potential example here is where people have more trust in organisations founded in their own country versus foreign organisations,

- Adults are more afraid of risks to their children than risks to themselves.

For an excellent review of risk perception see Ropeik D. and Gray, G. (2002) *Risk – A Practical Guide for Deciding What's Really Safe and What's Really Dangerous in the World Around You.*

So, where does the borderline between acceptable, tolerable, and intolerable risk lie in terms of annual risk of fatality?

The U.K. Health and Safety Executive carried out an excellent study on this subject, which is published in a document Reducing Risks, Protecting People H.S.E 2001. This study is affectionately called 'R2P2'. This was an extension of their first work in 1988 to establish the tolerability of risks from nuclear power stations. Others have had similar evaluations but with the same broad conclusions. The objective of this study is to enable decisions on risks based on whether those risks are:

- Acceptable
- Tolerable
- Unacceptable

In this context, 'tolerable' does not mean 'acceptable'. It refers instead to a willingness by society as a whole to live with a risk to secure certain benefits. Society does so in the confidence that the risk is worth taking and that it is adequately controlled. Both the level of individual risks and the concerns in society caused by the activity must be considered.

A risk of 1 in 100000 per year or better is generally considered acceptable for a single fatality. A risk worse than 1 in 10000 is intolerable, and the band between is tolerable if risk has been made as low as reasonably practical (see later section on ALARP, Chapter 28). But there are other factors, including 'Dread', where a particular type of accident leads to particularly dreadful or unknown outcomes. Hence high levels are expected of the nuclear industry.

The most important questions for individuals incurring a risk are how it affects them, their families, and the things they value. Though people engage voluntarily in activities that often involve high risks, as a rule, they are far less tolerant of risks imposed on them and over which they have little control unless they consider the risks as negligible. Moreover, though they may be willing to live with a risk, they do not regard it as negligible. If the risk secures them or society certain benefits, they would want such risks to be kept low and adequately controlled.

R2P2 identified a growing need in society to assess the benefits brought about by industrial activity against potential undesirable side effects such as the risk of being maimed or killed or of environmental pollution. This need is particularly true for risks:

- which could lead to catastrophic consequences.
- where the results may be irreversible, e.g. the release of genetically modified organisms.
- which lead to inequalities because they affect some people more than others, such as those arising from the siting of a chemical plant or a waste disposal facility.
- which could pose a threat to future generations, such as toxic waste.

Note that two aspects may make a risk tolerable or intolerable:

- The probability that it will happen.
- The consequence of it happening.

Thus, the Deepwater Horizon oil spill in the Gulf of Mexico and the Fukushima Daichi nuclear power plant disaster, both discussed in Section 2 are considered Low Probability, High Consequence events.

A road tanker collision could be considered a Higher Probability, Lower Consequence event.

For a man-made catastrophe to occur, there must be a cause, maybe several causes. Since we generally design facilities and manage them to avoid catastrophe, something must have gone wrong with the means of prevention. Section 3 on Catastrophe Prevention and Why It Is Failing explains many such prevention methods and why they fail.

Sometimes the causes of catastrophes are remarkably similar. For example, for the tragic incidents at B.P. Texas City (2005) and Flixborough (1974) (see Chapters 6 and 7) the following causes are similar between both tragic events (Bahr, 2015, p. 72):

- Worker trailers located too close to hazardous operations
- Supervisor absent
- Process safety not a priority – more safety focus on occupational safety
- Strong financial pressures impacted operational safety
- Abnormal start-up
- Operator inattention
- Poor communication during shift handover
- Did not take advantage of opportunities to replace inefficient and more hazardous equipment
- Some instrumentation did not work
- Lack of follow-up or follow-through from previous incidents

From the space industry, we have the tragedies of the loss of Columbus (2003) and Challenger (1986) (Bahr, 2015, p76). The following causes are similar:

- Inadequate safety culture across NASA organisations and undervalue if system safety engineering and system safety engineers
- Working-level engineers identified safety issues but did not get an adequate hearing in front of more senior management – barrier to dissent
- Performance pressure on operations
- Significant budget pressures and dwindling resources
- Lack of independence of the safety organisation
- Misunderstanding and lack of adequate analysis of previous anomalies and incidents
- Engineers did not push their safety concerns due to fear of the impact on their careers

These observations lead us to ask if we have learned from such catastrophes, and if not, why not.

With the above considerations in mind, let us move on, in Section 2, to discuss some actual catastrophes.

Chapter 4

Fundamental Concepts, Protection Barriers, and Risk Exposure Consequences

FUNDAMENTAL CONCEPTS

Our objective in this book is to understand why, unfortunately, many methods and approaches to preventing catastrophes fail to be effective and propose some distinctly new strategies to overcome some of the pitfalls. So, please, bear with me! It is necessary to provide some fundamental understanding for the reader who is unfamiliar with safety management.

A risk activity with which we are all familiar is driving a car. There are many aspects of operating a vehicle that help prevent driving accidents, for example:

- Drivers are allowed to drive only if they have a driving licence.
- A driving licence is only provided if the driver is medically able to drive.
- A driver must be trained and pass a test to get a driving licence.
- There is a code of practice for all travellers on the roads. In the U.K., this is known as the 'Highway Code'.
- Roads have signs directing traffic, including speed limits according to the hazard posed by the area.
- Roads are maintained to a minimum standard.
- Police and traffic officers patrol traffic in order to help ensure that standards are maintained and may take action against drivers they consider driving dangerously.
- Driving under the influence of alcohol or drugs is not permitted, and impaired drivers may be prosecuted.
- Satellite navigation systems reduce distractions from needing to consult maps and help avoid congested areas.

If an accident occurs, then there are various means which might reduce the potential harm to the driver and others involved in the accident, such as:

- Seat Belts

DOI: 10.1201/9781003360759-5

- Air Bags
- Availability of hard shoulders or refuges on motorways where the vehicles involved in the accident can, if possible, pull away from the main carriageways to reduce the risk of further collisions.
- Rapid response from emergency services, ambulance, fire, and police which can reduce the severity of injuries.

There is an essential but straightforward concept of considerable importance in incident prevention. For any risk, there are barriers to preventing the incident from occurring. Some barriers prevent the incident from being worse once it has occurred. A helpful diagram is a 'Bow Tie'. The barriers preventing an undesirable event from happening are shown to the left of the intersection of the lines on the Bow Tie. The barriers which help reduce the consequences of the undesirable event are shown to the right.

BARRIER ANALYSIS AND BOW TIE DIAGRAMS

Bow tie analysis is used in a wide range of industries. The interested reader can readily find examples available on the internet in air transport, railways, and the chemical, nuclear, oil and gas processing industries. Bow Tie diagrams can be used to illustrate issues relating to the development of serious incidents. This book uses these diagrams frequently. The paragraphs below briefly describe the technique. Readers of this book would benefit from understanding the basis of Bow Tie diagrams before proceeding further, although this is not essential – the diagrams are reasonably intuitive. There are some excellent YouTube videos by CGE Risk to be found on the internet. These videos take only a few minutes to explain the technique. The Chemical Centre for Process Safety (CCPS) has a book, 'Bow Ties in Risk Management – A Concept Book for Process Safety', which is easy to read.

Figure 4.1 shows a basic Bow Tie relating to the hazards of Driving a Car. In the centre of the diagram is the hazard and an event we want to analyse, in this case, 'Loss of Control of the Car'. In Bow Tie terminology, this event is called the Top Event. There is a 'Threat' on the left-hand side, something that can lead directly to the Top Event. I have chosen only two threats for this example – Driver Loss of Attention and Tyre Blow Out. Additional threats such as 'Intoxicated Driver', 'Break Failure', and 'Slippery Road' should be added for a complete analysis. On the right-hand side are two consequences 'Injury or fatality to passengers or driver' and 'Collides with other car or pedestrian with injury or fatality'. For a complete analysis, other consequences might be considered, such as vehicle damage.

There are some barriers that prevent Loss of Control of the Vehicle. The listed threats do not necessarily lead to loss of control of the vehicle. For the

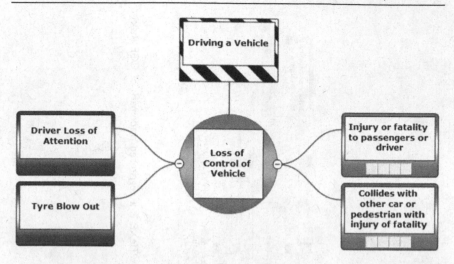

Figure 4.1 Basic Bow Tie – Losing Control of a Vehicle.

Source: Created using BowTieXP Software © CGE Risk Management Solutions B.V. 2021. BowTieXP is a registered trademark of CGE Risk Management Solutions B.V.

tyre blowout threat, regular checks of tyre condition can help prevent losing control of the vehicle. Defensive driving training in managing a tyre blowout situation could also help.

There are also barriers that prevent the loss of control of the vehicle from causing injury or fatality. Such features include seatbelts and airbags. In Bow Tie terminology, these are known as Mitigating Barriers. For the consequence of colliding with another car or pedestrian, we have no control, so in this case, the loss of control of the vehicle has no mitigating barriers (Figure 4.2).

There are some controls on the barriers which might make them more or less effective. Regular tyre checks can be mandated and audited, especially if the vehicle is a company vehicle. Defensive driving can be part of the employees' annual training plan, as was the case when I was working at Du Pont. In the above example, no barriers are considered for 'Driver Loss of Attention'; in more modern cars, there are driver alertness monitors of various types together with 'lane assist'. The company can mandate that vehicles be equipped with these and require limits on driving time (Figures 4.3 and 4.4).

In Section 2, several actual incidents are discussed where reference is made to hazards, threats, barriers, controls, and consequences. Barriers can be weak or strong. If the vehicle is not company controlled, as may be the case with your family vehicle, then the above barriers and controls could be weak – that is for you to consider!

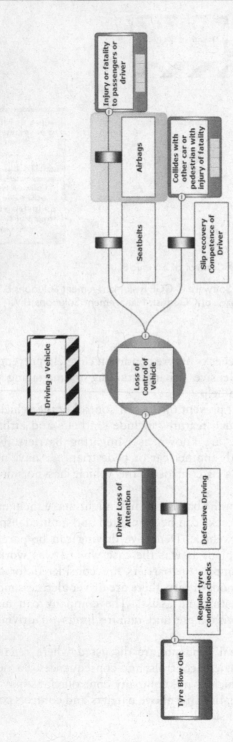

Figure 4.2 Basic Bow Tie – Losing control of Vehicle – Barriers.

Source: Created using BowTieXP Software © CGE Risk Management Solutions B.V. 2021. BowTieXP is a registered trademark of CGE Risk Management Solutions B.V.

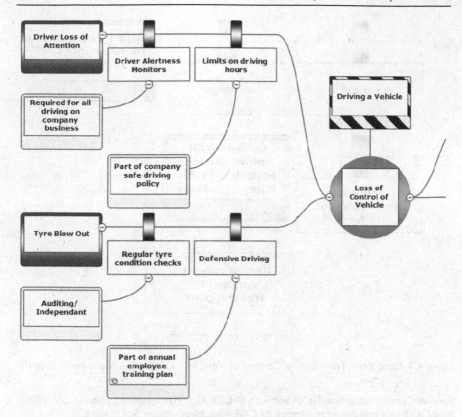

Figure 4.3 Basic Bow Tie – Losing Control of Vehicle – Controls – Left-Hand Side of Diagram.

Source: Created using BowTieXP Software © CGE Risk Management Solutions B.V. 2021. BowTieXP is a registered trademark of CGE Risk Management Solutions B.V.

The use of the Bow Tie technique is further expanded in Chapter 25.

CONSEQUENCES OF EXPOSURE TO A RISK

In the above driving example, the potential worst consequences are obvious. The people inside the car may be severely injured or die. If there is a collision between two cars, then the potential is double. But what about the case where one of the vehicles, or maybe both vehicles are carrying hazardous materials. What can we conclude about the potential consequences? Before considering in depth the hazards of particular industries in Section 2, I have given below a short review of the kinds of consequences that will be discussed, from a wide range of hazards, and how those consequences can be anticipated.

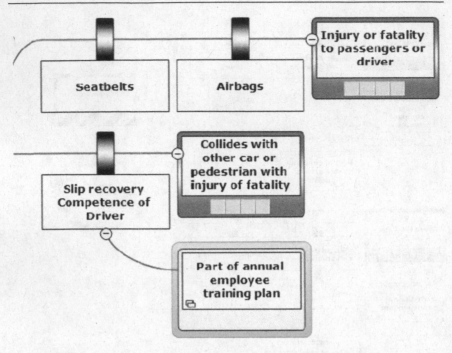

Figure 4.4 Basic Bow Tie – Losing Control of Vehicle – Controls – Right-Hand Side of Diagram.

Source: Created using BowTieXP Software © CGE Risk Management Solutions B.V. 2021. BowTieXP is a registered trademark of CGE Risk Management Solutions B.V.

Hazardous Material in Transport

A collision of a vehicle carrying hazardous materials might result in a fire, explosion, toxic cloud, or be damaging to people and the environment. It is useful to know the hazards of the materials on the road, and in particular any such vehicles going to or coming from premises in your area. Vehicles carrying hazardous material should be displaying a special plate. This can be valuable in considering the potential consequences of a vehicle incident, perhaps at the end of your street!

Materials are typically labelled according to the following types of hazards, each with a distinctive diamond as shown in Table 4.1. This table is from the U.S. DOT regulations, European requirements are similar:

Additionally, vehicles carrying hazardous material should have placards. In the U.K., these are known as HazChem Panels, ADR in mainland Europe. They are square with an orange background and bold black letters. These placards inform emergency responders of the hazards and nature of the material involved in the accident. In addition to the warning sign as indicated in Table 4.1, they include an Emergency Action Code and a U.N.

Table 4.1 Hazard Classes

Hazard Class	Hazard
1	Explosives
2	Gases
3	Flammable and Combustible liquids
4	Flammable Solids
5	Oxidising Substances, Organic Peroxide
6	Toxic Substances and Infections Substances
7	Radioactive Material
8	Corrosives (Liquids and Solids)
9	Miscellaneous Hazardous Materials

Table 4.2 HazChem Panels (U.K., Australia, Malaysia, New Zealand, India)

Material	Emergency Action Code	U.N. Number
Gasoline/Petrol	3 YE	1203
LPG	2YE or 2WE	1075
Ammonia	2TC	1005
Chlorine	2XE	1017

Number, plus a specialist advice telephone number. There is an equivalent in the USA and in most other countries (Table 4.2).

The emergency action codes tell the emergency responders what to do. For example:

3YE means use foam to suppress fire, use breathing apparatus, contain the spill from entering drain or watercourse, evacuation may be needed.

I found that it is quite difficult to look up the HazChem panel codes for any particular chemical. The Material Safety Datasheets sometimes gave them, sometimes not. There are inconsistencies, such as the one identified above for LPG. It looks like a sound system until you drill down.

Collision and Impact

Damage requires momentum, a combination of both speed and mass. It also depends on the properties of the two bodies colliding. In particular, the colliding bodies elasticity or ability to absorb energy by fragmentation. Buffers for trains have an elastic resistance created by heavy-duty springs or hydraulics. An automobile has a crush zone. This is provided so that some of the energy of the collision is absorbed in the fragmentation of the nose of the vehicle. If one of the vehicles is carrying hazardous materials, an impact may rupture the tank containing the material,

Gasoline tankers are particularly vulnerable in this respect since the tanks containing the gasoline are at atmospheric pressure and not designed as pressure vessels. I have had the unfortunate experience of seeing a gasoline road tanker leaking its contents on the roadside due to a multiple vehicle pile-up on a motorway. Propane tankers are less vulnerable because the tanks are more robust, being designed as pressure vessels.

Fire

If one of the vehicles involved in the collision carries flammable material such as gasoline, will there be a fire, and how bad would that fire be? Firstly, for there to be a fire, the vapour from the leaking material needs to find an ignition source. This ignition source is most likely to occur on a motorway from a running vehicle engine or an exasperated driver deciding to cool his/her nerves by smoking.

Switch off your engines and eliminate all potential sources of ignition.

If a fire does occur, will injuries occur, and at what distance?

The radiant heat from a fire can cause fatal injury without the person contacting the visible flames of the fire. The radiant heat load, together with its duration, are both critical in determining the effects. Later in this book, in Chapter 28, I will describe how you can easily calculate this radiant heat using free online software (ALOHA) from the U.S. EPA. No scientific background is necessary, just an enquiring mind. However, in order to demonstrate some of the difficulties engineers have in coming to firm conclusions about consequences, a glance at Table 4.3 is appropriate.

Table 4.3 Burn versus Thermal Dose Relationship

Harm Caused	Mean Dose $(kW/m^2)^{4/3}$ s	Comment
Pain	92	
Threshold for first-degree burn	105	
Threshold for second-degree burn	290	Pain induced will cause high motivation to escape, but any exposed skin will be very uncomfortable, making turning valves difficult or impossible.
Threshold for third-degree burn	1000	Consider casualty who cannot escape unaided
1–5% Fatalities	1000	
50% fatalities	2000	
100% fatalities	3500	

Note: derived from Health and Safety Laboratory (2004) Human Vulnerability to Thermal Radiation Offshore.

The units of mean dose merit explanation. kW is the rate at which energy is produced, otherwise known as power. kW/m^2 is the rate of that energy per unit area. S is the time of exposure. However, this is not linear – kW/m^2 are raised to the power (4/3), meaning that higher energies are disproportionately more damaging for the same duration – without reverting to our school mathematics, it is merely important to stress that the relationship is not simple. This complexity is why I explain this issue in some detail. Without needing to know how, you can anticipate that an engineer can calculate the radiant heat from a burning fire or, say, gasoline. That heat needs to be converted into potential radiant heat damage to human beings somewhere in the vicinity.

A pool fire of burning hydrocarbon emits radiant heat in the range of 50–150 kW/m^2. Let's choose 75 kW/m^2 by way of example. Exposure depends on the inverse of distance squared. Let's take an example of a person or object 5 metres from the fire. The local thermal radiation will be 1/25 of that at the source and therefore be around 3 kW/m^2. The duration before third-degree burns occurs is approximately 4 minutes. But as you can see from Table 4.3, there is a statistical probability of a fatality. The exposed person has a chance of death at this distance of between 1% and 5%.

When I wrote this book, I was reviewing a project plan for a flare stack that showed the maximum thermal radiation at structures being built at a distance to the flare of about 3 kW/m^2. The assumption for this to be satisfactory is that a person working on that structure can readily escape. In practice, the ability to escape will depend on the actual design of the adjacent structures, on the use of fire retardant clothing, on good housekeeping keeping escape routes clear, and on training to ensure that the persons potentially exposed recognises the hazard and escapes quickly.

Explosion

So, we have a vehicle collision involving a gasoline tanker which is leaking. We have established the potential for fire, but will there be an explosion?

There are various types of explosions. An explosion produces a pressure wave that can destroy buildings and equipment, as well as causing injury and fatalities. Damage also depends on the magnitude of the pressure wave and its duration. An explosion occurs due to a rapid expansion of gas and energy. Typically, this results from a vigorous chemical reaction such as burning or the ignition of a chemical/air mixture. Hence explosions are often, but not always, associated with fire. The explosion may also cause damage to tanks and vessels containing flammable materials, which results in the release of the contents of the tank and other fires.

Gasoline vaporises, producing gases that surround the vehicle and drift over adjacent vehicles. Further, a vehicle involved may be carrying compressed or pressurised gases such as propane. Most gases have lower and

Table 4.4 Dusts – Minimum Explosible Concentration

Dust	Minimum Explosible Concentration g/m^3
Coal	120
Sugar	45
Flour	50
Wood	15–200

Data derived from Dust minimum Explosible Concentration (www. dustwatch.com).

upper explosion limits, expressed as a concentration in air, between which an explosion might occur if the mixture meets an ignition source. For example, natural gas (primarily methane) has a Lower Explosive Level of 5%. If the concentration is less than 5%, there is too little oxygen to enable an explosion. Above 17% concentration natural gas and the mixture is too 'rich' and will not explode. Why is this important? A leak of flammable gas from a hole, say due to corrosion in a pipe or tank, will be concentrated at the hole and be above its upper explosive limit. Some distance from the hole and the gas will be below its lower explosive limit. It is only in between that an explosion can occur!

To complete the picture, the vehicle may be transporting finely divided solids such as sugar or flour. We do not consider these as hazardous. But with dusty materials, there is an explosion risk that is often neglected. Catastrophic dust explosions have occurred in sugar factories (see Chapter 12), wood factories, coal mines and coal handling processes, plastics factories, and flour handling processes. The minimum explosible concentration of some typical dusts is shown in Table 4.4.

Dust explosions due to vehicle collisions are extremely unlikely. I include the possibility here for completeness.

It is easy to visualise a cubic metre. There are very approximately five grams in a teaspoon. So, imagine dispersing five teaspoons of flour or sugar into a cubic metre – and you have an explosive mixture! All that is needed now is an ignition source. For dust explosions, this ignition source is often a spark from static electricity or hot conveyor bearings.

Acute Toxicity

Suppose the vehicle involved in the collision carries a toxic material, such as chlorine or ammonia. The tanks are strong because they are designed to be pressurised but leakages due to vehicle incidents have occurred.

Acute toxicity refers to adverse effects from exposure to a single dose of a toxic substance. The exposure might occur through breathing the substance, ingesting it in food, or by spillage of the material on the skin's surface. Consequences depend on the concentration of the toxic material to which a person has been exposed and the duration of exposure. Advisory bodies set control limits. These are often also legal limits. For example, the limit for chlorine is 0.5 ppm (parts per million) for 8 hours of exposure and one ppm for 15 minutes. Above these concentrations, varying degrees of irritation to the respiratory tract occurs. Above ten ppm, there is an immediate threat to life. Above 1000 ppm, or 1 part in 1000 is fatal within minutes. Chlorine is used in many locations and is present in chemicals that treat swimming pools. It can be inadvertently released in the wrong conditions, e.g. if these chemicals are heated or mixed with another chemical (see Chapter 15).

Chronic Toxicity

Suppose the vehicle carries a toxic material, which is not immediately toxic. Harm can occur with exposure over an extended period. Chronic toxicity is the development of adverse effects resulting from long-term exposure to toxic materials. Asbestos exposure is a prime example. Long-term exposure to asbestos fibres in the air leads to progressive lung damage. Lead poisoning from the exhaust gases of vehicle engines using leaded petrol is another historical example of exposure over the long term, causing severe health issues to many people. So, if the vehicle involved in the collision is carrying asbestos, there is no immediate cause for concern other than to minimise your exposure.

For more on chronic toxicity see Chapter 34 on Legacy Issues.

Why Is This Relevant?

Why start going into this 'science'? A significant part of improving the safety of hazardous facilities and operations is 'people pressure'. Applying this pressure is where you, the reader of this book, play a critical part. To exert that pressure, readers need to be informed and able to do simple 'ball park' estimates. Otherwise, we, the public, are too easily dismissed for our ignorance.

High-Risk Technologies and Their Catastrophes

Chapter 5

Introduction

This section discusses high-risk technologies, the kinds of catastrophic risks they cause and describes some catastrophes that have resulted from them.

HOW TO DEFINE A HIGH-RISK TECHNOLOGY?

Three methods characteristics of a high-risk technology have been included in this book:

1. Technologies where catastrophes are known to have occurred in the past somewhere in the world.
2. Where government regulations or government bodies have defined a facility or types of facility as high hazard.
3. Developing technologies where the extent of risk is yet to be fully understood.

This covers a broad range of sources of man-made catastrophes. Indeed whilst the book focusses on process industry examples, given the above definition I have also included catastrophes from other industries such as the aeronautical industry, military, and defence, together with consideration of pandemics and catastrophes which might occur in residential buildings.

HOW FREQUENTLY DO WE ENCOUNTER HIGH-RISK TECHNOLOGIES?

To develop, for the reader, some indication of occurrence of high hazard technologies, I started first with a list of high hazard installations that were available in the U.K. in 2018 under the Control of Major Accident Hazard Regulations (COMAH) – The equivalent of the Seveso regulations in the EU. This information was obtained by others under the U.K. Freedom of Information Act from the website www.whatdotheyknow.com. Whilst the information is not secret, a complete list seems no longer to be freely

DOI: 10.1201/9781003360759-7

Table 5.1 Spread of COMAH Installations by Type (U.K.)

Type	Number	%
Oil and Gas Terminals	222	28.4
Chemicals	284	31.6
Warehouse and Distribution	129	16.5
Alcoholic Beverages	78	9.9
Others	107	13.6
Total	784	100%

Note: Includes Upper and Lower Tier.

available. Currently, it is only by entering a postcode can a list of COMAH installations be provided, and only for facilities within 3 miles of your location. Based on the name of the installation and some knowledge of the companies involved, the following table shows the number and spread of such facilities (Table 5.1).

The prevalence of Oil and Gas and Chemicals is perhaps not unexpected. The inclusion of Alcoholic Beverages may be a surprise to readers. The risk arises because of the amount of ethyl alcohol in storage. As will be described later, there are real risks in the storage of alcoholic beverages, although I acknowledge that this is a relatively low risk.

The inclusion of warehouses and distribution centres may also be a surprise to readers. Warehouses store a wide variety of materials, some of which are hazardous either through flammability or toxicity. The hazards of warehouses are described in Chapter 13 on Process Industry Infrastructure.

However, to get a complete picture, we need to add some high-hazard facilities excluded from COMAH, primarily because they are covered by other legislation specific to the hazard. Also, the application of the COMAH regulations is based mainly on the volumes of hazardous materials stored and therefore would not include other hazardous operations such as air, marine, and railway transport.

So hazardous facilities which also need to be considered are:

- Nuclear
- Mining
- Air Transportation
- Railway Transportation
- Offshore oil and gas facilities

The U.K. Office of the Nuclear Regulator publishes details of licensed nuclear sites. It has 41 such locations.

Mining hazards are different from the types of hazards in industries considered so far and are primarily rockfall and explosions of gasses accumulated underground. There are some 1700 mineral workings in the U.K., but most of these are sand, gravel, and limestone surface workings with little hazardous potential, and it would create an incorrect picture to add all of these into our list of high hazard facilities.

Using the British Geological Surveys 'Directory of Mines and Quarries (2020)' it is possible to segregate some specific activities which may be high hazard:

- Coal – surface mined – 11
- Coal – Deep mined – 3
- Fluorspar (Calcium Fluoride) – 3
- Barytes (Barium Sulphate) – 3
- Lead – 2
- Gold – 2
- Tin – 1

Offshore oil and gas facilities are not shown in the COMAH list since they are covered by different legislation. There are approximately 100 offshore fields with structures in the North Sea (Wikipedia – List of Oil and Gas Fields in the North Sea).

Transportation risks are not focused on a particular geographical site – they are distributed along the flight path or railway line. There are approximately 230 airports or airstrips in the U.K., some very large like Heathrow and some small ones just for recreational flying. So, I have not attempted to include a number for transportation risks, I ask the readers to bear this exclusion in mind when reviewing the total picture.

In Table 5.2, we have added the known data for UK COMAH sites, nuclear sites, offshore structures, coal, and metal mining.

Table 5.2 Number of Hazardous Sites Estimated from Data

Type	Number	%	Section 2 Chapter
Chemicals	284	24.9	2.3
Oil and Gas Terminal	222	19.5	2.9
Warehouse and Distribution	129	13.1	2.9
Offshore oil and gas	100	10.1	2.2
Alcoholic Beverages	78	7.9	2.8
Nuclear	41	4.1	2.6 and 2.11
Metallurgical	35	3.4	2.10
Utilities	34	3.4	2.7
Mining – metals and coal	21	2.1	2.4
Other	38	3.8	
Total	982	100%	

This distribution will, of course, change from country to country and from region to region. Texas will have a higher proportion of oil and gas, as will Saudi Arabia, etc. Russia, Kazakhstan, and Australia will have a higher proportion of mining operations. The U.K. example at least identifies the spread of hazardous industries on a systematic basis, and other countries in Europe are likely to have a similar spread.

COMAH is the U.K. application of the Seveso regulations in use in the European Union. Similar data on the locations of Seveso sites can be discovered for other places in Europe. Appendix B shows maps displaying the density of Seveso sites in a selection of European countries. It seems that, at least in Europe, no one is far from a high-hazard facility but may well not know it. How far are you from your nearest hazardous facilities?

GUIDE TO SECTION 2

In each chapter of Section 2, the characteristics of each industry section are described. The kinds of hazards found in the industry section are discussed, along with the potential for a catastrophic event. There is also consideration of any progress made by the industry to reduce catastrophic potential. Several incident examples then follow, many of them catastrophes as defined by this book. Some other incidents are included, which were not catastrophes but demonstrate the potential for a disaster in the industry.

Finally, there is a conclusion where the potential for a future catastrophe in this industry is discussed. This is displayed in the form of a table summarise according to the following scale:

Not possible	Unlikely	Possible (50/50)	Likely, no more than one	Very likely/ maybe several
☐	☐	☐	☐	☐

The reader is asked to draw their own conclusions.

Chapter 6

Oil and Gas

OIL AND GAS UPSTREAM (Finding and Collecting Oil and Gas from Underground)

Crude oil and natural gas are found in naturally occurring underground reservoirs. These hydrocarbons are formed from the remains of plants and animals that died millions of years ago and have become trapped within geological features. Therefore, they are called fossil fuels. During the exploration phase, drilling rigs are used to discover if oil or gas is present in economic quantities.

Once a reservoir is proven, then it may be developed. Further wells are drilled using the above rigs, and production equipment is installed.

Significant risks occur from drilling oil wells. The most well-known involves a blowout. A blowout is the uncontrolled release of crude oil or natural gas from an oil well or gas well. Various control systems prevent should prevent this from occurring but as discussed below, these systems can fail.

'Blowout preventers' are used during drilling and reduce the risk of such an occurrence. However, failures do occur. Blowout preventers (BOP) are complex devices that can suffer mechanical failure, internal wear, external leaks of the hydraulic systems, loss of hydraulic power, corrosion, and erosion. During drilling, rods must be inserted through the blowout preventer. Although the blowout preventer is designed to shear such rods, failures have occurred to the shearing rams either because they failed to operate or did not have enough power to cut through the drilling rods. The BOP clearly should not be the only line of defence against blowout.

Oil drilling organisations have adopted the two-barrier minimum rule. In order to explain this rule, we need first to understand the concept of 'breaking of containment', a concept which we will come up against on many occasions in this book. Consider a simple tank containing a hazardous material. The material is contained. However, if we open a valve or remove sections of the vessel wall we are 'breaking containment'. The two-barrier rule simply states that for any work that must be done on a section of equipment such as pipework, which involves breaking containment,

DOI: 10.1201/9781003360759-8

there must be at least two barriers separating the fluids from the area of broken containment. The purpose behind specifying two barriers is to ensure redundancy. The risk of a total loss of containment is reduced if there are two independent isolations, as it is considered highly improbable that they would both fail simultaneously. (Wikipedia: Two Barrier Rule)

Notes from Trevor's Files

While carrying out a gas field survey, for insurance purposes, at a gas field on the bed of a now dry area of the Aral Sea, I was reassured that the organisation applied the two-barrier principle. Verification could be established by observing at the field the use of Blow Out Preventer and Drilling Mud, each of which provides a barrier during drilling of the well. Once the well is completed, the Blow Out Preventer is removed and replaced with a complicated valving system at the Well Head, known as a Christmas Tree. However, wells need to be maintained using a process known as 'well workover' from time to time. The wellbore may have become restricted with solids, or it may be necessary to perforate at a different level to extract the oil, or in this case, gas.

To safely enable this, the well is first pumped full of heavy mud (through the valving of the Christmas tree) to hold back the gas – the first barrier. And after the Christmas Tree is removed, a BOP is installed, completing the second barrier. All O.K. so far? Did you spot the problem? What about the time between taking off the Christmas Tree and installing the BOP. Where are the two barriers? There is only one at this point – the drilling mud.

The drilling manager explained that, whilst there was only one barrier, this is O.K. because the BOP is installed promptly. The time when only one barrier is in place is short.

Much discussion on this followed. My position – two barriers should be maintained at all times. The argument that the duration of a single barrier is short has some merit, but things do not always go to plan. What if there was an unforeseen delay in installing the BOP due to weather conditions, lost or damaged bolts, crane failure, etc. It is possible to maintain two barriers by inserting a plug through the bore of the Christmas Tree into the wellbore, prior to removing the Christmas Tree. These purpose-built plugs are in frequent use in the industry. After the BOP is installed, the plug (typically cement or polymer) needs to be drilled out. Other drillers elsewhere in the world confirmed that this is their practice – to install such plugs when preparing for workover. Unfortunately, I have not had the opportunity to return to the Aral Sea to discover if the gas field team has changed its practice. I suspect that the additional work of installing the plug was not considered necessary.

Drilling for oil or gas can be carried out onshore or offshore. The off-shore variant essentially copies the technology of the onshore rig, mounting it on a structure that can be floated to location and then tethered or 'jacked up' for the drilling operation. Rigs are mobile and move from place-to-place drilling well after well.

There are extensive hazards of fire and explosion resulting from the highly flammable hydrocarbons once the well is drilled into a hydrocarbon-containing layer underground. Fires and vapour cloud explosions can occur.

A further significant hazard results from the presence of highly toxic hydrogen sulphide often, but not always, found in the oil reservoir. Vented gases from an oil production facility are likely to include hydrogen sul-phide, and this presents a dangerous situation to people downwind of the vented gases. Crude oils with a high hydrogen sulphide concentration are called 'sour', and crude oils with low concentrations are termed 'sweet'. The differentiation between sour and sweet is generally considered 0.5%, so even sweet crude oils contain some hydrogen sulphide.

A further hazard arises from the crude oil itself, which, if it leaks, can cause tragic pollution events.

Depending on the geological location of the oil reservoir, the oil and water being extracted from underground may contain radioactive contaminants.

Special drilling muds are used when drilling a well. These muds are cir-culated through the wellbore to carry rock cuttings to the surface. They also lubricate and cool the drill bit. Drilling muds also control the well during drilling since they counterbalance the pressure of the hydrocarbons in the reservoir. This counterbalancing is often one of the two barriers necessary. Failure to properly counterbalance with drilling mud can result in a blowout. (See Deepwater Horizon and Montara incident examples below.)

Drilling muds contain a wide range of substances, many of which can be hazardous to human health and aquatic life.

The drill cuttings – the waste rock drilled out to make the well, can also be hazardous, although this is primarily of longer-term environmental concern. Historically drill cuttings have been disposed of on the seabed. Nowadays, they are usually separated and taken ashore for disposal at li-censed waste disposal sites.

Fracking deserves a special mention. Some deposits of oil and gas do not 'flow' when the oil or gas well is drilled. This is because the oil or gas is bound up within the rock formation and cannot flow under pressure to be captured by the oil well. Fracking is the process of drilling down into the underground rock layer where oil or gas is located, followed by a high-pressure water mixture directed at the rock to release the oil or gas inside. Water, sand, and chemicals are injected into the rock at high pressure. This activity ruptures the rock creating fissures that allow the oil or gas to flow out to the head of the well.

The extensive use of fracking in the U.S. has prompted environmental concerns. Many human activities such as mining, extraction of geothermal energy, dams, and oil and gas operations can affect local geological stresses and thus cause earthquakes. These earthquakes are typically of extremely low magnitude, not usually felt by human beings – 'micro earthquakes'. Such effects from human activity are called induced seismicity.

There have been cases where fracking has generated seismic events ranging from magnitude 1 to 4 on the Richter scale. These events are called 'anomalous induced seismic events' because they are unusual or inconsistent with what is expected. Seismic activity in the magnitude 3 to 4 range may cause detectable vibrations, similar to those felt when a heavy truck is driving by, but the event is unlikely to cause damage. (Oil and Gas Journal: About Fracking.)

The fracking industry suggests that pollution incidents result from poor practice rather than an inherently risky technique. As well as earth tremor concerns, environmentalists say potentially carcinogenic chemicals may escape during drilling and contaminate groundwater around the fracking site.

Notes from Trevor's Files: Mud Tank Alarms

Visiting a drilling rig in Pennsylvania, my attention was drawn to the repeated sound of a loud siren, which continued alarming on and off throughout the rig tour. After the field visit, in the quiet of a meeting room, I asked which of their alarm systems was the most important. Oh, oh, the participants exclaimed as if I should know – the mud tank level alarms, of course! The drillers then explained that a rapid rise or fall in mud tank level could be the early warning of a potential blowout as gases below displaced mud out of the wellbore, a characteristic known as a 'kick'.

So, I innocently asked if I could see this alarm. 'Why, did you not hear it' the drillers replied, 'it was sounding throughout your tour'. I then asked if they were aware of the dangers of alarm overload. Operators can become blind to the importance of alarms if they are constantly repeated and mixed with non-critical alarms. The risks of alarm overload are very well established and covered extensively by recognised sources of such as:

- International Society for Automation (ISA) issued standard ANSI/ISA-18.2-2009, "Management of Alarm Systems for Process Industries".
- Engineering Equipment and Materials Users' Association (EEMUA) standard 191 "Alarm Systems: A Guide to Design, Management and Procurement".

The International Electrotechnical Commission (IEC) – alarm management standard IEC-62682.

See also alarm overload comments in Chapter 21.

Astonished, my drilling companions explained that such mud tank level alarms were common to all drilling rigs; their alarm was no different. It was required to alarm loudly throughout the rig so that both the driller and the roustabouts could hear it.

So, every drilling rig has a critical alarm that typically sounds at frequencies likely to cause alarm overload. This practice is in contravention of accepted international standards. I challenge the oil and gas industry to recognise the implications of this well-known contributor to catastrophic events. How long before a blowout occurs with the incident investigation concluding that, with this level of alarms, it was 'an accident waiting to happen'.

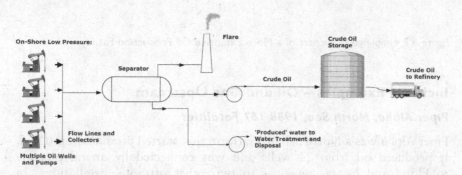

Figure 6.1 Simplified Flowchart of an Onshore Oil Production Facility.

From Figure 6.1, we can deduce that this oil field is probably 'sweet', i.e. low in hydrogen sulphide since the off-gas is being sent straight to flare where it is burnt as waste. The oil field is low pressure, meaning that 'nodding donkeys' or pump jacks are required to lift the oil in the well. The oil field is lean in gas, making gas recovery uneconomic, hence the flaring of the off-gas.

In Figure 6.2, a more complex oil production facility is shown. This field is higher pressure with pressurised well heads delivering oil from two different reservoirs feeding a separator. We can deduce that this is a sour oil field since there are H2S removal processes and sulphur biproduct for export. This is substantial gas content. The off-gas is treated and purified, forming a gas stream which is sold through the natural gas pipeline shown. Liquified Petroleum Gases are separated and sold employing a rail tank car filling facility.

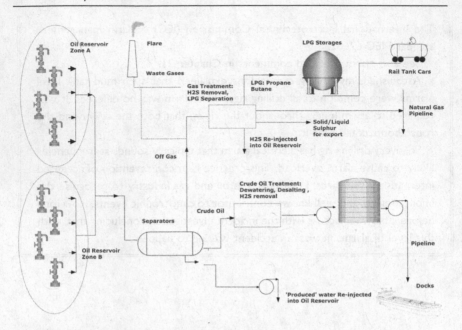

Figure 6.2 Simplified Flowchart of a More Extensive Oil Production Facility.

Incident Examples – Oil and Gas Upstream

Piper Alpha, North Sea, 1988 167 Fatalities

Piper Alpha was a North Sea oil platform that started production in 1976. It produced oil from 24 wells and was connected by an oil pipeline to Flotta and by gas pipelines to two other offshore installations. In July 1988, there was a massive leakage on Piper Alpha. This material ignited, causing an explosion which led to large oil fires. The heat from the burning oil and gas caused the rupture of one of the gas pipes exporting gas from another installation. A massive explosion and fireball then engulfed the Piper Alpha platform. All this took just 22 minutes. One hundred sixty-seven people died; 62 people survived (The Chemical Engineer (2018)).

The hazardous material involved is known as condensate, a liquid that, in some gas fields, comes along with the gas.

The inquiry concluded that the most likely cause of the first explosion was the release of a small amount of condensate over thirty seconds. This leak occurred from a pipe where a pressure safety relief valve had been removed as part of maintenance activity.

How did this happen? The initial cause involved two parallel pumps. Parallel pumps are quite common in industries where high reliability is needed. By installing two pumps in parallel, one pump can be taken off-line

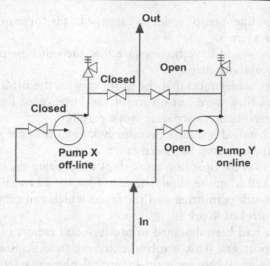

Figure 6.3 Parallel Pumps (Typical).

for maintenance without shutdown of the process. A typical installation of parallel pumps is shown in Figure 6.3. Pump X is shown as off-line and Pump Y as on-line.

Each of the pumps shown has a relief valve. Before the incident, the relief valve for the off-line pump had been removed for maintenance. It is understood that a blind flange had been installed temporarily in place of the valve, but it is assumed that this blind flange was not tightened up or pressure tested. See Figure 6.4.

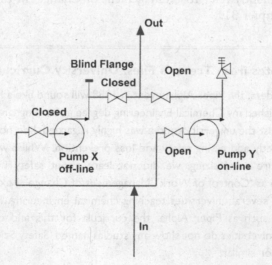

Figure 6.4 Parallel Pumps with Pressure Relief Valve Removed (Typical).

When the on-line pump suddenly stopped, the operators tried unsuccessfully to restart it.

The operators would have been aware that the other pump was out of commission for maintenance.

However, they understood that the problem with the off-line pump was not serious, but they were not aware of the absence of the relief valve, which was covered under a separate work permit.

The operators would not have necessarily known that the pressure relief valve for the off-line pump had been taken away.

It is believed that the operators took steps to reinstate the off-line pump. Condensate leaked from the blind flange. In Chapter 22, we will discuss the importance of work permitting and the faults which can occur with isolation under 'Control of Work'.

The platform had been designed to produce and export oil. It was later modified to export gas. This involved extensive modification of the platform. These modifications went on in several phases, starting with condensate separation and the production of gas of suitable quality for export via the gas pipeline, which was constructed to take gas from the platform.

The new facilities were located beside the control room. At this location, they were also under the electrical power, radio room, and accommodation modules. This is clearly a vulnerable area. When the condensate leaked, the vapour emitted found a source of ignition damage that occurred to these critical areas of the platform. The modifications do not seem to have been well thought through. (The Chemical Engineer, Loss Prevention Bulletin 261, June 2018, p. 3: Piper Alpha – What have we learned?)

We will frequently refer to 'Management of Change' in this book, particularly in Chapter 31.

Notes from Trevor's Files: University Curricula

To many readers, the Piper Alpha event in 1988 will sound like a long time in the past. I finished my Chemical Engineering degree at Birmingham University in 1974. Whilst the university course was highly regarded, I do not recall any lectures directly addressing safety and loss prevention. Whilst we did learn about pressure relief sizing, we did not learn about safety management systems such as 'Control of Work' 'Management of Change'. I examined the curricula for several universities teaching chemical engineering. Even today, after events such as Piper Alpha, the curricula for this and many other prestigious universities do not show any studies named 'Safety Science', 'Loss Prevention', or similar!

Deepwater Horizon, Gulf of Mexico, 2010, 11 Fatalities and Massive Environmental Issues

The Deepwater Horizon oil spill, also known as Macondo, was an industrial disaster that began on 20 April 2010, in the Gulf of Mexico. It is considered the largest marine oil spill in the history of the petroleum industry. The U.S. federal government estimated the total discharge at 4.9 million barrels (780,000 m^3) of oil, causing an environmental disaster. Eleven people died.

Much has been written about this disaster. The US Chemical Safety and Hazard Investigation Board investigated the incident. Their detailed report extends to some 525 pages distributed in 4 Volumes. Andrew Hopkins wrote 190 pages on the incident in 'Disastrous Decisions: the Human and Organisational Causes of the Gulf of Mexico Blowout'. A screenplay of over 100 minutes duration was produced, which covered, with some accuracy, around 24 hours of the most dramatic part of the tragedy.

In the pages that follow, I give a summary of the disaster and will refer to it frequently throughout the book. Learnings from the incident cover almost every aspect of safety management. Those learnings are applicable in virtually every sector of industry. It is important for the reader to have an open mind to the views of this author when reading about the incident.

Firstly, in my opinion, this is not a disaster caused by a 'rogue company' intent on cost-cutting at whatever expense to safety. I have done work, in the same era, with many of the major oil companies, including BP, the owner of the Macondo Prospect. From my sample, BP were at least as intense and genuine about safety as the other companies I worked with. Did they pursue cost reduction? For sure – and so does every commercial organisation. Were they looking for creative ways to become more efficient? Of course. It is what we expect. Did they do this at the expense of safety? I would suggest that their approach was on a par with other companies.

Secondly, this is not a disaster caused uniquely by deepwater drilling. This incident, perhaps on a somewhat smaller scale, could have occurred in any offshore oil platform and indeed on onshore wells also. Most of the operations and safety management procedures are just as required in a shallow-water rig. The location of the platform was not particularly arduous. Its location, just 41 miles off the coast of Louisiana, is just a short helicopter ride compared with journeys in the North Sea, some of which are over 100 miles. However, the Deepwater Horizon was certainly in deepwater – around 5000 ft (1500 metres), whereas a typical North Sea platform would be in approximately 500 ft (150 metres) or less.

Finally, this is not a disaster unique to oil and gas exploration and production. Major safety-critical pieces of equipment failed to perform as expected. Many aspects of the safety management system failed. It is disaster relevant to any organisation which has safety-critical equipment and relies on a safety management system to prevent catastrophes. The CSB in their investigation report writes, in the context of the problems with the BOP "The CSB provides its failure analysis of the BOP to spark a global re-examination of how industry is managing safety-critical elements as well as regulatory requirements and approaches used to ensure that these management practices are effective".

Safety critical elements can comprise hardware, people systems or software, or tasks whose failure could cause or contribute to a significant accident or is in place to limit the consequences of the accident.

The cost to BP in court fees, penalties, and clean-up costs has exceeded USD 60,000 million. Financial losses connected with the incident continued to impact BP over ten years later.

This event was a blowout during exploration in deepwater. After several failed efforts to contain the flow, the well was declared sealed on 19 September 2010. The Deepwater Horizon oil spill is regarded as one of the largest environmental disasters in history.

One of the causes of the incident was the failure of the Blow Out Preventer, which was located on the seafloor, to properly activate and shear through the rods which were within the borehole for drilling activities. Oil continued to flow from the seafloor for 87 days before the well was finally capped.

On the day of the incident, a team of Senior Managers landed on the rig to congratulate the crew on an excellent safety performance. This may sound ironic, but the management was concerned about occupational safety causing slips, trips, and falls. Process safety was not their focus. We will come across this issue of repeatedly mistaking good performance in occupational safety with process safety elsewhere in this book.

Hopkins (2012) points out that, whilst there was much focus on the BOP, this tool is only the last line of defence in preventing a blowout. There are many barriers preventing such a catastrophe (See Chapter 4), only one of which is the BOP. However, in detailing the barriers which failed, we also need to understand why they failed.

Notes from Trevor's Files: Blame

Unfortunately, there is a psychological reaction to a catastrophe of 'pinning the blame' onto someone or some organisation. Part of the reason for such a reaction to a disaster is stress relief. If we can attribute the problem to the organisation where the catastrophe occurred, we are relieved to conclude that it will not happen in other organisations, especially our own.

So, there are people who perceive BP to have been some kind of rogue organisation. Therefore, those who do so are relieved to conclude that such a tragic event would not happen in other oil companies with which they are familiar. It is also important to note that BP's subcontractors involved in the

> incident – Transocean and Halliburton – are experts in the field. I have also
> worked with them, and I have seen their dedication to safety.
>
> I am fortunate to have worked as a safety consultant and separately as an
> insurance risk engineer with most oil majors, both onshore and offshore. My
> perception is that BP was at least as dedicated to safety as the others, if not
> more so. Attribution of the incident to some rogue characteristic of BP does
> not help us learn.

The staff involved in drilling operations aboard the platform comprised (Table 6.1).

There were 126 people on board employed by 13 different companies. These included maintenance personnel, crane operators, catering, and laundry.

(U.S., Chemical Safety and Hazard Investigation Board – Investigation Report – Drilling Rig Explosion and Fire at the Macondo Well, Vol 1, p. 15).

It is not unusual for a drilling rig where many different contractors are involved in providing their own specialist services. Managing drilling operations is a serious contractor management challenge. We will come back to the issue of managing contractors in Chapter 20 on Human Factors. Post-incident, there was much focus on which contractor was responsible for what, since extensive legal cases and vast fines were to follow.

However, there were many others involved, most notably support staff to BP, Transocean, and Halliburton onshore. Unfortunately, the screenplay makes little reference to the importance of the onshore staff in the decision-making process, although, for the 24 hrs covered in the screenplay, apparently, there was little consultation with onshore.

> Whilst the screenplay is a reasonably fair representation of the disaster over the period covered, there are inaccuracies done for effect. The screenplay shows bubbles of gas coming from the seabed, as would be the case if there was a breach and there were fissures between the oil reservoir and the sea bed. This was not the case.

Table 6.1 Staff Involved in Drilling Activities on Deepwater Horizon

Position	Employer	Number
Well Site Leader	BP	2
Offshore Installation Manager (OIM)	Transocean	1
Drillers and Toolpushers	Transocean	9
Subsea Supervisor	Transocean	2
Cementer/Mudlogger	Halliburton/Sperry Sun	4

First, for the reader unfamiliar with oil well basics, just a little of the technology needs to be described. A drilling rig is used to drill a hole to determine if oil or gas are present at depths expected based on a geological study. The rig does not produce oil. The drilling platform is connected to the seafloor and from the seafloor to the bottom of the well by a long tube. Inside the tube another tube, typically the drill pipe, which is inserted to continue drilling down towards the target depth.

The drill pipe is inside the outer pipe and forms an annulus. A special mud is circulated, pumping down inside the drill pipe and back up in the annulus between the drill pipe and the outer pipe. Above the seafloor, this outer pipe is called a riser. Below the seafloor, it is called casing. Periodically, after drilling a section of the well, the drill pipe is removed, and a new section of casing is inserted into the freshly drilled wellbore. A highly simplified diagram is shown in Figure 6.5.

These casings decrease in size for mechanical reasons, with the largest being immediately below the seafloor (for simplicity, this detail is not shown in the diagram). At the time of the disaster, the well had been drilled,

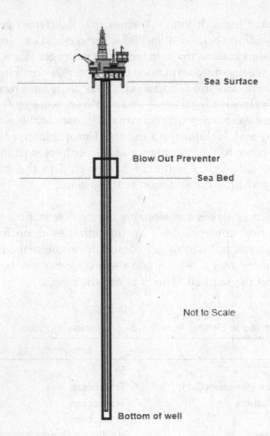

Figure 6.5 Drilling Basics.

oil had been found, and it was time to seal off the well. The drilling rig had another assignment to go to elsewhere in the gulf. Production of oil from the Macondo well would take place sometime later using different equipment. Therefore, the bottom of the well, which has no casing, had to be sealed with cement.

This is a regular operation in the temporary abandonment of wells. However, there are some obvious difficulties that are accentuated by deepwater operations. Imagine two straws, one a larger diameter than the other. Put the smaller straw inside the larger, to represent the drill pipe. Tie off the bottom of the larger straw to simulate the bottom of the well. If you now put some water into the inside pipe and pressurise it gently by blowing, the water will go down the inside straw and back up the larger straw. This represents mud circulation. Now put a blob of honey in the inside straw, followed by water, and see if you can blow just enough to have the honey settle out at the bottom of the two straws? Keep this picture in mind as we discuss the various events of the catastrophe.

As was well recognised, the completion of the well was behind schedule. There were undoubtedly managerial pressures to get the job done. What is not recognised in the descriptions of the tragedy that I have read so far is the pressure and desires of the personnel on the rig. My experience of many short trips offshore is that people want to get back home to their families as soon as possible. Life aboard a platform has many social deprivations. The people on board, even those who were fresh into a new rotation, probably wanted to get home at the earliest opportunity.

Cementing a well is a highly technical operation. Remember that the well is full of mud, which is holding back the pressure of the oil in the formation at the bottom of the wellbore. It is necessary to place the cement at the bottom of the well. The well is circulated with mud being pumped down through the inner casing and back up through the wall between the inner casing and the wall of the well. Sufficient cement is then pumped in to form the cement plug, followed by mud, and circulation continued until the cement plug is in place. Success in cementing is primarily judged by achieving 'full returns'. Completion of full returns is concluded when as much mud is returned as cement is pumped in. The engineers overseeing the cementing job saw 'full returns' and concluded that the cement plug had been successfully placed. The onshore team supported this conclusion based on the data received.

'Full returns' assumes that the cement had moved uniformly. From an outsiders perspective, I think it is easy to see that this is not necessarily the case. Channelling of the cement can occur on its long journey to the foot of the well. This particular cement had been formed by a relatively new process of foaming the cement with nitrogen which may or may not has made the cement more prone to channelling, or the foam might have destabilised. The purpose of the innovation was to reduce cementing time. Those making the decisions in the determination of the success of the cementing operation, together with those involved in judging the success of the well integrity test, were not informed as to experience so far in testing the nitrogen foamed cement. It had not in fact been proven under the conditions pertaining to the Macondo well. This failure is an example where Management of Change (see Chapter 31) was not correctly handled and points to another frequent cause of catastrophes.

There are also questions about the use of centralisers that keep the drill pipe centred within the wellbore when applying the cement. A decision had been made to use six centralisers versus the 21 originally planned. If the bottom of the well is 'offset' due to not being central, then uneven cement flow could result. However, there was evidence from previous operations that six centralisers were sufficient.

The engineers were affected by 'Confirmation Bias'. We will come back to Confirmation Bias decision-making failure many times in this book (see Chapter 20 on Human Factors). Confirmation bias is the tendency to search for, interpret, favour, and recall information in a way that confirms or supports one's prior beliefs or values. People tend to unconsciously select information that supports their views. They tend to ignore information that is not supportive of their opinion. People also tend to interpret ambiguous evidence as supporting their existing position. The effect is strongest for desired outcomes, emotionally charged issues, and deeply entrenched beliefs (Wikipedia: Confirmation Bias).

However, there is more to this incident than the well engineers' belief in success. They also knew that the well would be tested at the next stage. This test, the well integrity test, would show any deficiency in the cement job and detect other aspects regarding the integrity of the well.

The well integrity test was carried out by different personnel. Briefly, a well integrity test looks for the absence of a pressure rise when the well is not pressurised with mud. Remember that it is mud that was holding back the oil at the bottom of the well. When mud is partially replaced by lighter material, no pressure should be registered in the drill pipe since the bottom of the well is sealed, or should have been sealed, by the cement. However, the test did show a pressure rise. The rig personnel simply did not believe it! I have read many accounts of the integrity test, and so far, I have not rationalised why the first well integrity test was not considered valid. The best explanation I have is that the engineers thought that such pressure must have been somehow trapped. None of the experienced personnel on board had experienced a failed a well integrity test. The decision tree, which was available to them as a procedure on the test had no guidance on well integrity test failure. A further test was done using a smaller pipe connected at the BOP, known as the kill line. This showed no pressure rise. This result was chosen to be the correct one.

Following the incident, the question of the competence of people carrying out the test was raised. We will come back to competence as a means of failure elsewhere in this book. Drilling personnel are well trained, usually carry certificates verifying their highly technical training, and in my experience, are very smart. They are selected because of their technical skills and experience. I prefer the explanation that those carrying out the test had an unclear mental model of what was going on with regards to the hydrostatics of the test. The hydrostatics have been grossly simplified in this account. The CSB report includes an Appendix that describes in detail the considerations of the pressure test, and it is indeed a highly complicated affair. Other psychological factors such as 'risky shift and groupthink' also played a part (see Chapter 20 on Human Factors). It has been experimentally proven that groups are often more inclined to make risky decisions than individual group members would make when acting alone. 'Risky shift' can occur in environments where consensus management is promoted, as was the case here. Consultation with the onshore team was not apparently required at this stage.

At each step, however, those involved were partially reassured that there was a subsequent step where any failure would be detected. On the one hand, I would comment that many barriers to a catastrophe are good. This is known as defence-in-depth. However, here is the disadvantage that people involved at any particular stage develop a false sense of security about the reliability of subsequent barriers.

Well monitoring is carried out to assess that the amount of mud going into the well matches the amount returning through the annulus. See Notes from Trevor's Files on Mud Tank Alarms above for some further explanation. If the flows do not match, this indicates a failure with mud either being lost into the oil-bearing formation below or fluids (most likely oil and gas) entering the well. However, the rig was preparing to be moved to another location. Some of the mud and water used in the well integrity test was being offloaded into a ship waiting alongside. This operation meant that the all-important mud tank monitoring could not take place.

The above factors deprived the well abandonment process of three critical aspects:

- Cement plug test incorrectly judged successful
- Well integrity test incorrectly judged successful
- Well monitoring not possible

Oil and gas were indeed flowing into the well and later blew out onto the rig with the consequent disastrous explosion.

But there was another, final, line of defence – the Blow Out Preventer. There is certainly a suggestion that rig personnel and onshore engineers considered the BOP a highly reliable device.

However, some research done, through Transocean, had suggested that blowout preventers on deepwater rigs had a failure rate of 45%

(Hopkins, 2012, p. 59). It was also known that the BOP could shear only through normal straight drill pipe. Drill pipes must be joined and at the joint the metal can be twice the thickness. Furthermore, a BOP was not designed to deal with a blowout that was already at full throttle. The design assumption was that the crew would be monitoring the well at all times and that they would quickly recognise when control of the well had been lost and activate the BOP before full throttle blowout was occurring.

In fact, a BOP is not one but several barriers. For the Deepwater Horizon BOP there were the following means of arresting flow:

- Annular preventers (2), a rubber component designed to seal the annulus, the void between the drill pipe and the outer pipe.
- Pipe Rams (3), also a rubber component. Each ram was capable of sealing around particular diameter of drill pipe.
- Blind shear ram (1) designed to shear through the drill pipe if a drill pipe is in the hole (which was the case at the time of the incident)
- Casing shear ram (1) designed to shear through a section of casing is casing is being run. Whilst they are stronger, they cannot seal the well.

I understand that a Blow Out Preventer is not designed to stop a Blowout whilst it is in progress. The velocity and pressures of the oil are too much. Once hydrocarbons were detected on the rig floor, then annual preventers were first actuated but failed to stop the flow, probably due to erosion of the rubber components.

They then actuated the pipe ram, which did stop the flow temporarily. However, the Blind Shear Ram was then activated, either manually or automatically once the fire occurred on the platform. The BOP has an automatic system which, given loss of contact with the rig due to fire, rupture of the connecting cables, etc., the BOP Blind Shear Ram will automatically be activated. The Blind Shear ram was not rated to cut through the size of drill pipe in use. Furthermore, extensive examination of the BOP once it was retrieved from the seabed showed that the drill pipe was offset, bent into a position such that the Blind Shear Ram could not cut through it. The Blind Shear Ram did puncture the drill pipe accentuating the subsequent loss to the environment and may have damaged the seal achieved by the pipe ram. Furthermore, the CSB found that the BOP automatic activation system had been incorrectly wired.

This raises a critical question with regards to the operation and testing of safety-critical equipment. The BOP is just a representative of the many types of safety-critical equipment we will examine in many industries discussed in this book.

Once the hydrocarbon kick occurred, the rig operators had the opportunity to divert the uncontrollable flow overboard, reducing the chance of an onboard explosion. But they did not. Remembering that they had very little decision-making time, they channelled the flow to a mud/gas separator which, it was found, did not have the capacity for the volume of hydrocarbons being ejected from the well. I can sympathise with a reluctance to discharge the flow overboard, given that they wanted to prevent an environmental spill, and had no time to really assess that a spill of some size was now unavoidable. Had they done so it is likely that the size of the ultimate spill would have been considerably less.

We should also mention here that the emergency response plans devised in advance for the Macondo operation were inadequate. They would appear mainly written to satisfy the needs of the regulators. No blowout was conceived to be possible, and in any case, the rig was so far from the shore that any effect was judged not to be significant.

Eventually, a relief well was successfully intercepted the Deepwater Horizon well some five months later. The well was killed with heavy mud and the well capped. It is a sobering thought that if the relief well had missed its target, the spill could have continued even longer. In the next case study (Montara, see below) five well interceptions were attempted before success was achieved.

If we attempt to summarise this in a Bow Tie as described in Chapter 3, we can see the line-up of barriers which failed and below them the 'degradation' mechanism which caused them to fail.

This diagram is somewhat different from the diagram proposed by the CSB. I show only one threat and one outcome, strictly relating to this specific tragedy. Other threats and consequences should be shown in a complete analysis (Figures 6.6 and 6.7).

Referring to the chapters which follow in Section 3 on Catastrophe Prevention we see the above failings in the following as aspects of safety management (Table 6.2).

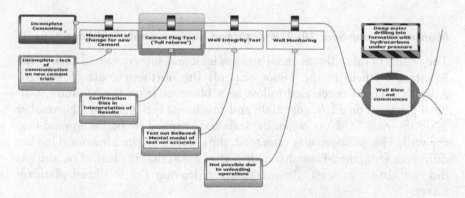

Figure 6.6 Deepwater Horizon Blow Out – Left-Hand Side of Bow Tie.

Source: Created using BowTieXP Software © CGE Risk Management Solutions B.V. 2021. BowTieXP is a registered trademark of CGE Risk Management Solutions B.V.

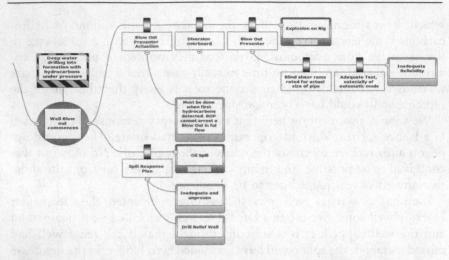

Figure 6.7 Deepwater Horizon Blow Out – Right-Hand Side of Bow Tie.

Source: Created using BowTieXP Software © CGE Risk Management Solutions B.V. 2021. BowTieXP is a registered trademark of CGE Risk Management Solutions B.V.

Note: Some commentators include the absence of a cement bond log as another ignored barrier. A cement bond log is an acoustic method in which an instrument is lowered into the well by a wireline to test the cement integrity. Hopkins (2012) points out that the cement bond log was only required if there were reasons to suspect failure of the cement, and since there were none, it was reasonable for the cement bond log to be omitted. The cement bond log is therefore not another independent barrier. A report by the National Academy of Engineering and National Research Council of the National Academies: Macondo Well: Deepwater Horizon Blowout: Lessons for Improving Offshore Drilling Safety also suggests that the Cement Bond Log would not have been able to extend to the depth required to identify problems with the cement.

Montara, Timor Sea, 2009, widespread environmental damage

The Montara oil spill was an oil and gas leak and subsequent oil slick in the Montara oil field in the Timor Sea, off the northern coast of Western Australia. The was released following a blowout from the Montara wellhead platform on 21 August 2009 and continued leaking until 3 November 2009 (in total 75 days), when the leak was stopped by pumping mud into the well. The wellbore was cemented, thus 'capping' the blowout. One big difference from the Macondo incident above was that the leaked oil and gas did not ignite, at least not immediately, leaving the wellhead platform intact.

Another difference, perhaps, is that the driller – PTTEP Australasia – is not well known! There was much-reduced media attention compared to Deepwater Horizon.

Table 6.2 Safety Management Failures (Deepwater Horizon Incident)

Aspect of Safety Management	Evidence of	See Chapter
Human Factors	Confirmation Bias, Risky Shift	20
Equipment Suitability and Reliability	Reliability of Blow Out Preventer Suitability of Centralisers	21
Operators	Routing of kick fluids to mud separator	19
Personnel Selection and Competence	Flawed decision making of crew, and decision makers on board and on shore False mental model of hydrostatics	20
Contractor Selection and Control of Contractors	Part played by contracted drilling and cementing services	20
Equipment Maintenance and Inspection	Lack of confidence in instruments	22
Assessing Risk Outcomes and Risk Reduction	BOP not treated as a safety-critical piece of equipment, especially with regards to automatic operation.	27
Emergency Response and Readiness	Flawed response to well kick and blow out Oil spill response plan incomplete	23
Governance of Safety	Lack of emphasis on process safety	31
Laws, Regulations, Standards	Insufficient evaluation by authorities of spill response	32

The blowout resulted in oil leaking until after the fifth attempt to plug the well from a remote platform. There was widespread environmental damage from the oil released.

The Australian government initiated a Commission of Enquiry. Many of the conclusions were similar to those of the Deepwater Horizon incident above, which occurred less than a year later. It is, therefore, significant that learnings from Montara had not been assimilated by the Deepwater Horizon engineers. The commission concluded that if the drilling company had followed the local authority's regulations or had adhered to its own well control standards, the incident, most likely, would not have occurred. (Wikipedia: Montara Oil Spill). The Australian Commission of Inquiry found that at the time, not one well control barrier complied with the drillers own Well Construction Standards (or, importantly, with sensible oilfield practice). Relevantly, a critical component, the casing shoe, which is positioned at the very bottom of the well, had not been pressure tested in accordance with the company's Well Construction Standards. In particular, the cement in the casing shoe was likely to have been defective. Multiple problems in undertaking the cement job should have caused a re-evaluation of the risks and the forward plan. These problems should have initiated a careful evaluation of what happened, the investigation of pressure testing and, most likely, remedial action. No such assessment was undertaken. The problems were not complicated or unsolvable, and the commission said the potential remedies were well known and not costly. Compounding the initial cementing problem was that, while the drilling operator chose two secondary well control barriers – pressure containing anti-corrosion caps – were programmed for installation, only one was ever installed.

It seems that the two-barrier principle was not fully implemented, and the leak occurred when the casing of a well failed.

Notes from Trevor's Files: Incident Sharing

It is recognised good practice to share incident information, as relevant, amongst an organisation and for any learnings considered and applied. In 2014 I visited an oil and gas production operation in the Middle East and asked about any learnings they had gleaned from the Montara incident. I was met by blank faces. Perhaps it was not considered relevant because the Montara was offshore, whereas the organisation I was visiting was onshore. Perhaps because PTTEP Australasia is not a well-recognised oil and gas company. It is all too easy to see the mistakes at both Montara and Macondo as being so 'stupid' as to be not possible in one's own organisation. This surely is a mistake!

Once the oil and gas are extracted, the crude oil and gas need to be transported to the facilities which use them. This aspect is covered in subsequent sections of this book in Chapter 9 on Marine Transport and in Chapter 13 on Process Industry infrastructure.

OIL AND GAS DOWNSTREAM (Taking the Crude Oil and Gas and Turning it into Commercial Products)

The oil and gas extracted from the wells are transported to an oil refinery or a gas plant. The materials are then processed into either fuels and/or materials used in the chemical industry.

Let's consider first the more straightforward process of gas purification. Natural gas is primarily methane. Gas from gas-only wells is typically wet and often contains valuable chemical components in addition to methane. It may also contain highly toxic hydrogen sulphide and mercury contaminants. The gas will be typically low pressure and cannot be piped much distance without compression. So, a typical gas plant dries the gas, removes toxic contaminants, and extracts high-value chemical components. The configuration required will depend on the geological make-up at the origin of the gas in the gas well. Some wells, such as the fracked gas wells of northern Pennsylvania, need to be dried and compressed. However, western

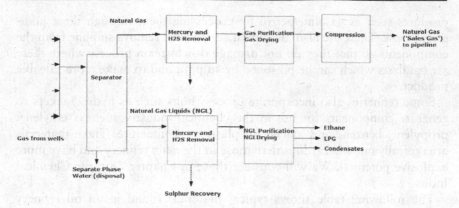

Figure 6.8 Schematic Diagram of a Gas Plant.

and southern Pennsylvania wells contain considerably higher value components such as ethane and merit more intense purification (Figure 6.8).

The hazards of natural gas plants, which might create a catastrophic situation, are those associated with large volumes of compressed flammable gas. Temperatures and pressures are not exceptionally high. Refrigeration is sometimes required to achieve the required separation, so temperatures significantly below zero Centigrade are not uncommon in some process streams.

An oil refinery normally has two distinct parts. The first comprises crude oil distillation and vacuum distillation.

Furnaces characterise these portions of the plant with stacks and wide ventilation ducts, which carry away the exhaust gases. Significant fuel gas is burnt to heat the incoming oil in these furnaces. The associated distillation columns are relatively short and "dumpy" and easily spotted when you pass by a refinery. In this section of the refinery, the many components of the crude oil are processed and separated into crude components streams including:

- Light gases which are later processed into LPGs
- Naphthas, which are later processed into gasoline
- Heavy components which are later processed into fuel oil, or further 'cracked' to make lighter more valuable components.

The second part of the refinery processes separates the many components of these streams to provide saleable products. These portions of the plant are characterised by tall, relatively thin distillation columns around which materials are circulated using pumps and recycle compressors. There are also reformers, crackers, and cokers which use heat and catalysts to break down the long chain molecules. These large molecules would otherwise end up as bitumen. The refinery produces a higher yield of more valuable products such as gasoline/petrol by processing them through these additional steps. There are also 'desulphurisers' which remove sulphur from the components so that they do not damage downstream process where there are catalysts which can be poisoned by sulphur and to make more saleable product.

Some refineries also incorporate process units such as hydrocrackers to generate components for use in the chemical industry such as ethylene, propylene, benzene and xylene for plastics manufacture. These materials are typically more hazardous than those in the main refinery and have more explosive potential. We will consider these in Chapter 7 on the Chemical Industry.

The following table shows typical materials found in an oil refinery processing (Table 6.3).

The auto-ignition temperature of all of these materials is between 250 and 600°C. So, in the event of a release impacting a very hot surface, a fire will occur. No spark is needed.

Ignition sources are controlled within a refinery, for example, smoking, use of static producing equipment, control of hot work, etc. (see Control of Work in Chapter 22).

Other materials are typically found in waste streams or as materials imported for processing components (Table 6.4).

The hydrotreaters and crackers use solid catalysts, typically in powder or pellet form. A catalyst is a material that enables a particular chemical reaction, but which is not itself consumed in the reaction. However, a catalyst degrades or becomes poisoned over time and needs to be replaced. The selection of catalysts is specific to each refinery. Many are not considered toxic but can release irritant dust containing trace toxic metals in the event of a fire or explosion. Catalysts used in a process called the Fluidised Catalytic Cracker (FCC) are relatively non-hazardous. In contrast, catalysts in Continuous Catalytic Reformers typically include platinum or rhenium metals.

Catalysts used in a process called hydrotreating, where hydrogen sulphide is removed, are typically Cobalt-Molybdenum catalysts (CoMo). Molybdenum is toxic, especially to ruminants, such as cattle and sheep. In the event of an environmental release, there is a concern that animal fodder with more than ten ppm of molybdenum would put most livestock at risk. Land used for ruminant farming around the refinery perimeter have where CoMo is used should ask about catalyst containment. What if the vessels containing CoMo would be over pressurised, do the pressure relief valves release to the atmosphere or via a separation vessel? The cost of baseline molybdenum measurements of the farmland and animals may be justified (Metalpedia).

Oil refineries vary enormously in complexity. The following schematic shows a relatively simple oil refinery. Most of the components shown are to be found at all refineries. An exception is the Coker, which might be

Table 6.3 Typical Materials to Be Found in an Oil Refinery

Material	Found in	Physical Properties	Comment on Fire or Toxicity Hazard
Crude Oil	Feed to the Crude Column	Liquid. Flash point below ambient.	Highly flammable: easily ignited by heat, sparks, or flames
Naphtha	An intermediate Crude Column Product, sometimes shipped to other refineries for further processing	Liquid. Flash point below Ambient.	Highly flammable: easily ignited by heat, sparks, or flames
Liquefied Petroleum Gases including Butane and Propane	Product of crude oil distillation and cracking processes. Further purified, then stored in pressurised spheres or bullets for export and sale	Gasses at atmospheric pressure. Flash point below Ambient.	Highly flammable: easily ignited by heat, sparks, or flames. Vapours heavier than air.
Hydrogen	Hydrogen manufacturing plant. By-product of reformer, used in hydrocracking.	Gas. Very low Minimum Ignition Energy	Highly flammable: easily ignited by heat, sparks, or flames. Vapours lighter than air.
Methane and fuel gas	Product of crude oil distillation. Normally burnt as fuel within the refinery,	Gasses at atmospheric pressure. Flash point below Ambient.	Highly flammable: easily ignited by heat, sparks, or flames. Vapours lighter than air.
Vacuum Gas Oil (VGO)	An intermediate heavy Crude Column Product, usually processed in a Coker of Fluid Catalytic Cracker	Combustible Class III liquid. Flash point above ambient.	Often surface temperatures of equipment processing VGO are above the Auto Ignition Temperature, so a leak of VGO spraying onto hot equipment is likely to result in a fire.
Gasoline/petrol	A product from refining used primarily as a fuel for vehicles	Flammable Class I B liquid. Flash point below ambient.	Can form explosive mixtures with air, especially if dispersed in droplet form.
Kerosene, diesel, jet fuel	A product from refining used primarily as a fuel for vehicles, locomotives, and airplanes'	Combustible Class II liquid. Flash point above ambient.	Safer from a fire perspective than gasoline, due to much higher flash point.
Fuel oils	A product from refining used primarily as a heating oil, also used as a fuel in	Combustible Class II or III liquids. Flash point significantly above ambient.	Hard to ignite, but as with VGO will can be a fire hazard if sprayed onto hot surfaces.
Methyl tert Butyl Ether (MTBE)	A material sometimes manufactured within the refinery for blending into gasoline. It is an anti-knock agent	Flammable Class I B. Flash point below ambient	Can form explosive mixtures with air, especially if dispersed in droplet form
Ethyl tert Butyl Ether (ETBE)	A material sometimes manufactured within the refinery for blending into gasoline. It is an anti-knock agent	Flammable Class I B. Flash point below ambient	Can form explosive mixtures with air, especially if dispersed in droplet form

Table 6.4 Other Materials to Be Found in Oil Refineries

Material	Found in	Physical Properties	Comment on Fire or Toxicity Hazard
Hydrogen Sulphide	Crude oil and oil refinery processing streams. It is an undesired component found in many crude oil sources	Gas	Highly toxic
Hydrofluoric acid (HF)	Sometimes used in process units known as Alkylation of "Alky"	Gas or liquid	Highly toxic and corrosive
Sulphuric Acid	Sometimes used in process units known as Alkylation of "Alky"	Liquid	Toxic and corrosive

replaced by other means of upgrading the heavy ends from the vacuum tower, such as Fluid Catalytic Cracker or Visbreaker (Figure 6.9).

The incident record of the oil and gas downstream industry has not been good in recent years. Marsh '100 Largest Losses in the Hydrocarbon Industry' 26th Edition 1974–2019 notes that, in terms of financial loss 'the past two years have been another turbulent period for the energy industry, with an unusually high number of large losses'. In the following pages, four major incidents are examined. The first is at a gas plant at Longford in Australia.

For Liquified Natural Gas see Chapter 13 on process safety infrastructure.

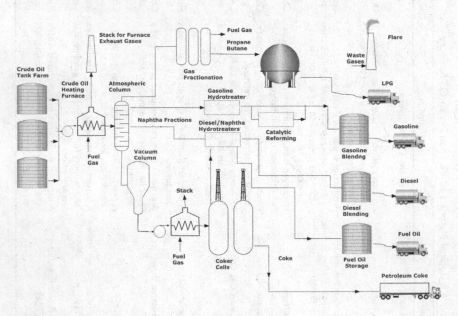

Figure 6.9 Simplified Flow Diagram of an Oil Refinery.

Incident Examples – Oil and Gas Downstream

Exxon Longford 1998 – Fire and Explosion with 2 Fatalities

Gas supplies to Australia's Victoria State were virtually shut down following an explosion and fire at this gas processing plant in 1998. The incident was caused by the rupture of a heat exchanger. This followed a process upset that was set in motion by the unintended, sudden shutdown of hot oil pumps. The loss of hot oil supply allowed some vessels to be chilled by cold oil, and when the hot oil was re-introduced to the heat exchanger, the vessel ruptured due to a brittle fracture. The release of gas exploded and continued to burn as a jet fire. The fire burned for two and a half days. The Longford Commission was set up to investigate. Operator error and improper training of employees were two of the major failures. A total of five explosions ripped through the gas plant. One hundred twenty workers were evacuated from the site, and police evacuated houses within a five-kilometre radius. The gas outage affected 1.4 million users state-wide and forced small and large businesses to shut down temporarily. For further explanation, see Hopkins excellent book (Hopkins (2000) *Lessons from Longford*).

BP Texas City 2005 – Fire and Explosion with 15 Fatalities

The incident occurred during the start-up of a distillation unit when the distillation tower was overfilled (CSB 2007, Refinery Explosion and Fire – BP Texas City). Pressure relief devices opened, resulting in a flammable liquid geyser from a blowdown stack that was not equipped with a flare. The release of flammables led to an explosion and fire. All fatalities occurred in or near temporary office trailers near the blowdown drum. A shelter-in-place order was issued that required 43,000 people in the surrounding communities to remain indoors. Houses were damaged as far away as three-quarters of a mile from the refinery. For further explanation, see Hopkins subsequent book (Hopkins (2008) *Failure to Learn*).

Underlying factors included:

- The distillation tower level indicator showed that the tower level was declining when in fact it was overfilling.
- Lack of supervisory oversight and technically trained personnel during the start-up.
- Operator fatigue from working 12 hr shifts for 29 or more consecutive days.
- Occupied trailers were sited too close to process units handling hazardous materials.

Note that the blowdown drum discharged to the atmosphere, not to a flare stack where it might have safely been burnt, depending on the capacity of the flare.

Chevron Richmond, CA 2012 – Fire and Explosion. Minor Injuries

The incident at the Chevron Refinery in Richmond, California, experienced a catastrophic pipe rupture of one of the Crude Oil distillation columns (CSB 2015 Chevron Richmond Refinery Fire). There were no fatalities. However, this incident has many valuable learnings, as well as having the potential to have caused more severe harm to people.

The rupture occurred in a pipework section referred to as the sidecut stream. The particular pipe was 8" in diameter and 52-inch long. The ruptured pipe released flammable light gas oil, which partially vaporised into a large, opaque vapour cloud engulfing employees. Eighteen of the employees safely escaped from the vapour cloud just before ignition; one employee, a refinery firefighter, was inside a fire engine that was caught within the fireball when the process fluid ignited. Because he was wearing full-body fire-fighting protective equipment, he could make his way through the flames to safety. Six employees suffered minor injuries during the incident and subsequent emergency response efforts.

The release, ignition, and subsequent burning of the hydrocarbon process fluid resulted in a large plume of vapour, particulates, and black smoke, which travelled across the surrounding area. This chain of events resulted in a Community Warning System alert. Surrounding communities were advised to stay indoors (shelter-in-place).

The rupture of the sidecut piping resulted from sulphidation corrosion. The component had become extremely thin. Sulphidation corrosion is a damage mechanism that causes thinning in iron-containing materials, such as steel, due to the reaction between sulphur compounds (which are found in crude oil) and iron at elevated temperatures. This damage mechanism causes pipe walls to gradually thin over time.

Virtually all crude oil feeds contain sulphur compounds; as a result, sulphidation corrosion is a damage mechanism present at every refinery that processes crude oil. Sulphidation corrosion can cause thinning to the point of pipe failure when not adequately monitored and controlled.

The sidecut piping involved was constructed of carbon steel. This metal is known to corrode at a much faster rate from sulphidation than other typical alternative materials of construction, such as higher chromium-containing steels. Furthermore, the vulnerability of a carbon steel pipe to sulphidation corrosion varies according to the silicon content. Carbon steel with a low silicon content being most vulnerable. The silicon content in carbon steel is variable. During the ten years prior to the incident, a small number of Chevron personnel with knowledge and understanding

Table 6.5 Safety Management Failures (Chevron Richmond Incident)

Aspect of Safety Management	Evidence of	See Chapter
Equipment Maintenance and Inspection	Assessment of criticality of pipe section replacement.	22
Equipment Suitability and Reliability	Use of Carbon Steel.	21
Emergency Response and Readiness	Emergency Response and Readiness – Misunderstanding of risks, working within hot zone	23
Governance of Safety	Consideration and prioritisation of inspection results and decisions relating to the replacement of the pipe	31

of sulphidation corrosion recommended on several occasions either a one-time inspection of every component within this particular sidecut piping circuit – known as 100% component inspection – or an upgrade of the material of construction of the sidecut piping. The recommendations were not implemented effectively, and the 52-inch component remained in service until it failed (U.S. Chemical Safety and Hazard Investigation Board 2015 Chevron Richmond Refinery Pipe Rupture and Fire).

Referring to the chapters which follow on Catastrophe Prevention, we see above failings in the following as aspects of safety management (Table 6.5).

Philadelphia Energy Solutions Refinery, 2019 – Fire and Explosion. Minor Injuries

The incident occurred in the refinery hydrofluoric acid alkylation unit. A ruptured pipework elbow is thought likely to be the initiating cause, but the incident is still under investigation at the time of writing. Fortunately, there were no injuries or fatalities and no toxic consequences to the neighbouring community. The explosion jettisoned several fragments, including a piece of metalwork weighing approximately 38,000 pounds. This fragment flew beyond the plant boundary and across an adjacent river. Approximately 5,200 pounds of hydrofluoric acid was released, of which about 2,000 pounds was contained by the water spray facility provided on the site as part of the unit emergency response provisions.

The potential for a much more severe community impact is clear. The nearest housing communities were less than 1 mile from the refinery and downtown Philadelphia less than 2 miles away. (U.S. Chemical Safety and Hazard Investigation Board (2019): Chemical Safety Board Releases Factual Update and New Animation Detailing the Events of the Massive Explosion and Fire at the PES Refinery in Philadelphia.)

CONCLUSIONS

In upstream oil and gas it seems, based on the above observations, there remains considerable potential for a major catastrophe, especially in the offshore industry. I don't see the issues raised by Deepwater Horizon and other incidents as being addressed.

Downstream operations have a more limited catastrophic potential. From the above observations and my experience in insurance work in the industry, I see that some progress is being made.

Upstream Oil and Gas – Potential for Future Catastrophes

Not possible	Unlikely	Possible (50/50)	Likely, no more than one	Very likely/ maybe several
☐	☐	☐	☐	☒

Downstream Oil and Gas – Potential for Future Catastrophes

Not possible	Unlikely	Possible (50/50)	Likely, no more than one	Very likely/ maybe several
☐	☐	☐	☒	☐

Chapter 7

Chemical Industry

BACKGROUND TO THE CHEMICAL INDUSTRY

Which chemicals? Indeed, the wide variety of chemicals produced makes this industry potentially significantly more hazardous than others. In order to assess this, the hazardous properties of each chemical produced, and the intermediates used to manufacture it, need to be understood. The technology used in manufacturing is typically very specific to the individual chemical, and any particular chemical may have only a few manufacturing plants globally. Compared with oil and gas the opportunities to share technology, including means to reduce risk, are somewhat reduced.

However, there are some common groups of chemical plants, and each would deserve a separate section in a more protracted publication. Some examples only are listed in Table 7.1.

The distribution of chemical sites listed as hazardous under the COMAH regulations, in terms of the number of installations in the U.K. is shown in Figure 7.1.

Unlike oil and gas, chemical manufacturing is diverse in terms of hazard processes and resultant risks. It is hard even to categorise, with 'Speciality', 63%, falling into no particular category.

Typical major hazards from such plants are fires, devastating explosions, and releases of highly toxic materials. Whereas catastrophic incidents in oil and gas upstream and downstream are confined mainly to destruction and loss of life within the plant boundary, major incidents in the chemical industry often extend to damage to the general public. This destructive power is due to the reactivity of materials used.

For those readers with a chemical background, oil and gas constituents are primarily saturated hydrocarbon chains. Unsaturated hydrocarbons are typically present in the products or process intermediates in the chemical industry. For the non-technical reader, it should be explained

DOI: 10.1201/9781003360759-9

Table 7.1 Chemical Industry – Types

Chemical Industry Group	Raw Materials	Products	Typical Product Uses	Occurrence
Petrochemicals	Petroleum derivatives – products of an oil refinery	Ethylene Propylene Vinyl Chloride Monomer Acrylonitrile Ammonia Ethylene Glycol Benzene, Toluene, Xylene	Polyethylene Polypropylene PVC Elastomers and textile fibres Fertilisers Pharmaceuticals	Commonly associated with an oil refinery and/or a polymer plant.
Cryogenic	Air	Oxygen, nitrogen, argon	Medical Chemical Industry Metallurgical	Located close to customers
Polymer Plants	Ethylene Propylene Vinyl Chloride Monomer Styrene	Polyethylene Polypropylene PVC Polystyrene	Packaging, Textiles	Commonly associated with a petrochemicals plant.
Ammonia and Fertilisers	Natural gas	Nitrogenous fertilisers including urea	Agricultural	Commonly associated with cheap and plentiful sources of natural gas
Household and Healthcare	Chlorine Salt Other Specialty Chemicals	Bleach Toothpaste Soap Cosmetics Detergents	General household and healthcare	Bleach-based industries are often close to sources of salt. Otherwise, the sites are not geographically restricted
Pharmaceuticals	Multiple pharmaceutical intermediates	Medicines	Medical and Health Care	Standalone can be small.
Textile Fibres	Acrylonitrile Ethylene Glycol Caprolactam	Acrylic Fibres, PET, Nylon	Textiles	Generally standalone factories
Elastomers	Butadiene Acrylonitrile Styrene	Synthetic Rubbers	Rubber products	Generally standalone factories
Speciality	Multiple	Multiple	Miscellaneous	Generally standalone

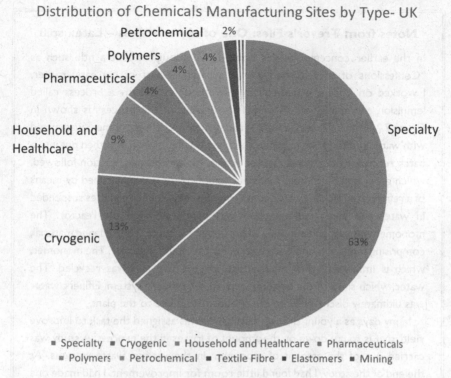

Distribution of Chemicals Manufacturing Sites by Type- UK

Petrochemical   2%
Polymers   4%
4%
Pharmaceuticals   4%
Household and Healthcare   9%
Specialty
Cryogenic   13%
63%

- Specialty - Cryogenic - Household and Healthcare - Pharmaceuticals
- Polymers - Petrochemical - Textile Fibre - Elastomers - Mining

Figure 7.1 Chemical Industry – Distribution in U.K.

that unsaturated carbons are much more reactive since they have one or more 'double bonds' which are always trying to find something to react with. Much of oil and gas processing involves separating hydrocarbon mixtures and eliminating undesired components. In contrast, chemical reactions are used in the chemical industry to produce new chemical molecules.

Highly toxic reagents are sometimes used in the chemical industry, presenting a more significant hazard for catastrophic effects extending beyond the fence line.

Chemical industry is often located close to the rivers, tidal estuaries, or the sea. The processes often require water. This water is not just for cooling; it is often mixed with the process chemicals to achieve the chemical reactions and quality of product required. After playing its part in the manufacturing process the water is then separated from residual chemicals and discharged into the plant's effluent stream, where it re-joins the river or tidal estuary. Complete separation of the chemicals from the water is rarely possible, so some contamination of the effluent water is normal, and subject to regulatory permitting.

Notes from Trevor's Files: One of my mistakes – Latex spill

In the earlier concepts for this book, I had thought about a title such as "Confessions of an Engineer"! For a substantial part of my early career, I worked on a plant manufacturing a synthetic rubber by a process called 'emulsion polymerisation'. A simplified diagram of the process is shown in Figure 7.2. In this process, a highly reactive chemical (a monomer) was mixed with water and soap-like materials to create an emulsion. It was then put into a batch reactor, agitated, and catalyst added. An exothermic reaction followed, which generated much heat. Constant temperature was maintained by means of a refrigerated jacket. A latex containing the small rubber particles suspended in water and unreacted monomer was produced from the reactor. The monomer was stripped out using steam under a vacuum. This stripped material, comprising monomer and water was collected in the decanter. The monomer, which is immiscible with water, floats to the top and was recycled. The water, which sinks to the bottom, went to the effluent system. Effluent water was ultimately discharged into the tidal estuary close to the plant.

In my days as a young process engineer, I was assigned the task to improve yield, that is to cut down on the amount of losses in the process. A study was carried out of the amount of monomer in the various waste streams. At the end of the study, I had found little room for improvement. I had made one vital mistake that had grave consequences some five years later. My mistake –

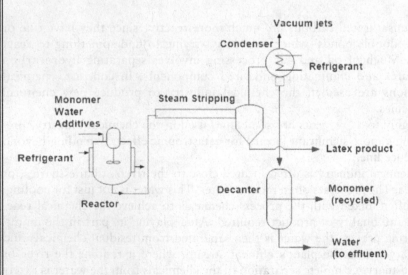

Figure 7.2 Latex Manufacture.

I had looked only at normal operation, not at the losses which can occur during abnormal operations.

After five years which included a totally different assignment in the U.S.A. in plant design, I returned to be plant manager of this same synthetic process.

One morning meeting, a laboratory result was reported to us, which showed very high levels of monomer in the combined effluent streams from the plant. The result we considered was just impossible. No one was aware of any process issues which might have caused such a loss. We just did not believe it! (Readers of Deepwater Horizon will recall how the drilling engineers just did not believe the negative pressure test – see Chapter 6). However, an engineer was assigned to investigate. Three days later, when I was having my lunch, the engineer passed by and announced that, according to his calculations, several tonnes of material had been lost. I recall vividly the shock at receiving this news. Of course, the authorities had to be informed. An in-depth investigation had to be concluded. The following months of my life were acutely stressful. As the manager of the plant, I was responsible, and my job was in jeopardy. I recall the upheaval to my personal life.

The investigation concluded that the condenser (see Figure 7.2) had frozen and been de-iced during the night preceding the high effluent analysis. This was a known procedure that occasionally had to be carried out. De-icing involved shutting down the stripping operation and applying hot water to the condenser jacket, an operation that required people to work on the building roof connecting hoses. An easier way was to maintain operation whilst temporarily shut off the refrigerant flow using the control panel. Whilst this was not in the operating procedures, it had, we suspect, occasionally been practised. So long as the duration of the shut-off is short, it probably worked fine, and I doubt that this was the first use of this unapproved procedure. However, we concluded that the refrigerant was shut off for an excessive time resulting in the decanter being overloaded and excessive monomer entering the effluent system.

There was no identified environmental consequence. But this can be no excuse. We had allowed excessive monomer to reach the estuary. There was much media attention. Media coverage was of a 'spill' and alluded to material being 'dumped', both terms implying that the incident was a deliberate act.

If only I had had the sense to consider abnormal operation in my yield studies five years before, maybe I could have put procedures in place to prevent this agonising event.

It is worth mentioning the measures we took to restore public confidence. I met with the local council. We had an open day in which local politicians came and asked any questions they wished. However, I doubt that full public trust was ever restored.

Petrochemical plants typically produce ethylene and propylene for consumption in other plants making plastics (polymers) such as polyethylene. A particular hazard of these chemicals is that they are highly reactive, and their release can lead to large explosions.

Chemical plants sometimes produce unique chemicals about which little is known, mainly acute and chronic toxic properties. Today we have legislation such as REACH (Registration, Evaluation, Authorisation and Restriction of Chemicals) in Europe and TSCA (Toxic Substances Control Act) in the U.S.A. This legislation requires extensive testing of new chemicals before their commercialisation, and today accelerated methods of predicting chronic effects are available.

Notes from Trevor's Files: TSCA

I was once made manager of a small speciality chemical manufacturing plant in the U.K. It was a recent acquisition by the company. One of the product lines used a small amount of an anti-bacterial agent. The previous operating company had used this component before the acquisition. Prior to acquisition, there is a process called due diligence in which the potential acquiring company has the opportunity to identify any issues with the plant. It had not been recognised, during due diligence, that the anti-bacterial component was not registered under Toxic Substances Control Act (TSCA). Its detailed chemistry was not understood. It was in fact, a mixture of chemicals that came as a biproduct of a coal chemicals process.

We exported our product to many customers, including plants in the U.S.A. It was not possible to immediately replace the anti-bacterial agent. Although an approved alternative was quickly identified, we were certified under ISO 8001. For good quality control reasons, ISO 9001 protects customers from changes in a manufacturing process until the client has tested the trial product in their process. It took about 18 months to see through this change, during which time the legality of what we were doing was questionable. The alternative, it would seem, would be to no longer supply our product. At that time, customers were reliant on our supply. There was no easy way out!

Serious incidents continue to occur in the chemical industry. In the following pages, seven major incidents are examined. The first is the most notorious and severe in terms of human suffering.

Incident Examples – Chemical Industry

Bhopal, India 1984

It is now 36 years since Bhopal, when the immediate death toll exceeded 2000 from the accidental release of a highly toxic chemical. Longer-term injuries and fatalities resulting from the exposure may have exceeded 500,000. It is considered the world's worst industrial disaster. Are we so much smarter now to be reassured that such a disaster will not happen again? The following later incident may cause you to doubt.

Just eight months after Bhopal, an incident occurred at a U.S. plant fairly similar to that at Bhopal, albeit involving a similar but slightly less hazardous material. The Occupational Safety and Health Administration (OSHA) conducted a careful examination and concluded that this was an 'accident waiting to happen'. OSHA cited hundreds of 'constant, wilful violations' at the plant. But this same plant had been previously inspected by OSHA and given a clean bill of health (Perrow, 1984 p359). I doubt that this is a rare occurrence where a government authority, insurance surveyor, third party auditor, or other entity entitled to examine an organisations facility has found them 'O.K.', only to judge them as inadequate after a tragic event. We will come back to this issue when discussing the value of audit in chapter 31.

At Bhopal, methyl isocyanate, a raw material for pesticides, was stored. The causes of the disaster are still disputed to this day. Such a dreadful accident was subject to heavy litigation between Union Carbide, the owner of the plant, the Indian government, and the U.S. government. What is not in doubt is that water entered the tank either by sabotage or misoperation. The tank, normally held under nitrogen pressure to prevent contamination and enable pumping, had been unable to sustain a nitrogen blanket for several months. (Willey R., 2014) demonstrates how to use an advanced technique called Layers of Protection Analysis (see Chapter 27) to analyse the incident. He concludes that the many layers of protection initially designed into the plant were sufficient but that in subsequent years most of these barriers to the incident were compromised.

Barriers to preventing the incident included:

- The tank involved had been overfilled, above the 50% maximum stated in the operating instructions. The rationale for the 50% had either been forgotten or not communicated. Perhaps the information was lost in the years since the plant was commissioned.
- There was no management of change procedure considering the consequences of the level limit change (see Chapter 31 on Management of Change).
- Many operators had been laid off; the plant was losing money, it could be that the value of training had practically disappeared.

- The refrigeration system installed to remove the exothermic heat of reaction within the tank had been disabled by plant management, probably as a cost-saving measure. This change also had not been reviewed by the Management of Change procedure.
- The plant had high-temperature and high-level sensors to alert personnel. There is no record of the manual intervention that could have followed in transferring material to another available tank. Other texts on the disaster question whether the instruments were working or previous failures led personnel not to believe the received data.
- There was a relief system that operated as planned. However, at the end of the relief pipework, the flare system that should have burnt off the venting gas was out of service, awaiting repair.
- Emergency response equipment was not adequate – the pressure hoses could not reach the plume, and the emergency response operators do not appear to have practised the scenario as part of their regular emergency response drills.

This particular tragedy is worth taking a look at in a simplified Bow Tie diagram (Figures 7.3 and 7.4).

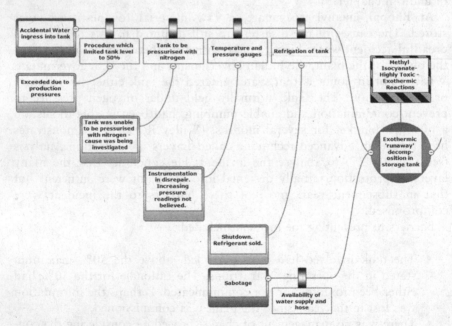

Figure 7.3 Bhopal – Left Hand Side of Bow Tie.

Source: Created using BowTieXP Software © CGE Risk Management Solutions B.V. 2021. BowTieXP is a registered trademark of CGE Risk Management Solutions B.V.

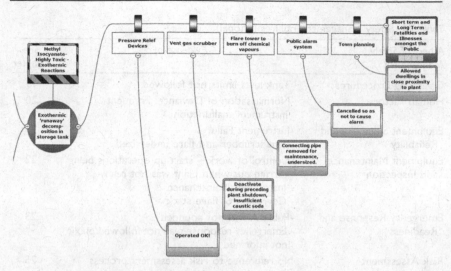

Figure 7.4 Bhopal – Right Hand Side Bow Tie.

Source: Created using BowTieXP Software © CGE Risk Management Solutions B.V. 2021. BowTieXP is a registered trademark of CGE Risk Management Solutions B.V.

There were plenty of barriers to the left and to the right of the Bow Tie. Any one of these might have eliminated or at least reduced the magnitude of the dreadful tragedy which occurred.

As a threat, I have also included sabotage, the cause favoured by the plant's majority U.S. shareholder. An interpretation as to why the company adhered to this explanation is that it absolved them of liability. I am not going to discount this possibility – but if it was sabotage, then it is still the case that the barriers to the right of the bow tie would have considerably diminished the magnitude of the disaster. Further, suppose the hazards of a facility are such that a disaster on such a proportion can occur. In that case, I believe that the operating organisation is obliged to take precautions against sabotage. No water connections should be available in the vicinity of the tanks. Water for clean down would be made available only by special work permit and supervision.

Referring to the chapters which follow on Catastrophe Prevention we see above failings in the following as aspects of safety management (Table 7.2).

It is for the reader to find his/her own level of comfort that these issues are all addressed elsewhere and that there will be no future Bhopal. I regret I am not convinced.

La Porte, USA 2014 – 4 fatalities

In November 2014, an incident occurred at Du Pont's chemical facility in La Porte, Texas. The incident involved the release of approximately

Table 7.2 Safety Management Failures (Bhopal)

Aspect of Safety Management	Evidence of	See Chapter
Operating Procedures	-Tank level limits not followed	19
Human Factors	-Normalisation of Deviance: Frequent instrument malfunction	20
Equipment Suitability and Reliability	-Instrument Failure -Vent scrubber and flare undersized	21
Equipment Maintenance and Inspection	-Control of work – start-up operations being carried out when plant was not ready - Instrument Maintenance - Corrosion of flare stack	22
Emergency Response and Readiness	-Public Alarm not sounded - Emergency response plan not followed/public not informed	23
Risk Assessment	No reference to risk assessment process during the investigation.	25
Governance of Safety	Management of Change not followed in removing of refrigeration	31
Consequence Analysis	Apparently not done, or consequences underestimated	28
Governance of Safety	No evidence of audit either by Indian company or by U.S. Owner	31
Laws, Regulations, Standards	Dwelling built too close to facility despite local regulations	32

24,000 pounds (10,900 kgs) of highly toxic methyl mercaptan. Four employees died. Analysis of the causes by the U.S. Chemical Safety Board points to some issues in common with the Bhopal incident.

Notes from Trevor's Files: Reputation of the Du Pont Company

This incident at Du Pont's La Porte plant motivated me most directly to write this book. I was an employee of the Du Pont Company for some 37 years. I had served the company as a chemical engineer, as a plant manager, a business director, and a management consultant. Du Pont had a strong reputation for good safety. For 14 of those years, I was employed by Du Pont, leading a team of consultants who were successfully helping organisations around the world improve their safety practices. I genuinely believe that Du Pont is an outstanding company in managing safety. Therefore, it was with great shock that I heard about this incident, which followed several other fatal incidents within the Du Pont of that era.

Unlike the Bhopal incident, however, the U.S. Chemical Safety and Hazard Investigation Board "Toxic Release at the Du Pont La Porte Chemical Facility (2019)" concluded that the causes of the release included inadequate engineering design.

The following description is paraphrased from the U.S. Chemical Safety Board (CSB) report:

The CSB investigation viewed the chain of implementation failures as starting with the flawed engineering design of the $20 million nitrogen oxides reduced scrubbed incinerator, a capital project implemented in 2011. The site had long-standing issues with vent piping to this incinerator. The design did not address liquid accumulation in the waste gas vent header piping to the incinerator. The engineers at the site could not fully resolve the liquid accumulation problem through hazard analyses or management of change reviews. Instead, daily instructions had been provided to operations personnel to drain liquid from these pipes. This draining operation was done inside the manufacturing building without explicitly addressing the potential safety hazards this action could pose to the workers. Operating instructions did not specify additional breathing protection for this task.

On the night of the incident, not realising the hydrate blockage in the methyl mercaptan feed piping was cleared, workers went to drain liquid from the waste gas vent piping. They did not know that high pressure in the waste gas vent piping was related to liquid flow through the feed piping and into the waste gas vent piping. Ineffective building ventilation had been identified during an internal company audit about five years before the incident. However, corrective action based on the audit recommendation was not taken. The ventilation design for the manufacturing building was based on flammability characteristics. It did not consider toxic chemical exposure hazards, even though the building contained two highly toxic materials, chlorine, and methyl mercaptan. Records indicated that the manufacturing building's dilution air ventilation design was based on providing sufficient ventilation to ensure that the concentration of flammable gases did not exceed 25% of the lower explosion limit. The building where the workers died was not equipped with an adequate toxic gas detection system to alert personnel to the presence of dangerous chemicals. First, the detector alarms were set well above safe exposure limits for workers. Second, the plant relied on verbal communication of alarms that automatically displayed on a continuously manned control board.

Finally, visual lights or audible alarms were not provided for the manufacturing building to warn workers of highly toxic gas concentrations inside it. When a release caused a detector to register a concentration above the alarm limit, the toxic gas detection system did not warn workers in the field about the potential leak and the need to evacuate. Among other factors, this detection system contributed to workers growing accustomed to smelling the methyl mercaptan odour in the unit.

Additionally, when the toxic gas detectors triggered alarms, personnel investigated potential methyl mercaptan leaks without using respiratory protection. Personnel normalised unsafe methyl mercaptan detection practices by using odour to detect the gas. This activity further degraded the importance or effectiveness of utilising instrumentation in response to alarms signalling potential toxic gas releases. During an unplanned shutdown shortly before the incident, water entered the methyl mercaptan feed piping and, due to the cold weather, formed a hydrate (an ice-like material) that plugged the piping and prevented workers from restarting the unit. A process hazard analysis had identified hydrate formation in this piping, but no recommendations had been satisfactorily implemented. When the hydrate formed, lacking safeguards to control the potential safety hazards associated with dissociating (breaking up) the hydrate (such as using heat tracing to prevent the hydrate from forming a solid inside the piping or developing a procedure to dissociate the hydrate safely), workers went into troubleshooting mode. Ineffective hazard management while troubleshooting the plugged methyl mercaptan feed piping formed yet another link in the chain. This situation allowed liquid methyl mercaptan to flow into the waste gas vent header piping toward the incinerator – a location where it was never intended to go. Workers dealt with the common problem of liquid accumulation in the waste gas vent header routinely by draining the liquid. They did this without an engineered solution or without ensuring the use of safety procedures or personal protective equipment. However, when the liquid drain valves were opened, flammable and highly toxic methyl mercaptan flowed onto the floor and filled the manufacturing building with toxic vapour.

There were also numerous issues with the emergency response. The site emergency response team received inadequate information on the emergency, particularly which chemicals were involved. They arrived with rescue gear suitable for a confined space rescue, not an ongoing chemical release, and lost time retrieving Self Contained Breathing Apparatus. Air monitoring was not carried out, at least partly because the shift supervisor was incapacitated. The assigned Incident Commander established an exclusion zone, the region into which only the Emergency Response team may venture with appropriate equipment and planning. An emergency call was made to the locality's emergency service, but again at the time of the emergency call, details of the chemical involved were not relayed.

Contributing to the incident were (with the chapters in Section 3 addressing these prevention methods) (Table 7.3).

The CSB further notes that the company subsequently closed the La Porte facility.

Table 7.3 Safety Management Failures (Du Pont La Porte)

Aspect of Safety Management	Evidence of	See Chapter
Equipment Suitability and Design	Ventilation, vent header, and design from flammable risks not toxic risks Insufficient toxic gas detectors	22
Governance of Safety	Deficiencies in auditing and follow-up of corrective actions Management of change not followed in implementing short-term workarounds for accumulated liquid drainage	31
Operators and Operating Practices	Procedures did not specify Self Contained Breathing Apparatus Communications between shifts were not adequate	19
Emergency Response Procedures	Communications delays over material involved.	23

Flixborough (1974), 28 Fatalities

Twenty-eight people were killed and 26 injured when an explosion occurred at the works, which produced an intermediate in the production of nylon. One of the six reactors, which were arranged in series, started to leak cyclohexane, and it was decided to shut down the plant to investigate. The reactor was found to have a very serious crack. It was then decided to install a bypass around this reactor and recommence production with the remaining five reactors. The connections to the reactors were 28", whereas the only available pipework was 20", but this itself does not seem to have been the cause of the subsequent catastrophe. Calculations had been done to confirm that the pipe was of sufficient diameter to take the required flow, and that it was capable of withstanding the pressure (as a straight pipe). The two flanges were at different heights, and the pipe needed to be welded to form a 'dog leg'. The bypass assembly was supported by a scaffolding structure intended to support the line and avoid straining the bellows installed between the pipe and each flange. The plant was restarted, and all appeared to be well, except for an unusually large nitrogen usage. This abnormality was being investigated at the time of the incident. The bypass line ruptured. The cloud of cyclohexane found an ignition source, and there was a massive vapour cloud explosion. The explosion destroyed the plant. Operational stress and the need to maintain production were part of the causes of this catastrophe (Lees, 2014, p. 453).

In many respects, the Flixborough tragedy was the birthplace of present-day process safety management thinking.

Seveso (1976)

One should not leave a description of incidents in the chemical industry without mention of 'Seveso', when the Seveso Regulations have impacted so much of industry in Europe. The Seveso regulations will get frequent mention in Section 3 of this book.

The incident took place in the town of Seveso, some 15 miles from Milan, in 1976. The plant produced the well-known antiseptic TCP (2,4,5 trichlorophenol). In the reaction process, a small quantity of the chemical known as Dioxin is produced. This component is normally a minor constituent of the waste products from the plant and was incinerated onsite. Dioxin is highly toxic. It can be taken into the body by ingestion, inhalation, or skin contact.

A batch of material was in progress, when the reaction was stopped for the weekend. Apparently, the batch of reactants was not properly cooled. It is understood steam was still in the reactor jacket. An exothermic reaction proceeded resulting in the bursting disk on the reactor blowing. Considerably more Dioxin is produced at higher reactor temperatures than at normal manufacturing temperatures. The consequences were not immediately apparent to those responsible for the plant. Communications between the plant and the local authorities became confused. Evacuation of the surrounding area took place some two weeks after the release.

It is important to emphasise that no deaths were attributable to the contamination. Several pregnant women who had been exposed had abortions (Lees, 2014, p. 457).

IQOXE (2020), Spain

This incident occurred whilst this book was in its infancy. I happen to be familiar with the industrial estate where this explosion occurred. The plant manufactured ethylene oxide, a material with well-known highly explosive properties. The material is used as an intermediate in the manufacture of detergents, solvents, and organic chemicals such as ethylene glycol. Three people died in the explosion. One of the people who died, who was inside a building 2.5 km from the plant which exploded, was fatally injured by flying debris.

Many other chemical industry incidents could be summarised here. For the interested reader, these can be readily found on the internet and the websites of the U.S. Chemical Safety Board and the European Process Safety Centre.

In conclusion, I summarise below a more recent chemical industry disaster. On 21 March 2019, a massive blast occurred at the Tianjiayi chemical plant in Xiangshui, China. Seventy-eight people were killed and over 600 injured, mostly members of the general public. The explosion is understood to have been sparked by a fire in the fertiliser factory.

CONCLUSIONS

In the chemical industry it seems, based on the above observations, that there remains a considerable potential for a major catastrophe. I don't see the issues raised by Bhopal and other incidents as being addressed.

Chemical Industry – Potential for Future Catastrophes

Not possible	Unlikely	Possible (50/50)	Likely, no more than one	Very likely/ maybe several
☐	☐	☐	☐	☒

Chapter 8

Mining and Mine Waste

INTRODUCTION TO MINING INDUSTRY

Mines are well known as hazardous places. Whether they are open cast or underground, the risk of rockfall is clear. Mines can become flooded. Most underground mines require extensive ventilation systems to get respirable air to the miners, and ventilation systems can fail. Many mines can accumulate toxic or flammable gases resulting in an explosion if ignited. Coal mines are especially vulnerable to underground explosions due to methane released from the coal during mining.

Hazards to the general public from mining operations are mostly found in the treatment of mining waste. Most mines concentrate the ore on site. Concentration of the ore involves crushing the rock into fine particles and then separating out the parts containing higher concentrations of the desired mineral using roasting and then either solution or flotation. Both methods require the rock particles to be suspended in water. The ore concentration process uses a lot of water, much of which is recycled. Unwanted ore particles are typically sent to a tailings dam area. Behind the tailings dam, a reservoir of wastewater accumulates, and the unwanted particles of the former ore, now depleted of the desired mineral, settle out. The water is then recycled from the tailings dam reservoir.

Some ore processing technologies use highly toxic materials. Gold mining, for example, typically uses highly poisonous sodium cyanide to dissolve the gold away from the ore. Many such gold mining facilities incorporate detoxification steps that reduce the effluent's cyanide but do not totally eliminate it.

Iron ores, however, are generally sufficiently rich in iron that they can be smelted directly after grinding.

Copper and chromium ores typically require concentration using sulphuric acid in autoclaves to dissolve the metal. Sulphuric acid is a highly corrosive material that poses some transportation risks.

A specific issue with mining concerns the extent of illegal, unregulated mining. When I wrote this chapter in November 2021, I read the sad story

DOI: 10.1201/9781003360759-10

of the loss of dozens of illegal jade miners missing after the collapse of a wall of an open cast mine in Myanmar. The illegal miners are trying as best to make a living. Some of these unregulated mines are allowed to exist by governments. Closing them because of safety concerns would deprive the miners of their livelihood.

Open Cast Mines

In principle, open-cast mines are less prone to catastrophes than underground mines. In an open cast mine, the overburden of unproductive soil and rock is cleared away, and successive layers of rock containing the desired minerals are quarried using explosives and excavators. Injuries can occur due to rock falls from the quarry sides and by misoperation of the process to deploy explosives used to break open the rock structure. Such incidents are mostly occupational safety issues, not catastrophes.

Flooding of open cast mines has catastrophic potential. In 1988 an opencast coal mine broke the banks of a river in the UK. The 300 ft breach resulted in water from the river pouring 200 feet down into the overcast workings – a massive waterfall. The downstream part of the river reversed flow somewhat amusingly, whilst the open cast mine was filling with water. Thankfully there were no fatalities. The lake formed by the flooded working remains (St. Aiden's Flood, Methley Archive).

Opencast mines don't just potentially flood from the top, as in the above example. They are often dug below the level of the water table. Opencast mines need to be constantly pumped to lift out the water which drains into the mine. Failure of the pumps leads to flooding. This water ingress would take some time and would not be catastrophic to personnel. They would have the time to evacuate. If the excavations penetrate a previously unidentified underground stream or a flooded mine working, the results are sudden and can be disastrous.

Underground Mines

Underground mines are particularly vulnerable to catastrophic events since many personnel work underground in restricted space. Catastrophes can occur from rockfall, leaving miners trapped inside. Most mines need to be continuously pumped out of water and can become flooded if the pumps fail or if the mine working penetrates a previously unidentified water source. The miners need a constant supply of oxygen provided by large ventilation systems, which can fail. Fires can occur with fatalities resulting both from the fire and the discharge of smoke.

In some cases, open cast mines preceded the development of underground mines. As the open cast mine became deeper and deeper as the miners worked through the ore body, underground mines were developed to reach the parts of the ore body which open cast could not reach. This is particularly the case with metallic ore where the ore body may be vertical. Coal deposits are more horizontal or inclined.

Gases, mainly methane, can be released from the rock during mining. This gas is a particularly high hazard in coal mines, and many underground explosions have occurred. Extensive ventilation is required to extract the methane (known to mine workers as Firedamp) released. This ventilation should have back-up from alternative ventilation shafts and machinery, but this is not always the case. Highly toxic gases including carbon monoxide (White Damp and Afterdamp – or the Silent Killer) and hydrogen sulphide (Stinkdamp) can also be released. The presence of these gases must be monitored. The accumulation of these gases is not always predictable, and it is feasible to drill into a pocket of toxic gas. Carbon dioxide (Black Damp), whilst not particularly toxic, displaces oxygen from the air and may cause death by suffocation.

Notes from Trevor's Files: Abai

The mines around Karaganda, Kazakhstan, are a rich source of coal that is particularly suitable for making coke used in steel making. Some mines have been the location of repeated significant incidents. I became involved as a safety consultant around 2009–2013. This assignment followed an accident at the Abai mine in 2008, killing 30 people.

It is, therefore, with particular sadness that I read of six fatalities at the same mine in November 2021.

Mining incidents are considered fairly common in former Soviet countries due to lax safety practices. I am not convinced this is a fair conclusion since I saw many good practices and respect for safety. However, there is a particular culture in these countries that severely limit any questioning of authority. We will return to this subject in Chapter 33 on Safety Culture. Risk assessments in which all levels need to think and discuss are almost impossible in such culture.

At the time of writing this book in November 2021 an explosion had occurred in a coal mine in the Kemerovo region of Russia. More than 50 people lost their lives (see Listvyazhnaya in Incident Examples below). I visited mines in the same area after the 2007 underground mine disaster, which killed 110 miners.

It should be emphasised, however, that the number of incidents and fatalities in mines in Russia is, in general, on a downward trend.

Tailings Dams

Tailings dams are embankments constructed to impound the tailings gen-erated in a mine processing plant. These tailings typically leave the mine ore extraction process as ground particles in a stream of water. The tailings need to be dewatered before they can be stored.

There is a considerable risk arising from the integrity of the tailings dam. The dam's collapse releases large quantities of water con-taining high concentrations of ore particles that separate out as a thick and unsightly mud. In many cases, the contents of the tailings reservoir are toxic.

In January 2019, a tailings dam collapsed near the town of Brumadinho resulting in the deaths of at least 248 people; further detail is given in the incident examples below.

Tailings dam failures regrettably occur far too frequently, often with disastrous results. Statistically, there is no evidence that the number of tailings dam incidents is reducing (Figure 8.1).

As shown in Figure 8.2 there is some indication that the total amount of effluent released is deteriorating.

Whereas better land-use planning might have been expected to reduce the number of fatalities from dam failures, this does not seem to be the trend. See Figure 8.3.

Graphs 8.1–8.3 were derived with the aid of 'Wise: Chronology of major tailings dam failures'. This reference is regularly updated with the latest incidents.

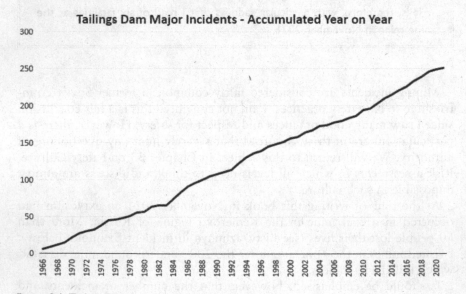

Figure 8.1   Tailing Dam Incident Frequency.

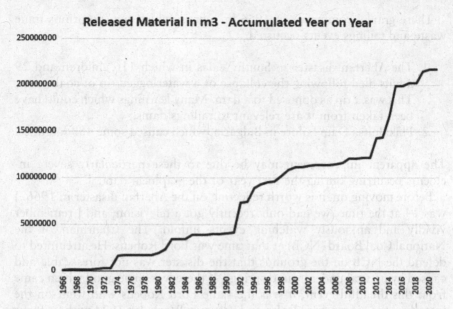

Figure 8.2 Releases from Tailings Dam Incidents.

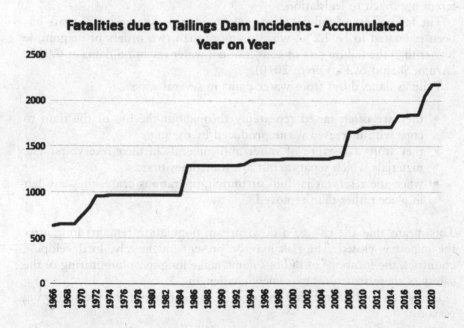

Figure 8.3 Fatalities due to Tailings Dam Incidents.

These graphs commence in 1966, a year in which two notorious mine waste and tailings events occurred:

1. The Aberfan disaster in South Wales in which 116 children and 29 adults died following the collapse of a waterlogged tip of coal waste. This was a tip as opposed to a dam. Many learnings which could have been taken from it are relevant to tailings dams.
2. Plakalnitsa Dam failure in Bulgaria, which caused some 488 fatalities.

The apparent improvement may be due to these particularly severe incidents occurring during the first year of the graphical data.

Before moving on, it is worth reflecting on the Aberfan disaster in 1966. I was 14 at the time, we had only recently got a television, and I remember vividly and anxiously watching events unfold. The Chairman of the National Coal Board (NCB) at that time was Lord Robens. He attempted to defend the NCB on the grounds that the disaster was not foreseeable and caused by natural underground springs. If there was some good which came from this dreadful event, it was the same Lord Robens who went on the introduce the pioneering Health and Safety at Work Act 1974 with its focus on self-regulation and risk assessment. I speculate that the Aberfan disaster fell heavily on his mind and led him to introduce such a dramatically different approach to legislation.

The failure rate of tailings dams worldwide over the past 100 years has been estimated to be 1.2%, which is more than two orders of magnitude higher than the failure rate of conventional water retention dams at 0.01% (Azam, Shahid & Li, Qiren. 2010).

Tailings dams differ from water dams in several ways:

- they are often raised repeatedly throughout the life of the dam to cope with increased waste produced by the mine.
- they store a mixture of water and minerals in their reservoirs. The materials which separate out are sometimes toxic.
- when the reservoir is 'full' or mining operations cease, they are left in place rather than removed.

This means that the risk to a downstream population remains long after the mine has closed. The risk may be present indefinitely. In developing countries, the locations of tailings dams make long-term monitoring of the safety of these structures extremely challenging.

Dams can and have failed for one or a combination of the following reasons:

- Overtopping caused by floods that exceed the capacity of the dam
- Deliberate acts of sabotage
- Structural failure of materials used in dam construction

- Movement and/or failure of the foundation supporting the dam
- Settlement and cracking of concrete or embankment dams
- Piping and internal erosion of soil in embankment dams
- Inadequate maintenance and upkeep

What Dam Construction Method Is Used?

Tailings dams are typically constructed mostly of waste rock.

There are three main types, as shown in Figure 8.4 – upstream (a), downstream (b), and centreline (c). Upstream constructions are generally regarded as the most susceptible to failure.

Upstream tailings dams are built progressively "upstream" of the starter dam. This is done mostly using rock and tailings materials (HR Wallingford, 2019, "A review of the risks posed by the failure of tailings dams").

Using tailings materials to build the dam reduces construction costs. However, an upstream dam is likely to be less stable. In particular, consider the case of an earthquake. Tailings materials may liquefy in earthquake conditions and lose their strength. In some countries, upstream tailings dams are not permitted.

Downstream tailings dams are built progressively 'downstream' of the starter dam. Centreline tailings dams are raised progressively whilst maintaining the original centreline of the starter dam. They often have an impervious layer on the upstream slope of the dam. Both these types of dams are considered more stable than upstream dams, especially in the event of an earthquake.

Methods for monitoring the 'health' of a tailings dam include:

- Measuring the hydraulic gradient. The tailings dam wall is typically composed of compacted rock and other mining debris. It is permeable and seeps water. During the dam's design, an acceptable hydraulic

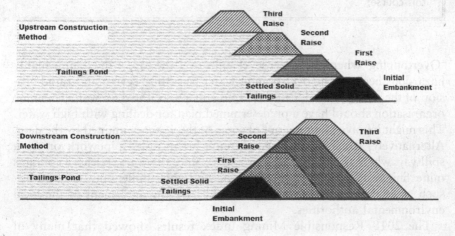

Figure 8.4 Types of Tailings Dams.

gradient should have been established. If the hydraulic gradient exceeds the design limit, then this is a clear warning signal requiring corrective action. The hydraulic gradient is typically measured using small wells ('piezometer' wells) distributed around the entire length of the tailings dam.

- GPS and optical monitoring. Potential early warning of dam failure may be detected from small deformations in the dam wall. Various new technologies including fibre optic sensors, GPS, drones, and satellite monitoring are now available to monitor dam wall deformations.

Notes from Trevor's Files: Ukraine Tailings Dam

We were working safety consultancy on mine and ore processing plant in Ukraine. My assignment was to review activities at the tailings dam. Standing at the top of the dam, an enormous lake of tailings was visible, and beneath me was a major river. The catastrophic potential was clear. The lady who managed the tailings facilities was very patient in explaining the many technical devices they had to ensure the dam's integrity. This included measuring the hydraulic gradient. Diagrams were produced showing the measured gradient as a red line versus that hydraulic gradient assumed in the design as a blue line. In places, the red line crossed the blue line. I had been told that the crossing of these lines was a clear indication of a problem. Fortuitously I had the opportunity afterwards to meet some directors of companies with tailings dams. I tried to explain that segmentation of a tailings facility would reduce the potential, since only one cell would be released in the event of a break. A view of the current tailings facility suggests that they have followed this course!

'Overtopping' where the water level spills over the top of the dam is highly hazardous because the overtopping water can cause severe and rapid erosion of the dam material, causing rapid collapse of the dam wall. The mine organisation should have a predetermined plan for dealing with high water. This might involve pumping out of some of the water contained by the dam. Alternatively, it can be released through pre-installed pipework or via a spillway, where these have been provided. Since the water released contains mine finings that are potentially toxic and environmentally undesirable, such an emergency response plan should have the agreement of the local environmental authorities.

The 2018 Responsible Mining Index results showed that many of the world's largest mining companies are not able to show that they are effectively addressing the risks of tailings dam failure. The 30 mining companies assessed in RMI 2018 scored an average of only 22% on tracking, reviewing, and acting to improve their tailings risk management. Fifteen of the 30 companies showed no evidence of keeping track of how effectively they are addressing these risks (Responsible Mining Foundation, 2018).

Many of the incidents that occurred in the past could have been prevented.

Whilst most mines have tailings dams; there are other solutions, notably dry stacking. In the latter method, the solid waste is dewatered to a point where it can be stacked as a solid. Dry stacking is generally significantly more costly than the tailings dam process. The company Vale in their sustainability reporting for 2020, have pledged, in the wake of the Brumadinho disaster (see Incident Examples), to increase the use of dry stacking from 40% of their tailings operations in 2014 to 70% in 2024. However, this does not cover the legacy issue – many tailings facilities have been abandoned once they were full. Vale undertakes to have these inspected and certified, and has set up a programme to 'decharacterise' the upstream tailings dams (see Figure 8.4). Such measures include downstream back-up dams. It is understood that this process will take a long time (Vale, SEC Form 20F, 2020).

There are many articles on tailings dam incidents, including efforts to prevent a recurrence. However, these seem to be based on a re-emphasis on inspection, geotechnical measurements, and accurate deformation measurement, plus enhanced emergency response plans should the tailings dam fail.

I remain highly concerned that a solution to ongoing tailings dam failures has not yet been found.

Incident Examples – Mining and Mine Waste

Buffalo Creek, U.S.A. – 1972 – 125 Fatalities

A coal slurry impoundment dam burst. The resulting flood unleashed approximately 500,000 m^3 of black wastewater, cresting over 9 metres high, upon the residents of 16 towns along the Buffalo Creek valley downstream of the dam. Out of a population of 5,000 people, 125 were killed, 1,121 were injured, and over 4,000 were left homeless. 507 houses were destroyed,

Dam 3 which was upstream of dams 1 and 2 failed first. The water from Dam 3 overwhelmed Dams 1 and 2. Dam 3 had been built on top of coal slurry sediment that had collected behind dams 1 and 2, instead of on solid bedrock. Dam 3 was approximately 79 m above a nearby town when it failed.

Officials from the mining company blamed the flooding on the substantial rain that fell prior to the dam breaking. However, subsequent investigations showed that the flooding was the result of improperly built coal waste dams (Buffalo Creek Flood, www.gem.wiki/Buffalo_Creek_Flood).

As mentioned above, six years earlier, a colliery waste tip at Aberfan, Wales, collapsed following a period of heavy rain. One hundred sixteen children died at a school that was engulfed by the waste, along with 28 adults. Some of the learnings from the Aberfan disaster may well have helped prevent Buffalo Creek. Still, it appears the mining company was only vaguely aware of the outcome of Aberfan (1966) and any potential applicability to Buffalo Creek. See The Buffalo Creak Disaster by Gerald M. Stern, an attorney with Arnold & Porter, concerning the subsequent court cases to establish liability and compensation (Stern (2008)).

Brumadinho Dam Disaster – 2019 – 248+ Fatalities

A tailings dam at an iron ore mine suffered a catastrophic failure. The mud hit the mine's administrative area. Further downstream at approximately 1 km from the mine, a small community was flooded. The mud reached the Paraopeba River which supplies water to one-third of the adjacent region.

The dam failure released 12 million m^3 of tailings and polluted over 300 kilometres of river. It is notable that, like many tailings dams incidents, this particular tailing facility was no longer active.

We will come back to this incident in Chapter 32 on Laws, Regulations, and Standards. Briefly, the owner of the mine employed an inspection service to certify the safety of tailings dams, including the Brumadinho dam. The employees of the inspection service found that the dam did not reach the necessary stability factor according to their calculation standards. Instead of refusing to issue a stability declaration, these employees looked for new calculation methods to achieve the desired result and certified the stability of the dam. As a result, neither the mine operators nor the authorities initiated effective stabilisation or evacuation measures (see also European Centre for Constitutional and Human Rights, "The Safety Business: TUV SUDS role in the Brumadinho dam failure in Brazil").

Referring to the chapters which follow on Catastrophe Prevention, we see above failings in the following as aspects of safety management (Table 8.1).

Soma Mine Disaster, Turkey 2014 – 301 Fatalities

An explosion at a coal mine in Soma, Turkey, caused an underground mine fire that burned for three days. Three hundred and one people were killed.

The blast occurred near a conveyor belt control unit. A fire started and the ceiling collapsed due to vibration. A high concentration of methane gas in the air helped the fire to spread, covering the main hallway of the mine. The power system shut down preventing the workers from escaping using

Table 8.1 Safety Management Failures (Brumadinho)

Aspect of Safety Management	Evidence of	See Chapter
Human Factors	Management of Contracted Inspection Company	20
Equipment Maintenance and Inspection	Dam Inspection and Certification Function	22
Emergency Response and Readiness	Warning and evacuation of people downstream	23
Governance of Safety	Absence of Audit Function	31
Legacy Issues	Disused dams and tailing facilities are no longer active	34

the elevators. Smoke covered all the main passageways. The workers were effectively trapped. One engineer suggested changing the ventilation direction to clear the smoke from the hallways to allow the rescue team to get inside. Unfortunately, his well-meaning advice causes the smoke to travel towards the trapped miners.

It is suggested that the Soma mine was not designed in the safest way. The conventional conveyor belts could have been replaced by semi-mechanised or fully mechanised equipment. The ventilation systems should have been upgraded as the mine grew. Apparently, the ventilation systems were not sufficient to minimise the carbon monoxide concentration in the accident location. The investigation revealed that the ventilation diagrams did not match their layout in the mine. The investigation also discovered 67 methane and carbon monoxide sensors which were either defective or not properly calibrated. This was a large mine with many such instruments (Soma Mine Disaster: Health and Safety in Coal Mining – U.K. Essays; Soma mine was death trap, report shows IndustriALL (industriall-union.org).

Listvyazhnaya Mine, Russia 2021 – 51 Fatalities

Whilst writing this book, a further dreadful mine disaster occurred. This time in Russia in the Kemerovo region. I visited this pleasant and wooded part of southwestern Siberia when working as a management consultant, although I was not at this particular mine.

In November 2021, 51 people were killed and dozens more injured. There were around 285 miners underground at the time of the explosion. The cause was an accumulation of methane which subsequently ignited (Hazardex). The incident provoked President Putin to make a personal statement. A preliminary report has revealed that management at the mine had attempted to hide details about high methane levels. These managers had even gone as far as to falsify methane sensor data. Three managers of the coal mine and two safety inspectors were reported to have been detained on suspicion of criminal negligence and breaking safety rules. Several of those who perished were members of the emergency rescue team who went into the mine to rescue miners trapped beneath. One of the rescuers who managed to escape suffered from carbon monoxide poisoning.

CONCLUSIONS

In the mining industry, it seems, based on the above observations, that there remains considerable potential for a major catastrophe. I don't see the issues raised by the Soma mine and other incidents as being addressed regarding underground mining. I do not see the issues being raised by Brumadinho and other mining tailings dams being adequately addressed globally. Here, there is a particular concern about legacy – older tailings workings that are no longer in use. In open-cast mining, the potential is less.

Open Cast Mining – Potential for Future Catastrophes

Not possible	Unlikely	Possible (50/50)	Likely, no more than one	Very likely/ maybe several
☐	☐	☒	☐	☐

Underground Mining – Potential for Future Catastrophes

Not possible	Unlikely	Possible (50/50)	Likely, no more than one	Very likely/ maybe several
☐	☐	☐	☐	☒

Tailings Dams – Potential for Future Catastrophes

Not possible	Unlikely	Possible (50/50)	Likely, no more than one	Very likely/ maybe several
☐	☐	☐	☐	☒

Chapter 9

Transportation Industry

AIR TRANSPORT

Airports and air travel need no description; they are familiar to almost all of us. The elements of substantial risk are obvious – heavy objects travelling at high speed propelled by burning fuel and flying in the air; held aloft only by virtue of their forward motion. Aircraft incidents are generally associated with large numbers of fatalities, mostly of passengers onboard.

However, much progress has been made in air transport safety, particularly since the 1950s, and air transport is now one of the safest transportation methods.

Comparisons of air transport versus travelling by car or foot (including both passengers and crew) are shown in Table 9.1.

The first column shows deaths per billion journeys. On this basis, air transport is approximately three times more likely to result in a fatality than travelling by car or foot. However, the second column shows deaths per billion hours of travel. On this basis, an hour of travel by foot is approximately seven times more likely to result in a fatality than air travel. An hour of travel by car is about four times more likely to result in a fatality than air travel.

Based on the third column, fatalities per kilometre, travelling by car is approximately 60 times more likely to result in a fatality. Travelling on foot is about 1000 times more likely to result in a fatality than travelling by air. Of course, this is not an entirely equivalent comparison. Most airline journeys cannot sensibly be completed on foot, and the number of air journeys that could sensibly be completed by car is limited. Air transport allows us to travel much further than would be practical by other techniques. It is rare for any aircraft incident to injure or kill people on the ground, either at the airport or in flight.

Almost all incidents relating to air transport are tracked in databases available online, for example, the Aviation Safety Network which covers incidents worldwide (https://aviation-safety.net/database/). Many incidents

DOI: 10.1201/9781003360759-11

Table 9.1 Deaths per 100

Type	Deaths per 1,000 million Journeys Taken		
	Per Number of Journeys	Per Number of Hours of Travel	Per Number of Kilometres Travelled
Foot	40	220	54.2
Car	40	130	3.1
Air	117	30.8	0.05

(Ref: Wikipedia – Transport Accidents).

are listed, but most incidents on the database do not involve actual injury and are considered near-miss, investigated to help eliminate potential serious incidents.

Typical aviation safety hazards are provided in Table 9.2.

Statistics show, however, that smaller aircraft, including helicopters are more accident-prone than larger commercial flights.

Why is there such a difference in fatal accidents between commercial airlines and smaller aircraft? Several factors can help explain this discrepancy. First, in any aircraft category, most accidents occur during take-offs and landings and a smaller aircraft needs to do more take-offs and

Table 9.2 Aviation Safety Hazards

Hazard	Description/Cause	Mitigation What Is Done to Prevent
Debris on runway	Parts falling from other aircraft	Periodic runway sweeps
Other Foreign Object Debris	Solids encountered in flight including hail, birds, dust including volcanic dust.	
Lightning	Exposure to lightning potential cannot be avoided for scheduled airlines servicing regions prone to lighting. Airliners are struck by lightning twice per year on average.	Aircraft withstand typical lightning strikes without damage. But what about positive lightning?
Ice and Snow		
Pilot Error and Fatigue		
Damage by Ground Support Equipment		
Runway Excursion/ Overrun/Incursion/ Confusion		Runway safety area – but only achievable in some newer airports
Suicidal actions		
Military Action		

landings to earn its keep. The most significant factors affecting flight safety are weather conditions along with pilot training, experience, and skill. Also, general aviation pilots aren't required to have as much training or experience as commercial pilots. Helicopter flights, in particular, are therefore more prone to accidents.

I spent some four years as a safety consultant visiting offshore oil platforms in the UK Sector of the North Sea. Any person going to a North Sea platform must have special training known as BOSIET (Basic Offshore Safety Induction and Emergency Training). A key component of this training is to be able to escape from a mock-up helicopter which has ditched and overturned in the training centre swimming pool. All offshore workers in the North Sea, where the water is typically below 4°C in winter, are required to wear survival suits. Nevertheless, occasional incidents do occur where helicopters have to ditch in the sea, sometimes with fatal consequences. In April 2009 16 people perished on returning from the Miller platform when the rotor detached from the main gearbox. Two other further forced ditchings occurred in the 2009–2013 period in the North Sea, but a controlled landing on water was thankfully achieved, the survival suits performed well, and all passengers and crew survived. (AAIB: Report 2/2011 - Aerospatiale (Eurocopter) AS332 L2 Super Puma, G-REDL, 1 April 2009; Report on the accidents to Eurocopter EC225 LP Super Puma G-REDW 34 nm east of Aberdeen, Scotland on 10 May 2012 and G-CHCN 32 nm southwest of Sumburgh, Shetland Islands on 22 October 2012).

Modern cockpit automation systems have been improving steadily for the past several decades. Accident rates have undeniably declined through the years with the rise in computerised systems. One of the consequences is pilot over-reliance on automation. American Airlines used a pilot training video back in 1997 that highlighted the issue. A Federal Aviation Administration (F.A.A.) study in 2011 found that many pilots were overly dependent on automated systems while flying and lacked hands-on flight experience.

To this day, even those pilots with extensive training and flight experience can be lulled into depending on their aeroplanes' computerised systems. Thus, they may not realise an emergency situation has developed until too late. Officials across the aviation industry agree that modern pilot training has pilots spending too much time learning computerised automated systems. It is suggested that they do not spend enough time manually flying aeroplanes. Pilots have become more like systems operators than traditional pilots. Many pilots simply aren't well-trained in hands-on flying techniques, especially those techniques that come into play when automated systems fail during a flight.

Unlike most other industries in this book, air transport safety responsibility is a combination of the airports, airlines, air traffic control, and state/global institutions.

This part on risk in air transport is divided into three sections:

- Airlines and aircraft (non-military)
- Airports and facilities
- Air Shows

Whilst I have no qualifications or experience in aircraft or airports, other than as a passenger, I spent many hours of research preparing this section. I also enlisted the help of my eldest son, who holds a degree in Air Transport Management and works in the field. This research included a review of the I.A.T.A. Operations Safety Audit (I.O.S.A.) standards manual and many online pilot and crew training texts and videos.

As an 'outsider', the airline industry has an impressive number of regulations, training, and reassuring certifications. This depth is, of course, assuring only all goes well. But what if a pilot or crew member gets somehow into flying the wrong aircraft? There seem to be many potential failure modes for an apparently robust system. As an example, consider the practice of naming a runway. Heathrow has two runways, so there would seem to be no danger in two aircraft, one landing on 09L and one taking off from 27R. Actually, 09L and 27R are both the same runway. The different labelling is used according to the direction in which it is being approached! (Figure 9.1).

Various approaches have been made to assessing the safety of airlines. However, there is no comprehensive assessment publicly available related to airline safety and use in most countries.

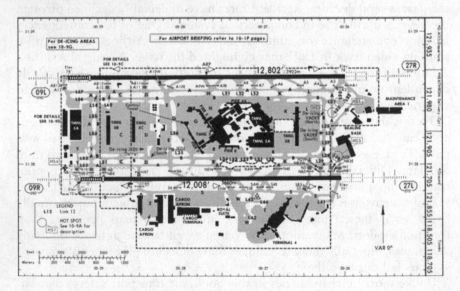

Figure 9.1 Heathrow Runway Diagram.

The website 'AirlineRatings' combines several features into one rating. This data provides an indicator of safety performance.

The 'Airline Ratings' website uses the following features:

1. I.O.S.A. Certification is issued following success in International Air Transport Association Operational Safety Audit.
2. E.U. Blacklist comprising airlines that are banned from entering the airspace of E.U. member states. Airlines are blacklisted for failing to meet E.U. regulatory oversight standards.
3. Absence of fatalities over the last 10 years.
4. Federal Aviation Authority endorsement. Similar to the E.U. Blacklist but applying to U.S.A. regulatory standards and prohibition from entering the airspace of the U.S.A.
5. International Civil Aviation Organisation (I.C.A.O.) eight audit parameters of the country of origin. The parameters comprise Legislation, Organization, Licensing, Operations, Airworthiness, Accident Investigation, Air Navigation Services, and Aerodromes.

Note that caution needs to be taken about interpreting an airlines performance on any one of these features. For example, an airline may have chosen not to participate in I.O.S.A. or I.C.A.O. audit, perhaps for financial or resource reasons. An airline that has not applied for routes in Europe or the U.S.A. may appear on the blacklist, possibly unfairly. The fatalities figures exclude terrorism, pilot suicide, and military action.

The details of the audit criteria are not readily available, but there are sufficient pointers on the I.A.T.A., I.C.A.O., and F.A.A. websites to produce a layman's checklist of questions. The I.A.T.A. Operations Safety Audit (I.O.S.A.) has a Standards Manual available for download (and runs to 433 pages)! It comprises the following sections:

- Section 1: Organization and Management System
- Section 2: Flight Operations
- Section 3: Operational Control and Flight Dispatch
- Section 4: Aircraft Engineering and Maintenance
- Section 5: Cabin Operations
- Section 6: Ground Handling Operations
- Section 7: Cargo Operations
- Section 8: Security Management

Much of the following is based on thoughts created by this manual.

It is worth enquiring if an airline might well appear on the E.U. or F.A.A. blacklists. The question should be asked both for the past and present. If an airline has been grounded, ask why and what has been done to rectify the situation.

Pre-1990 aircraft of the former Soviet Union are considered suspect. However, this does not mean that there is anything necessary to be concerned about in flying an aircraft with a Russian-sounding name! I have flown in quite a few.

Notes from Trevor's Files: Smoking Pilot

Many Russian companies have their own aviation department. A passenger wishing to visit a remote facility has no alternative but to use the company plane. We were to carry out a safety assessment at a remote steelworks in Russia and boarded the plane. There was no safety announcement. On landing, no disembarkation announcement. A strong smell of tobacco smoke emanated from the pilot's cabin. The pilots appeared coughing profusely and were the first to leave. Not a good start to our assessment!

Engine failure in flight is a significant risk. The consequences of an engine failure depend on the number of engines and engine reliability. However, aircraft can glide for substantial distances. For example, an aircraft at a cruising altitude of 36,000 feet could typically glide 70 miles to do an emergency landing at an airport.

Single-engine aircraft have limited commercial application and have in the past been constrained to operations only when there is good visibility, and during daytime. If the aircraft is flying over water, it should always be within gliding distance of land. However, advances in engine reliability have pushed the boundaries on this constraint. Nevertheless, catastrophic engine failures do occur. I would not choose to fly on such an aircraft for significant distances over water.

For multi-engine aircraft there is, of course, improved security. Before the 1980s, only three or four-engine aircraft were used for long flights over water, e.g. across the Atlantic Ocean. However, with improved aircraft reliability, procedures now allow two-engine aircraft to make such journeys. These procedures are known as ETOPS 'Extended Range Operation with Two-Engine Airplanes', jokingly referred to as 'Engines Turn or Passengers Swim'. Aircraft need to be ETOPS classified. Depending on the ETOPS rating, the aeroplane must always be within a specified number of minutes of flying time to an alternative airport. For example, the Boeing 777-300ER (Extended Range) has an ETOPS rating of 330 minutes – the current maximum. Next time you board an aeroplane, make it a point of interest to know if the flight is E.T.O.P., what the ETOPS rating of the aircraft is, and where the alternative airports are!

An E.T.O.P.s rating of 180 minutes gives much more flexibility over an E.T.O.P.S. rating of 60 minutes and only leaves a few inaccessible areas around the Antarctic.

Despite the improved reliability of modern aircraft with a high ETOPS rating, aircraft are still routed conservatively to have a closer safety margin to alternative airports. For the transatlantic routes, Goose Bay – Canada (Y.Y.R.), Kangerlussuaq – Greenland (S.F.J.), and Reykjavik – Iceland (KEF) are typical alternate airports.

Performance measures are a valuable indication of safety. Typical indicators in the operations or maintenance area include (Table 9.3).

Leasing is a common practice. There is, therefore, a risk that responsibilities will be confused. With 'wet lease', where the person owning the aircraft also provides one or more of the crew, it should be clear that the owner has the responsibility for aircraft airworthiness and the competence of the loaned crew. With 'dry lease', where no crew is provided, responsibilities seem even less clear.

The formation of ice on an aircraft is a very significant hazard. Where there is the likelihood of ice accumulation whilst the plane is on the ground, the aircraft should be inspected for visible ice and de-iced. Now the choice to de-ice results in delays and costs. Furthermore, environmental concerns limit the amount of de-icing fluid applied without creating a pollution issue at the airport. My observations suggest that de-icing practices are highly varied. Often, I have flown into and out of airports in Russia and Kazakhstan, where the temperatures are sub-zero in winter, with no de-icing taking place. Whilst this sounds questionable, there is less likelihood of icing where the air is cold and dry, as opposed to a freeze/thaw situation accompanied by snow as would be more likely in a Western European airport in winter.

There are various flight rules which are governed by the meteorological conditions and timing of a flight. Visual Flight Rules (VFR) requires strict adherence to precise rules regarding climatic conditions to ensure that pilots

Table 9.3 Performance Measures

Activity Area	Example
Flight operations	Take-off and landing tail strikes, unsatisfactory line or training evaluations
Operational Control	Flight diversions due to fuel
Engineering and maintenance	In-flight engine shutdowns, aircraft component/equipment failures
Cabin operations	Inadvertent slide deployments
Ground handling	Aircraft damages due to vehicles or equipment
Cargo Operations	Dangerous Goods Spills
Operational Security	Unauthorised interference or access events

have the necessary visual contact with hazards, routes, and airports. Typically, VFR-only certification indicates that the pilot is a student or restricted to pilots who fly small aircraft for their entertainment. The majority of aircraft incidents are with smaller aircraft. A distinct hazard arises when a VFR-certified pilot encounters unexpected climatic conditions. He must therefore transfer to Instrument Flight Rules (IFR), but what happens if the pilot is not trained or certified? See incident examples below for a tragic example.

Air shows require separate consideration since there have been several occasions when there have been fatalities and injuries amongst the general public during an air show. The most notable recent example is the Shoreham air show crash in the U.K. See Incident Examples below.

Hofstede (Hofstede, 1980, p. 115) offers an interesting insight into aviation safety. He raises the issue of how different cultures work together. Hofstede identifies a 1994 article written by executives in the safety section of the then Boeing Commercial Airlines Group. They demonstrated a clear positive correlation with a measure he calls power distance index and a negative correlation with an index he calls individualism (see Chapter 31 on Safety Culture and an explanation of Hofstede's cultural dimensions). Suppose the pilot and co-pilot relationship is hierarchical rather than one of shared responsibility. In that case, this may lead to a co-pilot not correcting errors made by the pilot. Standard procedures in the cockpit assume two-way communication between pilot and co-pilot. For an example of this kind of failure, see the 2010 Air India crash in Table 9.4. My eldest son also suggested the same explanation when discussing some of these incident examples.

Incident Examples – Airports and Aircraft

Here is a table of some selected major aircraft and airport incidents, along with at least one of the causes of the incident.

Shoreham Air Show Crash, 2015 0 11 Fatalities

In 2015 11 people were killed at an air show at Shoreham, England. The former military aircraft, which was privately owned, failed to complete a loop manoeuvre and crashed, hitting vehicles on the main road. The pilot survived. The subsequent investigation concluded that the crash resulted from pilot error.

The public generally trusts that the relevant authorities, pilots, and airports competently manage risks. The Air Accidents Investigation Branch (Aircraft Accident Report 1/2017) revealed several irregularities which point to the kinds of questions the general public might genuinely ask in advance of such an air show:

Table 9.4 Incident Examples

	Incident	Location	Fatalities	Probably Cause	Post Incident Improvements
2020	Ukraine Airlines PS752	Nr Sabashar	176	Military action, probably a mistake	
2019	Boeing 737 Max	Ethiopia	157	Software design faults leading to conflicts with pilots' intentions. Incomplete pilot training in the software is also suspected.	Grounding of Boeing 737 Max. Redesign of software. More rigorous design oversight by federal authorities is also anticipated.
2018	Boeing 737 Max	Indonesia	189	As above	See above
2016	Germanwings 4U9525	Southern France	150	Human Factors: Deliberate and planned suicidal actions by co-pilot	Improved medical evaluations of pilots. Improved cockpit procedures so that one pilot is not left alone when the other pilot visits the toilet.
2016	EgyptAir Flight 804	Mediterranean	66	Disputed	
2015	Metrojet Flight 9268 (from) Sharm-el-Sheik	Northern Sinai	224	Terrorism. Suspected by an airport cargo handler.	Improved airport security, and screening of airport staff.
2014	Indonesia Airlines QZ8501	Java Sea	162	Malfunction of a rudder control mechanism combined with unauthorised resetting of circuit breakers in-flight by one of the pilots. Communication issues between Captain and First Officer.	Improved training in recovery from critical situations.
2014	Malaysian Airline Flight 17	Ukraine	298	Military action, probably a mistake.	

(Continued)

TABLE 9.4 (continued)

	Incident	Location	Fatalities	Probably Cause	Post Incident Improvements
2010	Air India Express Flight 812	Mangalore	158	Human factors – incorrect flightpath. Captain failed to discontinue unstabilised approach despite calls from First Officer to 'go around'. Overran runway which led to plane falling down cliffs. Insufficient rest was considered a contributing cause.	Extended runway again at Mangalore. Improved rest discipline.
2001	11 September terrorist attacks	New York, USA	2996 both passengers and surface inhabitants	Terrorism	Improved security screening equipment and security. Improved cockpit security.
2001	American Airlines 587	New York, USA	265 including five people on the ground	Mechanical failure of vertical stabiliser due to excessive rudder pedal inputs. Simulator training did not match actual roll conditions.	Pilot training programme improved.
1996	Charkhi Dadri Mid-Air Collision	Charkhi Dadri, India	349	Mid Air Collision – misunderstanding over altitude plus some linguistic issues.	For the Delhi Airport: Separation of inbound and outbound Aircraft through the creation of 'air corridors' Installation of a secondary air-traffic control radar for aircraft altitude data Mandatory collision avoidance equipment on commercial aircraft operating in Indian airspace

Year	Flight	Location	Fatalities	Cause	Outcome
1996	ValuJet Flight 592	Florida, USA	110	Hazardous Materials in cargo hold	Smoke Detection and Fire Suppression in Cargo Hold
1994	China Airlines Flight 140	Japan	265	Stall on descent due to conflict between crew commands and autopilot	Software upgrades and improved training
1985	Japan Airlines Flight 123	Tokyo, Japan	524	Faulty repair following tail strike incident	Faulty maintenance procedure discontinued. Airplane model temporarily withdrawn from service.
1979	American Airlines Flight 191	Chicago, USA	273 plus 2 people on the ground	Maintenance issues leading to uncommented stall followed by detachment of on engine during take-off	
1978	Air India Flight 855	Nr Mumbai, India	213	Human factors – loss of situational awareness when one of the Attitude Indicators failed.	
1977	Tenerife Airport Incident	Tenerife, Spain	583	Runway collision in fog. Communications errors with A.T.C.	Improved Communication Protocols. Installation of ground-based radar so that controller can see aircraft on a screen even if not visually.
1974	Eremenonville Air Disaster - Turkish Airlines Flight 981	Paris, France	346	Incorrectly secured cargo door	Redesign of cargo door.

Compiled with the aid of Wikipedia: Aviation Accidents and Incidents.

1. The aerobatic manoeuvre commenced at a height lower than the pilot's authorised minimum for aerobatics, at an airspeed below his stated minimum, and proceeded with less than maximum thrust.

2. Although it was possible to abort the manoeuvre safely at the top of the manoeuvre, it appears that the pilot did not recognise that the aircraft was too low. Several credible explanations include not reading the altimeter due to workload, distractions, or visual limitations, misreading the altimeter, or incorrectly recalling the minimum height required at the apex.

3. The investigation found that the guidance concerning the minimum height at which aerobatic manoeuvres may be commenced was either not applied consistently or is unclear.

4. There was evidence that other pilots did not always check or perceive correctly that the required height has been achieved at the apex of the manoeuvres.

5. Training and assessment procedures in place at the time of the incident did not prepare the pilot fully for the conduct of the relevant escape manoeuvres in that particular aircraft.

6. Defects and exceedances of the aircraft's operational limits were not reported to the maintenance organisation.

7. Mandatory requirements of its Airworthiness Approval Note had not been met.

8. During prolonged periods of inactivity, the aircraft's engine had not been preserved per the approved maintenance schedule.

9. Whilst the aircraft had been issued with a Permit to Fly and its Certificate of Validity was in date, issues identified by the investigation indicated that the aircraft was no longer compliant with the permit's requirements.

10. Parties involved in the planning, conduct, and regulatory oversight of the flying display did not have formal safety management systems to identify and manage the hazards and risks.

11. There was a lack of clarity about who owned which risk. Who was responsible for the safety of the flying display, the aircraft, and the public outside of the display?

12. The regulator believed the organisers of flying displays owned the risk. The organiser believed that the regulator would not have issued Permission for the display if it had not been satisfied with the safety of the event.

13. The aircraft operator's pilots believed that the organiser had gained approval for overflight of congested areas, which was otherwise prohibited for that aircraft; whereas the display organiser believed that it was the responsibility of the operator or the pilot to fly the aircraft in a manner appropriate to the constraints of the display site.

Following the incident, the Civil Aviation Authority published 'Actions that impact on U.K. civil air displays', raising the requirements for civil air displays. These requirements include a comprehensive risk assessment. It states that "it is the responsibility of those organising the air display to identify the risks that the air display may pose, consider how those risks can be managed and mitigated and then inform those who may be affected by the air display".

The new requirements include consideration of the proximity of congested areas.

There is an expectation that the organisers will appoint a Fight Display Director who is the central point of contact.

All air displays are performed inside an imaginary three-dimensional box, with a crowd line, a hard base, and the lowest allowable altitude during the display. The geographical limits of the display are designed to avoid built-up areas. The size of these airport-specific boxes was increased after Shoreham.

These additional requirements have precluded some air shows and reduced spectator excitement.

Referring to the chapters which follow on Catastrophe Prevention, we see above failings in the following as aspects of safety management (Table 9.5).

Incident Involving VFR – Death of High-Profile Footballer

In 2019 a pilot and passenger set off from Nantes airport on a trip to Cardiff. The high-profile footballer player had joined Cardiff football team.

The investigation identified the following causal factors:

1. The pilot lost control of the aircraft during a manually flown turn, which was probably initiated to remain in or regain Visual Meteorological Conditions (VMC).
2. The aircraft subsequently suffered an in-flight break-up while manoeuvring at an airspeed significantly more than its design manoeuvring speed.
3. The pilot was probably affected by carbon monoxide poisoning.

A loss of control was made more likely because the flight was not conducted following the safety standards applicable to commercial operations. The flight was operated under Visual Flight Rules (VFR) at night in poor weather conditions. The report suggests that the pilot did not have training in night flying and lacked recent practice in instrument flying. There was no carbon monoxide detector with an active warning in the aircraft, which might have alerted the pilot to the presence of carbon monoxide in time for him to take mitigating action. It is alleged that this was an unlicensed charter flight. The aircraft had an airworthiness certificate from 1984, which is OK as long as routine maintenance and inspection programmes

Table 9.5 Safety Management Failures (Shoreham Air Crash)

Aspect of Safety Management	Evidence of	See Chapter
Operator and Operating Procedures	Pilot did not follow rules on height and speed, guidance was not clear	19
Human Factors	Training was inconsistent for this particular kind of aircraft Pilot did not recognise that the plane was too low	20
Equipment Maintenance and Inspection	Defects not reported, Airworthiness Approval not complete. During periods of inactivity, maintenance schedule not followed.	22
Risk Assessment	No formal safety management systems in place to identify and manager the hazards and risk	25
Governance of Safety	Various parties in disagreement about who was responsible for what	31

were completed. At first sight, this was apparently the case. However, the inspection was carried out in the context of private and not commercial flying. Furthermore, the report alleges, the altimeter and transponder calibration had not been carried out, and the Certificate of Release to Service specifically stated that because of this, no IFR flight was permitted (summary of Aircraft Accidents Investigation Branch 'Report on the accident to Piper PA-46–310P Malibu, N265DB' 1/2020).

So, this appears to be an incident of major pilot culpability. But not so fast. CAP 667, Review of General Aviation Fatal Accidents 1985–199475, discussed the major types of General Aviation fatal accidents, including loss of control in Instrument Meteorological Conditions (I.M.C.). This report by the Civil Aviation Authority found that:

- Three-quarters of the pilots involved were attempting to fly in I.M.C. when not qualified to do so.
- More than two-thirds of the pilots were flying outside the privileges of their licence, often leading to structural break-up.
- Almost two-thirds continued flight into adverse weather, and more than half were thought likely to have suffered from disorientation.
- Almost a quarter experienced some kind of technical failure.

So, such activity is commonplace. See Chapter 33 on Just Culture, and Chapter 20 on Human Factors and Normalisation of Deviance.

Commonplace failures should not be blamed on the individual, but on the system that is managing them.

RAF Nimrod MR2 Aircraft XV230, Afghanistan (2006) – 12 Fatalities

Acts of war, whilst indeed a man-made catastrophe, are not within the scope of this book. Incidents involving military aircraft were therefore not included in my research. However, the case of a reconnaissance aircraft that was lost over Afghanistan deserves note. The aircraft suffered a catastrophic mid-air fire. The aircraft was a total loss, and the 12 crew onboard perished. No enemy fire was involved. Although the aircraft was lost in enemy territory, photographs were able to be taken of the wreckage before the combat search and rescue team had to be airlifted out of the location. It is clear that a failure in the aircraft itself caused the fire. The incident was the subject of an RAF Board of Inquiry and subsequently an independent review (Haddon-Cave,2009, An Independent Review into the Broader Issues Surrounding the loss of the RAF Nimrod MR2 Aircraft XV230 in Afghanistan in 2006). We will return to the review in Chapter 32 on Laws, Regulations, and Standards since the review brings to light issues with the 'Safety Case' methodology which is discussed in that chapter.

In summary, the Nimrod aircraft was an extensive modification of the de Havilland Comet, one of the world's first commercial jet airliners. In order to operate as a long-range reconnaissance aircraft, one of the modifications involved enabling some of the jet engines to be shut down in flight to conserve fuel. A cross-feed duct was designed to enable hot air to be directed to engines on either side of the aircraft so that engines could be restarted in flight. This duct became very hot in use and passed close by some fuel lines in a compartment that had no means of fire detection or suppression. A leak, perhaps due to joint failure, resulted in the fire which escalated. This was an issue apparently not identified in the safety case.

MARINE TRANSPORT

Marine transport may not, at first sight, appear as a particularly hazardous industry. At least, it did not seem that way until I read Charles Perrow's book 'Normal Accidents' (1984). Perrow analysed incidents in a wide variety of industries and concluded, Marine Transport:

> Collisions, groundings, and tanker explosions occur for no good reason .... There would seem to be every reason for the accident rate to decline, instead of rising, as it has since ships are equipped with technological marvels from collision avoidance devices to satellite navigation devices ... and are increasingly subject to national and international regulation.

When I was writing this section of this book in April 2021, the Suez Canal had been blocked for over a week when one of the largest container ships in the world ran aground and swung across the canal, blocking the entire width. It was said at the time that high winds and a sandstorm caused the incident. As a frequent visitor to the Middle East, my experience is that high winds and sandstorms are not unusual. That such conditions would bring about the complete closure of the canal for a week seems somewhat incredible, and one might have thought that modern technology would enable navigation in poor visibility. Fortunately, no lives were put at risk in this event; but losses to the Egyptian state and the insurance industry are likely to run into hundreds of millions of dollars.

The primary risk in marine transport is a collision with another ship or grounding on rocks. This event could be anywhere at sea or close to land and could involve cruise ships carrying thousands of people or chemical tankers carrying toxic or explosive materials.

Recent trends are hard to assess given statistics, including damage by wars, hurricanes, hijackings, and migrant ships. Perrow stated that there were 71,129 ships in service worldwide in 1979, and that 400 of these were lost in that year. This data gives a risk of loss of 5.6 per 1,000 ships per year. A better measure, as provided by Allianz in their A.G.C.S. Safety Shipping Review 2020 is an accident rate per ton-mile. Allianz reported that this accident rate increased 74% over a decade.

An analysis of the effectiveness of incident prevention methods between the aviation industry and the marine industry shows some key differences (Abdushkour et al. 2018).

A ship has an assigned 'Order of the Watch' (OOW). The OOW makes the decisions that need to be taken on the navigational bridge and is the primary person responsible for collision avoidance. The performance of the OOW is subject to debilitating factors such as lack of sleep, high workload, stress, noise levels, variable levels of experience, mental health issues, and missing home and family. The ergonomics of the navigational bridge, which varies according to the age and design of the ship, maybe a performance-enhancing or a debilitating factor.

Clear procedures and navigational aids can help the OOWs situational awareness and reduce the potential for human error.

There are marine 'Rules of the Road' known as COLREGs (Collision Regulations). These comprise a set of rules provided to the OOWs in order to help them assist the situations where they encounter other marine traffic. These encounters include crossing, head-on or overtaking. Nevertheless, making decisions requires complete understanding and interpretation of the whole situation around the ship. The rules guide the OOW about the suggested actions to avoid collision with other vessels, and give prohibited actions.

These rules and their interpretation need to be well understood by the OOW to avoid any conflict situations at sea. Three conditions of vessel conflict have been identified that cover all possible collision situations at sea, according to the International Maritime Organisation (2005). The collision situations are Overtaking, Head-on, and Crossing:

- Overtaking situation – Any vessel approaching the other from the stern is an overtaking vessel, and she shall keep clear of the vessel being overtaken.
- Head-on situation – Any vessel meeting the other on a reciprocal or near reciprocal course in a head-on situation, where both vessels shall alter course to starboard side so they pass port to port.
- Crossing situation – Any vessel on a crossing course with another, where the risk of collision exists, is in a crossing situation. The vessel seeing the other on her starboard side shall keep clear and avoid passing ahead of the other vessel if the circumstances would otherwise allow it.

So far, so good. We have an internationally respected organisation – the International Maritime Organisation (I.M.O.) specifying these and many other rules to avoid collisions in like manner to the aircraft industry organisations. However, there is considerable subjectivity and uncertainty of COLREG regulations (Abdushkour et al. 2018). Firstly, these regulations have not stopped accidents from happening. The rules are, to a degree, somewhat subjective in nature. Sometimes they do not inform the OOW about the exact action to take, instead, it leaves the decision to OOW to decide. Some phrases used include: 'If the circumstances of the case permit', 'In ample time', and 'If there is sufficient sea room'. These cases are subject to the interpretation of the situation. The COLREG does not inform the OOW about the magnitude or the time to take actions. The judgement is left to the experience of the OOW. These judgements cause dangerously subjective decisions to be made.

The navigational aids and equipment on a ship's bridge, which are used to assist the OOW in understanding the situation around the ship and conducting the navigational watch, include:

- Ship's conning display unit
- Weather monitoring unit
- Automatic Identification System (A.I.S.)
- Radar, X and S bands/Automatic Radar Plotting Aid (ARPA)
- Electronic Chart Display and Information System (ECDIS)
- Global Positioning System (G.P.S.)
- V.H.F. for external communication
- Echo sounder

The O.O.W. must collect all this information, analyse it and provide the most appropriate decision for the ship's safety.

Ships Carrying Hazardous Materials

The transportation by ship of hazardous materials has particularly cata-strophic potential, given the possibility that the dangerous materials could be released in the event of a ship collision. In recent years, the ship transportation of Liquified Petroleum Gases (LPG) and Liquified Natural Gas (LNG) has grown considerably. LPGs are kept liquid using pressure. Should the pressurised storage vessel be punctured, the contents will ra-pidly vaporise, creating a large vapour cloud. LNG is liquefied by cooling to below 143°C at the supply terminal and maintained liquid by insula-tion during shipment. Heat lost through the insulation causes boil-off of some LNG. This gas is typically used in the ship's boilers.

There are also installations where a Floating Liquified Natural Gas (FLNG) terminal is positioned in preference to build an LNG terminal on land. This FLNG is a modified LNG carrier which is permanently an-chored. It is equipped with equipment to vaporise the LNG, using heat transfer from seawater, and connected to the natural gas main onshore. See chapter 13 on process safety infrastructure for the dangers of 'roll-over' during LNG storage.

Accidents to such carriers are rare. I examined 500 reports available from the Marine Accident Investigation Board (M.A.I.B.). The M.A.I.B. com-municates and investigates marine incidents around the world. Only 13 involved hazardous cargo, and most of these were near-miss, anchor drag, and groundings where there was no release of dangerous material. There were two incidents of dockside release. One of these was a release of toxic and flammable styrene monomer was due to a faulty valve, presumably corroded, which released a large amount of vapour. Another was due to a defective sample point, resulting in a large release of LPG. Another inter-esting incident involved a cargo believed to be non-hazardous – 'incinerator bottom ash'. However, two explosions occurred due to hydrogen being released from the payload of incinerator bottom ash when the material came in contact with water. This consequence is an issue of inadequate risk assessment.

Incident Examples – Ships

Costa Concordia (2012) – 32 Fatalities

One of the most notorious recent marine incidents was that of the Italian cruise ship Costa Concordia which ran aground, capsized, and sank in January 2012. Thirty-two people out of the 3206 passengers and 1023 crew members on board died, and a further 157 were injured. Subsequently, the ship's captain was found guilty of manslaughter, causing a maritime accident and abandoning his ship.

The investigation concluded that the ship was sailing too close to the coastline in a poorly lit shore area. The ship's captain had planned to pass at an unsafe distance at night at high speed. Apparently, the objective was to perform a maritime 'salute', a common practice that included the cruise ship sounding its horn; the Concordia had performed several in the past.

The captain ordered a course change, who had more than seven years of experience, but the Indonesian helmsman steered the boat in the opposite direction, probably due to language issues. It reportedly took 13 seconds to correct the manoeuvre. The boat's bow ultimately swung clear, but the stern collided with the rocks.

Confusion on the bridge resulted in conflicting orders, but the damage had been done: the Concordia's port (left) side had suffered a 174-foot (53-metre) tear. An assessment of the damage revealed that five compartments, including the engine room, were flooding, and the ship soon lost power. In addition, with neither the engines nor rudder functioning, the ship could not be steered. The drifting ship eventually ran aground near the shore. The Italian Coast Guard called the Concordia, having been alerted by a passenger on her mobile phone. The captain, however, downplayed the damage, only noting that the vessel had experienced a blackout. Some 10 minutes later the coast guard contacted the ship again, and at this time the crew admitted that the vessel was taking on water. The captain's only request was for tugboats.

According to reports, approximately 15 minutes after the first rescue boat arrived, the captain finally ordered the Concordia to be abandoned, though lifeboats had already been launched. The captain and crew abandoned the ship leaving approximately 300 people still on the ailing vessel. A coast guard captain called the Costa Concordia captain, who was in a lifeboat with other Concordia officers and ordered him to return to the ship to oversee the evacuation. He refused. However, the rescue operations included 25 patrol boats, 14 merchant vessels, and numerous helicopters by this time. By early morning, 4,194 people were evacuated from Concordia. Thirty-two people died in the disaster, and the last body was not recovered until November 2014.

(Ministry of Infrastructures and Transports (Italy) Marine Casualties Investigative Body Costa Concordia, Report on the safety technical investigation) together with (Wikipedia – Costa Concordia).

Referring to the chapters which follow on Catastrophe Prevention we see above failings in the following as aspects of safety management (Table 9.6).

Dona Paz, Philippines, 1987, 4386 fatalities

A collision occurred between the passenger ship Dona Paz and an oil tanker named Vector.

Travelling to the Philippine capital, the vessel was seriously overcrowded, with at least 2,000 passengers not listed on the ship's manifest. It has also been claimed that the ship did not have a radio and that the lifejackets were locked away. However, official blame was directed at Vector, which was found to be unseaworthy and operating without a license, look-out, or

Table 9.6 Safety Management Failures (Costa Concordia)

Aspect of Safety Management	Evidence of	See Chapter
Human Factors	Normalisation of Deviance – deviating from course for the maritime salute was common practice	20
	Misunderstandings due to the language difficulty	
	Captain downplayed the damage to the Coastguard	
Emergency Response	The emergency response was delayed	23

qualified master. With an estimated death toll of 4,386 people and only 25 survivors. It remains the deadliest peacetime maritime disaster in history.

Marchioness, UK, 1986 51 Fatalities

The Marchioness disaster was a collision between two vessels on the River Thames in London in the early hours of 20 August 1989, which resulted in the deaths of 51 people. The pleasure steamer sank after being hit twice by the dredger Bowbelle.

The steamer had been hired for the evening for a birthday party and had about 130 people on board, four of whom were crew and bar staff. Both vessels were heading downstream, against the tide, the dredger travelling faster than the smaller vessel. It is likely that the dredger struck the steamer from the rear, causing the latter to turn to port, where she was hit again, then pushed along, turning over and being pushed under the dredger's bow. It took just thirty seconds for Marchioness to sink; 24 bodies were found within the ship when it was raised.

A formal inquiry in 2000 concluded that "The basic cause of the collision is clear. There was poor look-out on both vessels. Neither vessel saw the other in time to take action to avoid the collision".

In each vessel, the maintenance of a good look-out was seriously hampered as a result of design so that visibility from the wheelhouse was seriously restricted and in neither vessel were sufficient steps taken to overcome this difficulty. A further major contributory factor was the failure of Marchioness to keep to the starboard side of the channel: this is linked to the failure of look-out in that her Skipper plainly thought that the channel was clear.

The collision and the subsequent reports led to increased safety measures on the Thames. Four new lifeboat stations were installed on the river. (MAIB Report into the collision between the Passenger Launch Marchioness and MV Bowbelle, together with Wikipedia: Marchioness Disaster).

Transhuron (1974) – Close Call

Even though there was no loss of life or environmental damage, this final example was selected because the incident demonstrates some of the complexity of maritime systems and how that complexity contributes to an incident. For a detailed and enthralling description see Perrow C. (1984): Normal Accidents: Living with High-Risk Technologies.

The Transhuron was operating in the Arabian sea. When the ship had been reconditioned, air conditioning directly under the propulsion switchboard. Whilst it was recognised that the pipework should not be in the vicinity of switchboards, this piping was separated by steel floor from the switchboard and ran to a nearby condenser.

After installation, engineers found that they needed a bypass valve. An iron nipple was installed on a bronze condenser head to hold a gauge. The resulting modifications resulted in a dissimilarity in metals which slowly accelerated corrosion. Eventually, a connection failed, spraying water into the propulsion switchboard through an opening in the deck through which cables from the switchboard passed. The switchboard shorted out, and a large fire resulted. The carbon dioxide fire suppressant systems did not work when turned on. Hand-held fire extinguishers were eventually used to extinguish the fire.

The ship's master sent an urgent message to his company, located in New York, requesting a tug. The message was sent through the nearest relay in India. The ship had no propulsion system and was drifting in the Arabian Sea. No reply was received, nor after four further requests over the next seven hours. The next morning the Transhuron radioed that they were drifting in heavy 10-foot seas through frequent rain squalls, 23 miles from an island with only emergency generator power.

The radio operator in India then informed the ship that his station did not recognise urgent messages; all messages went in on a routine first-in, first-out basis. The operator simply deleted the word 'urgent' that headed all messages. The New York Company, still not fully up to date, did not immediately respond to the emergency, other than by asking many questions about finding the cause, and saying nothing about getting a tug. Meanwhile, passing ships had offered assistance since he was expecting a tug to be despatched by his home company. This approach is questionable in hindsight. The captain is fully in charge of the ship. At 11 am he sent out a distress signal. The closest ship responding was 110 miles distant. Opportunities to drop anchor at appropriate potential anchorages were overlooked. A more urgent distress signal found a ship 45 miles away, and this ship sped toward the Transhuron. However, the tow line fired from their line throwing gun fell short, and a seaman had been injured when firing the gun – they, therefore, needed to take him to hospital. They pleaded with the ship to give them a short tow since they were now just clear of a nearby island. This time it was decided that the Transhuron

Table 9.7 Safety Management Failures (Transhuron)

Aspect of Safety Management	Evidence of	See Chapter
Equipment Suitability and Reliability	Design failures – location of gauge and materials used	22
Equipment Maintenance and Inspection	Gauge and nipple not serviced Carbon dioxide system failed	23/24
Operators	Switchboard should have been de-energised immediately. Captain failed to request a tow for passing ships sooner than he did.	21
Emergency Response	Failure of the operators in New York to respond promptly	23

would use its line throwing gun. At this point, the ship struck the bottom. The ship started to break up and lose its cargo of fuel oil.

An Indian naval vessel appeared and boarded the ship. After making an underwater survey, it was determined that the ship was lost. The master and four officers decided to stay on board. The requested tug finally arrived. When the Transhuron started to move in heavy seas, the officers prepared to abandon ship. There were further difficulties in launching the lifeboats. Fortunately, all officers were saved, and the cargo was successfully pumped out. It might have been so much worse.

Perrow lists the following contributors to the accident, to which I have added the relevant chapter numbers in the section of this book (Table 9.7).

Ennerdale (2006) L.P.G. Leak

There was a leak from a gas carrier Ennerdale whilst cargo sampling operations were taking place alongside the Fawley Marine Terminal. The leak was sealed 29 hrs later after an estimated 66 tonnes of propane had been lost to the atmosphere. Fortunately, there was no ignition and no damage, but the incident demonstrates the potential for a catastrophe involving gas-carrying ships (M.A.I.B. Report 10/2007).

During the sampling operation, the sampling valve assembly came off in the hands of the cargo surveyor, and a release of propane from the open connection commenced. The emergency shutdown valves (ESD) were activated, but the leak continued, and it was clear that the ESD valve was not completely shut. The emergency services were called and doused the ship with water. The port of Southampton was closed to all traffic, and all ships at the terminal were evacuated. The leak was eventually sealed using a sealing compound typically used to make temporary repairs to leaking pipework.

Table 9.8 Safety Management Failures (Ennerdale)

Aspect of Safety Management	Evidence of	See Chapter
Governance of Safety	Management of Change – regarding the use of the drain as a sample point	31
Equipment Suitability and Reliability	Regarding the sample point	21
Risk Assessment	The only other barrier to a release if the sample point failed was the ESD system which did not work.	25
Equipment Maintenance and Inspection	The ESD system did not seem to be routinely tested and proper operation of the valves verified	22

The drain valve used for sampling had never been designed as a sampling point. It involved a screwed fitting that became partly unscrewed whilst orientating the nozzle in the desired position for connecting the sampling container. Subsequent investigation of the ESD valve showed that it had been jammed open by a small burr. There were no records of when the ESD valves were tested.

Referring to the chapters which follow on Catastrophe Prevention we see above failings in the following as aspects of safety management (Table 9.8).

RAIL TRANSPORT

Rail incidents are generally well remembered. Consider the nature of rail travel – Massive vehicles confined to tracks travelling as fast as possible to get passengers or goods to their destination in the quickest possible time.

Railway safety is improving. The 'Report on Railway Safety and Interoperability in the E.U. 2020' published by the European Agency for Railways comments:

> The railway safety level of the Union railway system remains high; it is actually one of the highest worldwide. In a multi-modal comparison, rail appears as the safest mode of land transport in the E.U., with the fatality rate for passengers gradually approaching that for aircraft onboard passengers. The safety level in terms of fatal accident rate has improved continuously since 1990, with an average annual reduction of more than 5%. Major accidents resulting in five or more fatalities have become rare: only two such accidents occurred in the last two years.

The E.U. report continues:

> After the exceptional year 2018, with no single major accident re-
> corded, two such accidents occurred in 2019. An overall downward
> trend has been observed since 1988, whereas the rate of improvement
> has been 'softening' over the past two decades. There were on average
> 13 major railway accidents each year during the 1990s; this figure has
> now reduced to an average of eight accidents per year in the 2000s and
> four in the 2010s. Estimated for the most recent past years, major
> accidents occur after train runs more than one billion kilometres.
>
> The fatality risk for a train passenger is one-fourth of the risk for a bus/
> coach passenger, but almost twice as high as that for commercial air-
> craft passenger. The use of individual transport means, such as pas-
> senger car carries a substantially higher fatality risk: car occupants have
> almost 50 times higher likelihood of dying compared to train passenger
> travelling over the same distance. The fatality risk for an average train
> passenger is now about 0.05 fatalities per billion passenger-kilometres,
> making it comparatively the safest mode of land transport in the E.U.

However, the data is not so encouraging for some other countries.
Furthermore, incidents involving Dangerous Goods in the E.U. shows an
adverse trend.

If we exclude suicide fatalities, most fatalities on railway premises are
from accidents to persons. Fatalities from level-crossing accidents account
for 29% of fatalities. Fatalities from collisions and derailments represent
less than 2% of all railway fatalities. People strictly internal to railway
operation (passengers, employees, and other persons) represent only 3% of
people killed on E.U. railways, and the latter are mostly occupational health
and safety incidents as opposed to those which constitute a catastrophe.

In 2018, on average more than seven suicides were recorded every day on
E.U. railways, totalling 2,637. Sadly, suicides represent 75% of all fatalities
on railways and, together with the unauthorised person fatalities, constitute
an overwhelming 91% of all fatalities occurring within the railway system.
However, suicide does not fall into our classification of major catastrophes,
and we will not consider suicides here.

The E.U. report identifies precursors (contributory causes) to incidents as
follows:

- Broken wheels and axles
- Wrong-side signalling failures
- Signals passed at danger
- Train buckles
- Broken rails.

As I researched this topic, I became convinced by the statistics that train safety was improving. Then on 8 April 2021, I read:

> A busy passenger train carrying 490 people has derailed in a tunnel in eastern Taiwan, killing at least 50 and injuring dozens more … at least 69 survivors were being treated in several hospitals.

From the perspective of catastrophe, the issue of derailment is of major concern, especially if the train is carrying hazardous materials. See incident examples below. Not having much railway background, I initially considered derailment an inevitable risk over which there can be little control. Amongst the causes of derailments are

- Earthworks eroded or wasted away due to flooding of rivers and streams crossing or running parallel to the railway.
- Subsidence due to water accumulation and high-water level due to insufficient drainage.
- Collapse of bridges and tunnels.
- Frost heave (the upwards swelling of earth during freezing conditions caused by an increasing presence of ice)
- Ruptures and excessive wear of main rails. Track twist. Heat buckles.
- Broken or missing rail fastenings.
- Point 'geometry' failures, e.g. a component of the point system becomes bent.
- Failure of signalling, e.g. ambiguous signalling information or points being allowed to operate whilst a train is passing or located on top of the point.
- Ruptures of axles and wheels, and suspensions.
- Wagon frame twist, failure of wagon load-bearing elements or buffers.
- Brakes fail or do not release and overheat the wheels.
- Improper loading such that the load on the axles is seriously skewed (Table 9.9).

Potential preventative measures are shown in Table 9.9.

This list was compiled from the Det Norske Veritas (2011) report 'Assessment of freight train derailment risk reduction measures'. Only examples are shown requiring little technical knowledge of railway operation.

It seems that there is a lot that could be done to reduce derailments, but many of the potential measures have not yet been adopted broadly.

Many hazardous installations have railways as a means of importing materials for processing and exporting products. These include oil and gas operations, chemical works, power stations, and mines. Such large-scale rolling stock moving in amongst highly hazardous chemical processing equipment is a clear danger and a potential precursor to a disaster. Railway

Table 9.9 Potential Preventative Measures to Avoid Train Derailments

Preventative Measure	Extent of Adoption
Check rails to prevent derailments, in particular at sharp curves and at points.	Widely used at points, limited use at sharp curves
Track and flange lubrication to reduce rail flange friction.	Partial
Ground penetration radars to survey conditions of track bed superstructure.	Partial
Interlocking of points operation whilst the track is operated.	Utilised except in shunting yards.
Hot axle and bearing journal detectors	Partial
Acoustic bearing monitoring equipment	Partial
Hot wheel and hot brake detectors	Partial
Dragging object and derailment detectors	
Ultrasonic rail inspection of track at sufficient frequency to detect rail cracks	Widely utilised, frequency varies.
Action limits for track twist	Lack of consistency between intervention limits
Onboard lubrication of locomotive flanges to be able to provide track//flange contact lubrication.	Widely applied.
Secure breaking gear so that if braking components become loose, they do not fall to the ground risking causing a derailment.	Partial
Axle inspections	Widely
Qualifications of registered person for loading	Partial
Prescribed Braking actions to suit difficult track geometries, steep slopes, etc.	Partial

sidings operations are carried out at a slow speed. The siding is typically located well away from the hazardous processing equipment. Nevertheless, railway locomotives are an ignition source. The loading and unloading of rail tankers create many risks including:

- Unloading hose leaks or breaks during unloading, potentially through train movement or wear and tear.
- Vapours emitted during tank filling which drift to an ignition source.

The HSE Report (2003) 'Transport Fatal accidents and FN-Curves' shows an interesting comparison between rail, road, and air transportation expressed as a means of displaying risk magnitude known as an FN curve. We will return to the analysis of FN Curves in Chapter 28. The FN curves can be plotted against the background of tolerable and intolerable risk we discussed in Chapter 3. This study was used in the assessment of whether Advanced Train Protection (ATP) was merited over the less costly Train Protection and Warning System (TPWS).

For catastrophe prevention in tunnels see separate paragraphs at the end of this chapter 'Road and Rail Tunnels'.

Note that in Europe there is a legal obligation known as R.I.D. (Regulation concerning the International Carriage of Dangerous Goods by Rail).

Incident Examples – Rail Transportation

Viareggio Train Derailment (2009) – 32 Fatalities

32 people were killed and 26 injured when a freight train derailed. This was followed by a flash fire fuelled by the escape of liquified petroleum gas (LPG) from one of the tank cars. A total of 14 LPG tank cars were carrying approximately 46 tonnes of LPG each.

The cause of the derailment was reported to have been the rupture of one of the wheel axes of one of the wagons. Such wagons are required to undergo X-ray inspection of the wheel axes every five years, and it appears that this had been completed but did not reveal a fatigue fracture. However, the investigation focussed more on the consequences. It seems to me that a derailment must be assumed as a possible scenario in any railway safety analysis.

There was no immediate ignition. The engine drivers had time to shut down the engine and reach an escape point. The ignition point for the gathering cloud of vapour was probably a passing car.

Narratives on the incident appear to me to be sparse in terms of measures to prevent recurrence. LPG is transported in large quantities on railways. Derailments have a significant probability.

Perhaps consequence assessment of heavy gas releases in urban areas may contribute to better mitigation and emergency planning (Wikipedia: Viareggio train derailment, and Journal of Loss Prevention in the Process Industries (2011) The Viareggio LPG accident).

Lac-Megantic (2013), Canada

'Forty-two people dead following train derailment' read the news headlines on 6 July 2013. Where? Not in China, but in Quebec, Canada where we might expect highly developed standards of safety. The cause was a runaway train that had been parked on a hill. The train was composed of locomotives and 72 tanker cars carrying about 8 million litres of crude oil from the oil fields of North Dakota, bound for the refineries of New Brunswick. The train was deliberately parked with the engines running to maintain the air-braking system. However, not enough handbrakes were applied, and the driver left the train improperly secured on a downward slope. After the driver left for the night, a fire broke out on the train's main locomotive. Firefighters responded and put out the fire. They also switched off the train engine. The train's air brakes were now no longer operable because the engine was shut off. The train began moving towards the town of Lac-Megantic because not enough hand brakes had been applied. The train reached a speed of over 100 km/h, and eventually derailed as it approached the centre of the town. In this beautiful tourist resort at Lac-Megantic, who could have conceived of such a potential tragedy? Interestingly the driver admitted to not applying enough handbrakes but insisted that this was regular practice (see Normalisation of Deviance – Chapter 20 on Human Factors).

Paddington (1999), London, UK

The Ladbroke Grove rail accident, near Paddington in London, killed 31 people and injured 523 (Wikipedia – List of rail accident in the United Kingdom). The collision was between a Turbo (a slower stopping train) and a high-speed train. Whilst the driver of one of the trains (the Turbo) passed a signal at danger (SPAD), there were many other issues, including the fact that the particular signal was hard to see. Signal siting reviews should have been caried out, especially where there has been concern. These reviews are intended to identify these and take corrective action.

In the aftermath of the disaster, Automatic Train Protection (ATP) benefits, previously rejected on cost grounds, were much discussed and subsequently fitted on some lines. I recall the incident vividly. It severely damaged public confidence in the privatised railway system at the time.

I will go into some more detail on this accident, primarily because we will refer back to these in Section 3 on Catastrophe Prevention and Why It Is Failing.

I extract the following from Lord Cullen's report (HSE, 2000, The Ladbroke Grove Rail Inquiry) with some simplifications since the precise details are not relevant for readers of this book.

With regards to the train driver:

> The evidence of train drivers, supported by that of independent experts, demonstrated that there was persisting difficulty in the sighting of the signals which formed part of the re-signalling scheme between Paddington Station and Ladbroke Grove, and in particular those on gantry 8. SN109 had been passed at danger on eight occasions since August 1993. The driver of the Turbo had recently passed out as a driver. It was acknowledged by the train operator that he had not been instructed directly about signals which had been passed at danger (SPADs), and the assessment of his route knowledge did not specifically cover the section between Paddington Station and Ladbroke Grove.

The ergonomics and reasonable response time of the signallers have been widely analysed (for example Stanton, N, Baber, C, (2008) 'Modelling of human alarm handling response times: a case study of the Ladbroke Grove rail accident in the UK'). We will come back to this in Chapter 21 on Equipment Suitability and Reliability.

With regards to the signal personnel:

> It is clear that for a period of time after being alerted to the fact that the Turbo had passed a signal at danger, the signaller did not take action as he was expecting the driver to stop within the 200 yards overlap beyond the signal involved,

(since this was his previous experience – see Normalisation of Deviance in Chapter 20).

"In addition to the putting back of the next signal to red, an emergency stop message was sent to the Turbo by means of its cab secure radio".

Lord Cullen continues

> The evidence of the signaller and other members of the staff at the I.E.C.C. (Integrated Electronic Control Centre) indicated that there was a serious under-rating of the risks involved in SPADs, a failure to realise the importance of immediate and direct communication with the driver where that was possible, and a dangerously complacent attitude to SPADs as being simply a matter of driver error.

With regards to the fitting of Advanced Train Protection:

> Since it was highly probable that the crash would not have happened if the Turbo had been fitted with Automatic Train Protection (ATP) in connection with the pilot system which was in operation in the Great Western Zone, the Inquiry examined the basis for the decision of the train operator that their trains should not be so fitted. The conclusion of a previous analysis was that the fitting of ATP was not justified as the costs outweighed the benefits. … . I conclude that the decision not to proceed with ATP but to install (the less expensive) Train Protection Warning System (TPWS) was reasonable.

Neither ATP nor TPWS were present on the affected track.

Referring to the chapters which follow on Catastrophe Prevention we see above failings in the following as aspects of safety management (Table 9.10).

Carmont (2020), Aberdeenshire, Scotland – 2 Fatalities

A passenger train derailed after striking a landslip. Two people died. Unusually heavy rainfall led to a landslip. The investigation, which has not been concluded at the time of writing this book, focussed on:

Table 9.10 Safety Management Failures (Paddington)

Aspect of Safety Management	Evidence of	See Chapter
Operators and Operating Procedures	Train driver did not stop on red Train driver not trained on specific 'Signals Passed and Danger' issues	19
Human Factors	Too many alarms at control room, alarms not responded to in time	20
Equipment Maintenance and Inspection	Signal inspections not carried out or actioned.	22
Governance of Safety	Not enough attention paid to SPAD metrics	22

- the sequence of events and the actions of those involved
- the operating procedures applied
- the management of earthworks and drainage in this area, including recent inspections and risk assessments
- the general management of earthworks and drainage and associated procedures designed to manage the risk of extreme weather events
- the behaviour of the train during, and following the derailment
- the consequences of the derailment and a review of the damage caused to the rolling stock
- underlying management factors
- actions taken in response to previous safety recommendations

(from the UK Government Website, news – passenger train derailment near Carmont).

I have included this lesser but recent example to demonstrate that the risk of derailments still exists.

ROAD TRANSPORT

Compared to air, rail, and ship, the lesser size of an individual vehicle and its contents make road transport less prone to major catastrophes. Road incidents are very significant in the number of individual fatalities, which don't classify as major catastrophes under the definition used in this book. Even though massive motorway pile-ups of 100–200 vehicles occur the number of fatalities and serious injuries is low by comparison with the catastrophic incidents we have been reviewing. For example, in 1997, a pile-up near my current home close to Bromsgrove in the U.K. involved 160 vehicles. There were just three fatalities and sixty injuries. The most significant pile-up occurred in Sao Paulo in 2011 and involved 300 vehicles, killing one and injuring thirty.

Most dangers of a catastrophic potential are incidents involving road tankers of gasoline, LPG, or toxic chemicals. However, these road tankers are generally well constructed, and serious releases of hazardous contents are rare. Terrorism in this context is considered in Chapter 40.

There is a separate and emerging catastrophic potential with regard to the road transportation of liquified natural gas (LNG). As a material LNG is unusual, as it is liquified by using very low temperatures. It is retained in the liquified state simply by good insulation. Heat penetrating the insulation causes a small amount of the natural gas to boil off. Unlike LPGs the LNG cannot be contained by pressure and the road tanker of LNG simply vents the natural gas. This venting creates a risk that flammable gas will continue to boil off from the vehicle's vent in the event of a breakdown of the road tanker. Natural gas is continuously venting. Whenever the vehicle is not moving, and if the space around the vehicle is confined, natural gas vapours could accumulate to explosive proportions. See also Chapter 17 on New and Emerging Technologies.

ROAD AND RAIL TUNNELS

Tunnels are a particular point of vulnerability. Even without hazardous cargo, there is a risk from the total ventilation failure in a long tunnel. If a train or vehicle were to stop in a long tunnel and ventilation failed, oxygen in the air in the tunnel would eventually be consumed. If there is a fire, the tunnel rapidly fills with smoke, running a severe risk to the driver and passengers.

The Channel Tunnel connecting England and France comprises three tunnels, one for train movement in each direction and a separate service tunnel that provides access in emergencies. The two locomotive tunnels are connected periodically by 'piston relief ducts'. These ducts equalise the tunnel air pressures as trains move up and down. Fires have occurred notably in 1996 and 2008. Prior to 1996, the established emergency procedure was for the train on fire to continue forwards to the end of the tunnel. This procedure was changed to that of stopping the train and escaping. Both fires involved Heavy Goods vehicles, and in the 2008 incident the train was carrying some pharmaceuticals, albeit in a small volume (Wikipedia: The Channel Tunnel Fire).

Dangerous goods are allowed to be carried through the Channel Tunnel. There are some restrictions to the amount of some explosive materials. (Le Shuttle – Eurotunnel's dangerous goods guide 2021).

The Mont Blanc Road tunnel between France and Italy suffered a fire in 1999 in which 39 people died.

A transport truck carrying a 'non-hazardous' cargo of flour and margarine caught fire in the tunnel. Note that there had been 16 other truck fires in the tunnel of the previous 35 years in which the driver was able to stop a deal with the fire on the spot. The Mont Blanc Tunnel is a single shaft with an up lane and a down lane. During the 1999 fire, the driver was alerted to the fire and stopped in an attempt to deal with the fire but was driven back when the truck exploded in flames. The ventilation system in the tunnel drove toxic smoke back down the tunnel faster than anyone could run to safety. The tunnel was too narrow for larger trucks to turn around. The fire melted the wiring and plunged the tunnel into darkness. Some of the fire crews themselves were trapped and had to shelter in emergency fire cubicles which were provided every 600 metres. The fire burned for 53 hours and reached temperatures of 1000°C. The tunnel remained closed for three years (Wikipedia: Mont Blanc tunnel fire).

Manslaughter charges were brought against 13 defendants ranging from fines to suspended prison sentences, the largest of which was brought against the head of security at the tunnel (see Chapter 33 Social Structures and the role of punishment).

In neither tunnel disaster could I find significant measures taken to prevent a recurrence. Indeed, some restructuring of emergency response was made in both cases. This seems another case where we really don't have an answer to catastrophe prevention.

CONCLUSIONS

Air Transport – Potential for Future Catastrophes

With the Boeing 737 Max software-related crashes and the Germanwings suicide incidents particularly in mind, I find the potential for future catastrophes as follows:

Not possible	Unlikely	Possible (50/50)	Likely, no more than one	Very likely/ maybe several
☐	☐	☐	☒	☐

Marine Transport – Potential for Future Catastrophes

It seems to me that the issues with the marine industry are solvable. I find the potential for future catastrophes as follows:

Not possible	Unlikely	Possible (50/50)	Likely, no more than one	Very likely/ maybe several
☐	☐	☐	☒	☐

Rail Transport – Potential for Future Catastrophes

With the incidents involving highly hazardous cargo particularly in mind, I find the potential for future catastrophes as follows:

Not possible	Unlikely	Possible (50/50)	Likely, no more than one	Very likely/ maybe several
☐	☐	☐	☐	☒

Road Transport – Potential for Future Catastrophes

Potential is low, except for transportation in tunnels. I tend to think that the emergency response in tunnels has been improved, therefore I find the potential for future catastrophes as follows:

Not possible	Unlikely	Possible (50/50)	Likely, no more than one	Very likely/ maybe several
☐	☐	☐	☒	☐

Chapter 10

Utilities – Nuclear Power Generation

NUCLEAR POWER – WHY THE FEAR?

The nuclear industry, especially power generation, is probably the most widely associated with catastrophes. Not because they have had many but because the consequences are so insidious. Radiation from nuclear incidents can cause dreadful diseases, sometimes many months after exposure to the radiation. The incidents that have occurred are almost household names – who has not heard of Chornobyl?

Note: in this book the Ukrainian spelling Chornobyl will be used in preference to the more common English translation of Chernobyl.

The HSE first researched and published "The Tolerability of Risk from Nuclear Power Stations" in 1988. This work precedes the more general 'R2P2' HSE (2001) *Reducing Risks, Protecting People*. Faced with an increasing concern, expressed in relation to planning applications and public enquiries into the construction of nuclear power plants, the HSE sought to explain these by comparing nuclear risks versus other risks. This analysis is probably the first time the word 'tolerable' was used in relation to industrial risk. The document explains that risk from nuclear installations is less tolerable (in statistical terms) because of radiation:

- Can harm without being felt.
- Is capable of causing cancer.
- Can harm unborn children.

This, together with the fact that most people do not understand radiation and therefore dread it more than other risks.

At the time of writing this book, there were approximately 440 commercial nuclear reactors worldwide, of which 93 were located in the USA and 56 in France.

It is not the place in this text to describe the variations in the technology of a nuclear reactor involved in power generation. It is sufficient for our purposes to describe a nuclear reactor as a place where the radioactive fuel

DOI: 10.1201/9781003360759-12

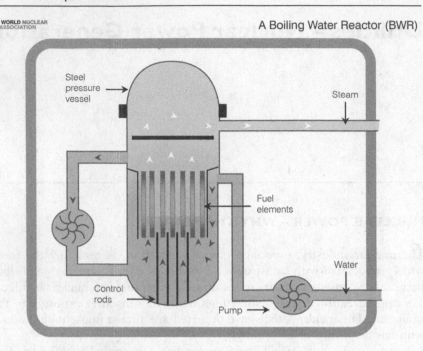

WORLD NUCLEAR
ASSOCIATION

A Boiling Water Reactor (BWR)

Steel
pressure
vessel

Steam

Fuel
elements

Water

Control
rods

Pump

Figure 10.1 Diagram of a Boiling Water Reactor (World Nuclear Organisation, Boiling Water Reactor).

(typically enriched uranium) releases its energy. One type of nuclear reactor is shown in Figure 10.1.

The key to the nuclear reaction is the release of neutrons due to radioactive decay. Released neutrons may then collide with other uranium atoms, splitting them and releasing more neutrons and heat. The rate at which the fuel releases heat can be varied by the control rods, which are typically made of boron and absorb neutrons. The control rods are raised and lowered inside the reactor to speed up or slow the reaction. The moderator slows the speed of the neutrons. The moderator may be in the form of rods, probably graphite, or simply water.

There is an immediate issue of deep concern with the control of nuclear reactors. The reaction cannot be stopped instantaneously, as you can switch off the ignition in a car. For a period after the control rods are inserted, neutrons will continue to split uranium atoms creating heat. The speed at which the reaction can be stopped depends on the reactor's design. Even once shutdown, the fuel continues to generate heat which must be removed to avoid damaging the core. Otherwise, a meltdown may occur, and containment of the radioactive material is lost.

Nuclear power plants require electricity to power pumps and motors, control panels, and lighting just as any industrial plant. Most importantly, pumps and compressors are needed to pump the coolant around the nuclear reactor. If the reactor goes into emergency shutdown, then sufficient electricity is needed to ensure that the cooling is maintained until the reactor is in a safe condition. This length of time varies considerably according to reactor design. If coolant is not supplied, then a meltdown occurs. The most severe scenario is that the molten reactor fuel can melt its way through the foundations of the reactor containment. The release of radiation could be huge. Thankfully total meltdown of this type has never occurred, but we have come close at both the incidents described below. Therefore, nuclear power plants have emergency electricity generators. Mostly these are powered by diesel, which are designed to continue to provide electricity sufficient to ensure a safe shutdown. Emergency diesel generators are found in industry for all sorts of purposes and are known to have reliability issues (see Chapter 23 on Emergency Response Readiness).

In addition to the radiation contained in the nuclear reactor, many nuclear installations have spent fuel pools. Fuel rods discharged from the reactor are held in big pools of water which absorb the neutrons. These require a constant supply of fresh water and cooling. Without this the water level would reduce to below the level of the fuel rods and the fuel rods would overheat releasing their contents.

As an engineer, I have strong faith in my profession. Whilst I never worked in the nuclear industry, apart from a short summer job when I was at university, I always wanted to be a part of it, partly because of its immense complexity and challenges. I assumed that those in the nuclear industry were highly competent. Perrow (2007) points to words used by the U.S. Nuclear Regulatory Commission in 1984 that the Commission had "made a tacit but incorrect assumption that there was any uniform level of industry and licensee competence".

Many issues have occurred that question whether inspections were properly and fully carried out at nuclear facilities. See Chapter 22 – Equipment Maintenance and Inspection.

There remains, therefore, a question as to the degree to which current plants have planned and implemented safety improvements. The Carnegie Institute for International Peace makes this comment:

> To date, there have been three severe accidents at civilian nuclear power plants. Two of these led to significant radiation releases, which averages out to about one major release every seven thousand five hundred years of reactor operation. The International Atomic Energy Agency's (IAEA's) International Nuclear Safety Group believes that major releases of radiation from existing nuclear power plants should occur about fifteen times less frequently if best practices are implemented. Indeed, an improvement on this scale is probably necessary for nuclear power to gain widespread social and political acceptance. (Carnegie Institute for International Peace, 2012)

I am a supporter of nuclear power. As the pressure to eliminate fossil fuels grows, so too will the need to turn to nuclear power. An interesting development is the construction of mini reactors (Rolls Royce 2018 Small Modular Reactors). Small Modular Reactors are open to factory-made module manufacture and on-site assembly. At first, the concept of lots of small reactors may look like it increases the risk of a nuclear failure. However, the idea here is that they will be produced largely identically, reducing infant failures and enabling improvements between identical reactors. This concept aligns with 'shrinking the targets' promoted by Perrow (2007) as a method of catastrophe risk reduction (see Chapter 44).

FUTURE DEVELOPMENT IN NUCLEAR REACTORS

Bill Gates in How to Avoid a Climate Disaster – The Solutions We Have and the Breakthroughs We Need (2021) points out that the high-profile accidents at Three Mile Island, Chornobyl, and Fukushima Daiichi have put a spotlight on the risks of nuclear power. He argues that whilst real problems led to those disasters, instead of getting to work and solving those problems, we just stopped advancing in the field, at least in the U.S.A. Amongst the solutions he is promoting is the sodium fast reactor. A demonstration plant is to be constructed at the site of two coal units in the U.S.A., which are due to discontinuing operations. The reactor design uses liquid sodium as a coolant. It includes a molten salt-based energy storage system; the latter being used alongside renewable energy sources by storing and despatching energy in response to intermittent generation from other sources. Using sodium as a coolant allows several design trade-offs that reduce the plant's cost and complexity. The higher boiling point of sodium means it can operate at atmospheric pressure without expensive pressure vessels. Sodium's more efficient heat transfer capabilities allow for less equipment and a smaller plant (The Chemical Engineer, January 2022).

Some readers will immediately recognise increased risks here from molten sodium, which spontaneously catches fire when in contact with air or water. Apparently, the nuclear industry has prior experience with sodium fast reactors, chiefly in Russia. We also have bad experiences with early sodium fast reactors (see Incident examples below – Monju Reactor (1995)).

Incident Examples – Nuclear Power

Our purpose in including incident examples is to look for reasons why these incidents occurred and what safety methodology failed to prevent the incident. Hence the following two incident examples have been abbreviated to focus only on selected safety failures meriting emphasis.

Fukushima Daiichi (2011) – Release of Radioactive Materials

An earthquake occurred, which was followed by a tsunami. The earthquake caused the automatic shutdown of the nuclear reactors. Failure of electrical supply also occurred, and the emergency diesel generators for the site automatically started as they were designed to do. Using power from the emergency generators, cooling of the reactors could be maintained by pumping water through the reactor core.

However, the following tsunami caused flooding, and the emergency generators then failed. The resultant loss of cooling led to the meltdown of three nuclear reactors, three hydrogen explosions from the hydrogen emitted, and radioactive contamination released. The hydrogen is a consequence of water reacting with metals which typically encase the fuel rods. High temperatures and radiation accelerate this reaction. Radiation was released into the atmosphere, and large amounts of contaminated waste were passed into the Pacific Ocean.

To better understand the failings in safety management, we will go into this incident in some depth, referring to methods of safety management discussed in Section 3, which might, if carried out better, have limited the extent of this incident.

The incident was initiated by a magnitude 9.0 earthquake centred in the Pacific Ocean about 80 kilometres east of the city of Sendai at 14:46 on 11 March 2011. Elsewhere in this book, I have not included a timeline since this is largely irrelevant to the purposes of this book. However, since 4 Units at the site experienced incidents emitting substantial radiation from the same root cause, I have reconstructed a timeline for all the affected units in Table 10.1. This event was initiated by the largest earthquake ever detected in Japan and is the fourth largest detected worldwide since records began in 1900. The reactors facilities along the Japanese coast were equipped with seismometers to detect earthquakes and automatically initiate a shutdown. The seismometers worked as designed, and 11 nuclear reactors at four nuclear power plants on the East Coast of Japan went into emergency shutdown (known in the industry as a SCRAM). The earthquake itself did little damage to the nuclear power stations in the region. 'Walk down inspections' carried out immediately after the earthquake showed damage only to a tiny portion of equipment with low earthquake resistance, which did not affect the safety of the nuclear power reactors.

Collapse of electricity transmission towers did occur throughout the region, and there was a complete loss of external electrical power supplies. Emergency generators automatically started as programmed.

(The Atomic Energy Society of Japan, (2014) The Fukushima Daiichi Nuclear Accident – Final Report of the AESJ Investigation Committee).

So far, all went reasonably well, until just under one hour after the earthquake, a tsunami attacked the Western Coast of Japan. The Fukushima Daiichi power plant suffered widespread flooding. Most functions at power

Table 10.1 Timeline of Events at Fukushima Daiichi

Event	Date and Time Unit 1	Date and Time Unit 2	Date and Time Unit 3	Date and Time Unit 4	Notes
Earthquake	11 March 14:46				
SCRAM Automatic Shutdown	With earthquake				Unit 4 already shutdown for maintenance
Start-up of Isolation Condenser (IC)(Unit 1) and Reactor Core Isolation Cooling (RCIC)(Units 2,3)	With SCRAM				
Tsunami	11 March 15:30				
Diesel Generators Lost	With tsunami	With tsunami		With tsunami	
Seawater Pumps Lost	With tsunami	With tsunami		With tsunami	
AC Power Lost	With tsunami	With tsunami	Not lost		
DC Power Lost	With tsunami	With tsunami	Not lost		
Monitoring and control instruments Lost	With tsunami	With tsunami			
IC (Unit 1) and RCIC (Unit 2)	Lost with DC Power	Lost or could not be monitored	Manually started at 18:00		
Water level dropped below top of active fuel	11 March 15:30	14 March 17:00	13 March 17:00		
Reactor Pressure Vessel Damaged/Core damage	12 March 02:30	14 March 19:20	13 March 10:40		
Alternative water injection commenced using a fire engine	12 March 04:00	14 March 19:14	13 March 09:25		
RCIC Lost (Unit 3)			12 March 11:36		
High-Pressure Coolant Injection System Started (Unit 3)			12 March 12:35		
HPCI Stopped (Unit 3). Safety relief devices opened.			13 March 02:42		
Primary Containment Vessel Opened/Vented	12 March 14:00	Unsuccessful	13 March 09:20		At unit 1 using temporary compressor, Unit 3 probably by rupture disc
Hydrogen explosion	12 March 15:36	None	14 March 11:01	15 March 06:12	

panels were lost. AC power was lost to Units 1 through 5 and DC power to units 1 and 2 and 4.

Units 5 and 6 were spared fuel meltdown, at least partially because they shared power and were shutdown for inspection.

The original design basis tsunami height was 3.12 m established in 1966 for Unit 1 at Fukushima Daiichi. This height was chosen because a 1960 earthquake off the coast of Chile created a tsunami of that height on the Fukushima coast (Carnegie Institute for Peace (2012) Why Fukushima was preventable).

TEPCO decided to locate the seawater intake buildings at 4 metres above sea level and the main plant buildings at the top of a slope 10 metres above sea level.

In 2002, on the basis of a new methodology for assessing tsunami safety developed by the Japan Society of Civil Engineers, TEPCO re-valuated the tsunami hazard and adopted a revised design-basis tsunami height of 5.7 metres (AESJ Report p. 65). However, it does not seem that any action was initiated at the Fukushima Daiichi site in response to this revised guidance.

Tsunami heights coming ashore at the time of the catastrophe have been estimated at 13.1 metres, plus a 'run up' along the slope to the plant reaching 14–15 metres and in a few places 17 metres (AESJ Report and Carnegie Institute, 2013). The argument is made that a tsunami height of 14 m was not a credible event that could have been reasonably planned. I disagree. Historically there have been tsunamis around the world with higher wave heights. I have had the good fortune to see a tree in Hilo in Hawaii on which the height of actual tsunamis is indicated. In 1946 the island was devastated by a 14 m tsunami and in 1960 by a 10 m tsunami.

In the last century, there had been eight tsunamis in the Japan region with maximum amplitudes above 10 metres (some much more), these having arisen from earthquakes of magnitude 7.7 to 8.4, on average one every 12 years. Those in 1983 and 1993 were the most recent affecting Japan, with maximum heights 14.5 metres and 31 metres respectively, both in-duced by magnitude 7.7 earthquakes. This 2011 earthquake was magnitude 9 (World Nuclear Organisation: Safety of Nuclear Reactors).

Hence, I suggest that the selection of a 5.7 m tsunami wall was too low, given the consequences, that this should have been foreseen in the basic safety scoping for the project.

There seem to be failings here in Table 10.2.

The inundation by the tsunami flooded the diesel-driven emergency generators, all of which failed due to flooding. Each reactor unit had two emergency generators. Without any electrical power, cooling water could not be pumped around the reactor, and the sequence of events leading to the meltdown of the reactor commenced.

Table 10.2 Safety Management Failures (Fukushima Daiichi)

Aspect of Safety Management	Evidence of	See Chapter
Equipment Suitability	Construction of Tsunami defenses and location of equipment	21
Governance of Safety	Risk Acceptance and the Management of Change	31
Laws, Regulations, Standards	The Nuclear and Industrial Safety Agency not sufficiently independent of the operator and from government to push for upgrades	32
Legacy Issues	New data became available, on tsunami protection, but it was not retrospectively applied to older plant.	32

It is suggested that had the warnings with regards to tsunami height been recognised the following measures could have been taken:

- Moving emergency diesel generators and other emergency power sources to higher ground on the plant site.
- Establishing watertight connections between emergency power supplies and the plant.
- Building dikes and seawalls to protect against a severe tsunami.
- Installing emergency power equipment and cooling pumps in dedicated, bunkered, watertight buildings or compartments.
- Assuring the seawater-supply infrastructure is robust and/or providing additional reliable sources to serve as the plants' ultimate heat sink.

(Carnegie Endowment for International Peace (2012)) (Figure 10.2).

Alternative or additional heat sinks to the pumped seawater are conceivable. In addition to emergency diesel generators located inside the plant and above the level of the postulated design-basis tsunami, a Taiwanese reactor has two gas turbine electricity generators available at an elevation of 22 metres. In the case of the loss of the primary heat sink, for emergency cooling, two water reservoirs were additionally installed at an elevation of 62 metres ((Carnegie Endowment for International Peace (2012)).

There were also potential issues between the state regulator (NISA) and the plant owner (TEPCO), in that the regulator was not totally independent. As a government body, they understandably had an interest in maintaining viable the nuclear generation business, and once the design had been authorised NISA is also considered at fault. (See Chapter 32 on the role of local regulators.)

There were some additional features installed at Fukushima Daiichi which might have reduced the magnitude of the catastrophe. These are

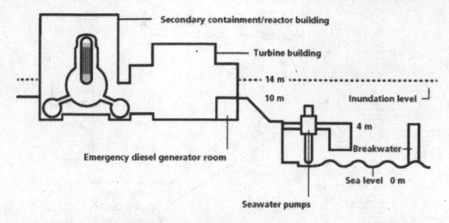

Figure 10.2 Simplified Diagram of a Fukushima Daiichi Reactor (Carnegie Institute for Peace (2012)).

rarely mentioned in commonplace accounts of the incident. The following details are extracted from the Final Report of the AESJ Investigation Committee.

The failure of both the generators to each reactor is a 'common mode failure'. See Chapter 27 for more on common mode failures. The common mode failure of flooding affecting all the generators was foreseeable. Other common mode failures are shown in Figures 10.3 and 10.4, connected with the loss of electricity.

Focussing first on Unit 1, the DC electrical supply failed as well as the AC supply. Now the DC supply is used to power several interlock valves as well as power instruments. It is usual in industry to have 'Uninterruptible Power Supply' (UPS) back-up for critical interlocks and for the control panel instrument display, even when the plant has emergency diesel generators on hand. A UPS is essentially a set of batteries installed for the emergency, which are routinely held in good, charged condition by a UPS control station. Batteries were eventually connected some six hours after the earthquake and five hours after power was lost due to the tsunami. It is understood that regular 12-volt batteries such as car batteries were brought in for this purpose. It was not possible to monitor the critical water level inside the reactor during this time. Retrospectively it has been calculated that the water level dropped below the top of the fuel approximately four hours after the tsunami.

Unit 1 featured an isolation condenser where steam boiling off from the reactor would be automatically condensed in the event of a failure, in this case when steam was not passing to the turbine following the SCRAM. The isolation condenser worked fine after the SCRAM until the tsunami hit, and then the interlocked valves failed closed on loss of DC supply. As identified by the AESJ p143 the early restart of the Isolation Condenser would have eliminated or at least significantly reduced the loss of water, which was

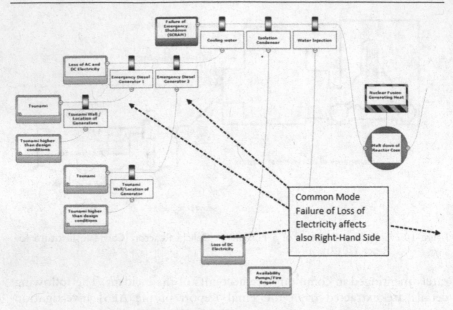

Figure 10.3 Simplified Left-Hand Side Bow-Tie for Melt Down Scenario – Unit 1.

Source: Created using BowTieXP Software © CGE Risk Management Solutions B.V. 2021. BowTieXP is a registered trademark of CGE Risk Management Solutions B.V.

boiling off as steam and being vented through the Reactor Pressure Vessel relief valve. Had the operators been appropriately trained, they could have connected a battery to open the interlocked valves. This action was subsequently carried out, but too late to prevent the meltdown. The isolation condenser, of which there are two, has its own supply tank of cooling water. Each condenser could have been operated alternately, and the cooling water tanks topped up with water from a fire engine.

Once battery power was available, the water level could be read but showed an unexpectedly high level. Subsequent investigation revealed that the water level had dropped below the fuel level, but the resulting water evaporation affected the water level instrument. In my opinion, this failure mode in such a critical instrument should have been predicted.

Water injection using fire engines was commenced some 12 hrs after the tsunami hit. It took considerable time for the fire engines to start water injection partly due to the difficulty in finding the inlet nozzle to which the fire engines should be connected.

It was considered necessary to vent the containment vessel. Since AC power and compressed air were not available, two valves had to be manually opened. However, one of the valves could not be accessed due to the high radiation doses being monitored. The valve was eventually opened using a small generator, and the pressure in the containment vessel declined.

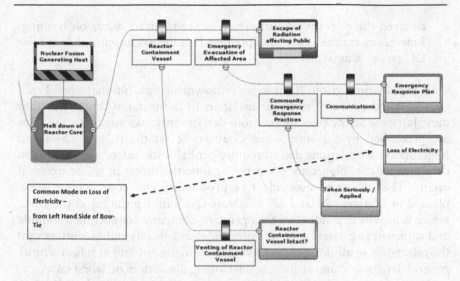

Figure 10.4 Simplified Left-Hand Side Bow-Tie for Melt Down Scenario – Unit 1.

Source: Created using BowTieXP Software © CGE Risk Management Solutions B.V. 2021. BowTieXP is a registered trademark of CGE Risk Management Solutions B.V.

Some 25 hours after the tsunami, the Unit 1 reactor building exploded. Hydrogen released by the reaction of the fuel rods with water under the extreme heat inside the reactor led to an accumulation of an explosive mixture with air.

Units 2 and 3 were equipped with reactor core isolation cooling systems using turbine-driven pumps.

The operator started these following the earthquake and SCRAM, and operation was satisfactory until the tsunami hit. Similar problems occurred compared to Unit 1. Both units experienced a core meltdown, and Unit 3 had a hydrogen explosion. It is thought that Unit 2 was spared a hydrogen explosion due to a blow-out panel that had been displaced by the explosion on unit 1.

Unit 4 was in a shutdown condition for maintenance. However, it also experienced a hydrogen explosion which was subsequently believed due to hydrogen migrating through ducts from unit 3. The attention at Unit 4 was on the spent fuel pool. Since the emergency diesel generators had been lost, cooling was also lost, and there was a danger that water in the pool would overheat, exposing the spent fuel. Unfortunately, this did also occur, although this was not until the end of March.

As noted by the Atomic Energy Society of Japan Report on this incident

Nobody had considered that measures against the loss of all AC and all DC power were actually required. Reactor personnel had never received the relevant education and never received hands-on training. Emergency manuals had been premised on the assumption that at least DC power was available.

One explanation given ((Carnegie Endowment for International Peace (2012)) for the lack of response in Japan to the tsunami threat was that most Japanese safety rules follow from deterministic assessments – the rules are prescribed by the state – see Chapter 32 on the role of Laws and Regulations. Regulations do not require probabilistic safety assessments to demonstrate that plants are protected against the threat of severe external events. The subject of assessment of probability and consequence is explained in Chapters 26 and 27. Contrast this with legislation in the U.K., which is heavily dependent on the operating company conducting a suitable and sufficient risk assessment. Japanese nuclear officials and executives said the reluctance of authorities to re-evaluate tsunami risk might reflect a more general Japanese cultural bias against open discussion of worst-case scenarios or contingencies for which Japanese society and its authorities may be unprepared. These kinds of cultural differences we will return to again in Chapter 33 on Social Structures and Culture.

Chornobyl (1986) – Upwards of 30 Fatalities, Release of Radioactive Materials

Whereas Fukushima Daiichi severely tested measures to contain radiation, Chornobyl involved an intense fire without sufficient provision for containment. The Chornobyl accident destroyed the reactor, and its burning contents dispersed radionuclides far and wide. This tragically meant severe consequences, with 56 people killed, 28 of whom died within weeks from radiation exposure. It also caused radiation sickness in a further 200–300 staff and firefighters, and contaminated large areas of Belarus, Ukraine, Russia, and beyond. It is estimated that at least 5% of the total radioactive material in the Chornobyl 4 reactor core was released from the plant, due to the lack of any containment structure. Most of this was deposited as dust closed in the adjacent land. Some was carried by wind over a wide area (World Nuclear Organisation, Safety of Nuclear Power Reactors).

For an excellent review of the Chornobyl incident see Gerstein M. & Ellsberg, M. (2008) 'Flirting with Disaster – Why Accidents are Rarely Accidental'.

The Chornobyl disaster occurred during a test of its back-up power-generating safety system. Engineers had recognised that, if external power was lost, the reactor's back-up diesel generators required a 30-second delay before they would be able to supply sufficient power. However, they speculated that with some modifications the residual momentum from the turbines would be sufficient to bridge the gap. A test was needed to confirm this, and it was during the conduct of this test that the Chornobyl incident occurred. It is worth emphasising that just because a test is being carried out to improve safety, this does not mean the test itself is safe!!!

The test required that the reactor operates at low power. The turbine would then be allowed to spin down, and the electricity generated by the slowing turbine would be used to power the cooling pumps. The test could therefore only be carried out when low reactor output was required, and finding such opportunities was rare. On the day of the test, a suitable opportunity had been planned, but an unexpected surge in demand led to the postponement of the test.

The test was initially planned for dayshift, where the operators had been prepared for the test. Now the trial commenced late at night with a crew which had not been prepared.

The reactor fell to a much lower level of power than planned. The rapid decline in power had released a significant amount of xenon gas as a natural by-product of the reaction creating a phenomenon known as xenon poisoning. Since this gas absorbs neutrons restoration of power was impeded. A further issue was that the water in the reactor had fallen below its boiling point. In its liquid state water is a better neutron absorber and was also impeding restoring reactor power. It is understood that the operators response was to withdraw many of the reactor's control rods. This was a violation of procedure. The reactor was now unstable. Small variations in flow and temperature dramatically affected power levels due to the 'positive void coefficient', which was an unfortunate characteristic of this reactor. An increase in steam bubbles is accompanied by an increase in core reactivity.

Further seemingly desperate attempts were made to restore the reactor, which included bypassing a low water level interlock. Apparently, the spinning turbine powered the pumps as planned. However, in the reactor's low power state, the gradual reduction of pumping strength as the turbine spun down produced a rapidly increasing level of reactor activity. In response to the operator pressed the shutdown button (SCRAM) to drive the control rods down and stop the reaction completely.

When the SCRAM button was pressed, the insertion of control rods into the reactor core began. The control rod insertion mechanism moved the rods slowly as designed. However, there was an inherent issue with the design of the control rods. There was a graphite neutron moderator section attached to the end of each control rod, by design, in order to boost reactor output by displacing water when the control rod section had been fully withdrawn from the reactor (Wikipedia: Chornobyl disaster).

The emergency SCRAM initially increased the reaction rate in the lower part of the core. This behaviour was discovered when the initial insertion of control rods in another RBMK reactor in 1983 induced a power spike. Countermeasures were not implemented in response to the 1983 observation. A subsequent UKAEA investigative report (INSAG-7) later stated, "Apparently, there was a widespread view that the conditions under which the positive scram effect would be important would never occur".

Local overheating is believed to have occurred. The reactors power spiked at many times its designed full power. This caused the uranium fuel to fracture into small pieces and for the water to vaporise explosively. The resultant steam explosion burst some of the reactor's pressure tubes that contained the fuel, ruptured the steel container around the high-temperature graphite, and lifted the shield on top of the reactor core. The explosion engulfed the reactor building itself and started graphite fires, as high-temperature radioactive debris was thrown from the core.

It would be easy to blame everything on the unit operator for his error in bringing the power down to 1%. Yet, it is fair to ask why the building was so ill-equipped to handle an explosion. And why were there no backup measures within the design of the building in case of fire

Whilst the Gerstein and Ellsberg (2008) account mentions only the operator, I think it is highly unlikely that an operator would be allowed to conduct a test alone. I personally, as an engineer, have spent many night shifts supervising tests. Some accounts more accurately suggest that the deputy chief – engineer of the entire Chornobyl Nuclear Power Plant was present to oversee and direct the test (Wikipedia: Chornobyl Disaster).

A dramatised screenplay version (Chornobyl 1986) portrays a senior engineer in the control room adopting a dominant attitude towards the operators, who tried to question the engineer's direction. I don't know if the source of the screenplays information is correct, but it sounds accurate. If the operators were alone in the conduct of the test, then that would be totally unacceptable.

Whilst this screenplay includes some elements which may be either fictitious or over-dramatised, I consider it to be on the whole, well researched. For the record, one example of fiction created for effect concerns the plant operators who were sent beneath the plant to open drain valves to release accumulated water, in circumstances which were foreseen as undoubtedly leading to their death. In practice, they all survived, a fact which you can only detect by reading a script displayed as the film ends. A tragedy for me in this film is that despite, for the most part, being very appropriate and educational in disaster prevention, it leads the audience to the conclusion that this was a Soviet Union-era accident that could not have happened elsewhere. Having lived in the former Soviet Union, there are some cultural aspects where I might agree – for example, the initial secrecy. Incident investigation in the former Soviet Union is consumed with finding the guilty person, and once identified the investigation is largely uninterested in finding out other useful facts. Fear of punishment leads to secrecy. The lack of means of containment is also seen as Soviet Union cost-cutting. Maybe.

Other aspects have a lot in common with incidents which we report elsewhere in this book:

- Pressure to complete the safety test, which had been deferred before.

- On the planned day, a further delay due to unforeseen circumstances relating to the consumer load, which switched the performance of the test from dayshift (which had been trained, or at least briefed) to nightshift which had not been briefed. A tired senior specialist engineer responsible for the conduct of the test switched to night shift to see the test to its conclusion.
- Written instructions which were unclear.
- Unwillingness by the operators to question the senior engineer.
- Rejection by the senior engineer of any questions or hesitation by the operators.

I would ask those who think these bullet points are a problem of the Soviet Union to reflect on their own experience in questioning authority in a stressed situation. The consequences may not be as severe, but I suggest you envisage how it happened.

In defence of the former Soviet Union, the manner in which they were able to organise the massive emergency response effort with many volunteers putting themselves at considerable risk is quite simply astounding.

Looking at the measures taken following the explosion, firefighters initially attempted to use regular fire-fighting equipment. It seems that they had not been informed about the risks of radiation, or they put their own lives at risk in an attempt to deal with the emergency. Some died soon afterwards from radiation exposure. Subsequently, about 200–300 tonnes of water per hour were injected into the intact half of the reactor using the auxiliary feedwater pumps. This was stopped after half a day owing to the danger of the water flooding units 1 and 2. From the second to tenth day after the accident, some 5000 tonnes of boron, dolomite, sand, clay, and lead were dropped onto the burning core by helicopter in an effort to extinguish the blaze and limit the release of radioactive particles (World Nuclear Association: Chornobyl Disaster). No doubt, the helicopter pilots' actions were heroic since they required the helicopter to fly close above the roof of the reactor that was open due to the explosion.

The next task was cleaning up the radioactivity at the site so that the remaining three reactors could be restarted, and the damaged reactor shielded more permanently. About 200,000 people ('liquidators') from all over the Soviet Union were involved in the recovery and clean-up between 1986 and 1987. They received high doses of radiation, averaging around 100 millisieverts (mSv). Some 20,000 liquidators received about 250 mSv, with a few receiving approximately 500 mSv. Later, the number of liquidators swelled to over 600,000 but most received only low radiation doses. The highest doses were received by about 1000 emergency workers and on-site personnel during the first day of the accident.

Many people put themselves at risk to contribute to the clean-up effort.

Table 10.3 Safety Management Failures (Chornobyl)

Aspect of Safety Management	Evidence of	See Chapter
Equipment Suitability	The plant was not designed to safety standards in effect and incorporated unsafe features	21
Risk Assessment	Inadequate safety analysis was performed Insufficient attention to independent safety review	25–27
Operators and Operating Procedures	Operating procedures not founded satisfactorily following safety analyses. The operators did not adequately understand safety aspects of the plant	19
Governance of safety	Operators did not sufficiently respect formal requirements of operational and test procedures	31
Laws, Regulations, Standards	The regulatory regime was insufficient to effectively counter pressures for production	32
Safety Culture	There was a general lack of safety culture in nuclear matters at the national level as well as locally	33

Several prosecutions followed the catastrophe, one of which was the deputy chief engineer who supervised the test. He was found guilty "of criminal mismanagement of potentially explosive enterprises" and sentenced to ten years imprisonment – of which he would serve three – for the role that his oversight of the experiment played in the ensuing accident (Wikipedia – Chornobyl).

INSAG-7 (The Chornobyl accident: INSAG-7: a report by the International Nuclear Safety Advisory Group) found several contributing factors to the incident. I have added reference to the corresponding safety management elements described in Section 3 of this book. Some of these contributing factors are shown in Table 10.3.

Monju (1995) Japan – Fire

This reactor was a sodium-cooled fast reactor in Japan. An accident occurred in December 1995 in which several hundred kilogrammes of sodium leaked, causing a major fire. There was a subsequent cover-up of the extent of the accident, which delayed its start-up. There was a further incident involving dropped machinery, which resulted in a further shutdown of the reactor. The reactor was subsequently closed (Wikipedia: Monju Nuclear Power Plant).

CONCLUSIONS

Nuclear Power Generation – Potential for Future Catastrophes

With the incidents involving Fukushima Daiichi incident particularly in mind, I am not convinced that there is enough being put in place to avoid other similar incidents. However, I tend to think that incidents of the dimension of Chornobyl, where the reactor went critical for a very short time, are unlikely to happen in the future. I suggest that the potential for future catastrophes as follows:

Not possible	Unlikely	Possible (50/50)	Likely, no more than one	Very likely/ maybe several
☐	☐	☐	☒	☐

Chapter 11

Utilities – Other

FOSSIL FUEL POWER GENERATION, HYDROELECTRIC POWER, WATER, WASTEWATER, WASTE DISPOSAL, INCINERATORS

Compared to nuclear power generation, mining, chemicals manufacture, and oil and gas, the utilities covered below have a comparatively low catastrophic potential. Nevertheless, some interesting hazards are lurking in these relatively benign facilities, as we shall see.

Power Generation Using Fossil Fuels

Fossil fuel power generation can be by coal from mines, fuel oil from oil refineries, or natural gas from underground gas reserves, extracted and compressed for transport by pipeline to the power generation facility. All such power generation involves furnaces that burn fossil fuel, boilers where water is turned into steam under pressure, and turbines where the steam pressure is converted into the turbine's rotary motion, which generates the electricity. There is, therefore, the potential for an explosion of the furnace or disintegration of the pressurised boiler. However, such events are unlikely to reach catastrophic proportions. Incidents associated with fossil fuel power generation are generally related to the supply of fuel. These hazards are either already covered in oil and gas, or mining or will be covered in Chapter 13 on process industry infrastructure, including pipelines.

Water, Wastewater, and Effluent Treatment

Dams are also constructed for water supply amongst other purposes such as hydroelectric power generation and mine tailings treatment (see Chapter 8 on Mines and Mine Waste). Most catastrophes associated with water are due to dams. However, other smaller-scale catastrophes have occurred primarily with personnel entering tanks and simple drowning incidents.

DOI: 10.1201/9781003360759-13

Water and wastewater treatment plants often use chemicals stored at the treatment plant, including chlorine and hydrochloric acid. Such hazardous substances are prone to leakage or mishandling. Effluent treatment plants often release small quantities of toxic gases which can accumulate within enclosed areas.

Many of these facilities operate remotely and are not constantly manned, which increases the potential for incidents.

Entry into sewerage systems has caused a number of fatalities. People entering the sewer without proper respiratory equipment have been overcome by toxic gases. The sewer can also act as a means of unintended distribution of hazardous materials which have been spilled (see Guadalajara in the incident examples below).

In liquid and waste disposal, materials are often poorly identified, and sometimes materials are sent for waste disposal without proper risk assessment.

Hydroelectric Power

Hydroelectric power involves the harnessing of large quantities of water behind dams. The catastrophic risk comes from the dam's failure with the risk of flood in the valleys and plains below. Dams can be exposed to additional stress due to heavy rainfall. They may also degrade in time.

Waste Disposal and Waste Recycling

Waste disposal is intrinsically hazardous because of the lack of control over the constituents of the waste. Waste disposal sites are notorious for fires. As shown below, some are serious.

In the U.K., in 2013, there were 298 fires, 248 in regulated sites, and 50 in unregulated sites (Chief Fire Officers Association – Waste management and Recycling Fires).

Fumes and smoke from a waste disposal centre should always be considered toxic since they come from poorly defined burning materials. The number of waste disposal facilities and their throughput is increasing. This growth is a healthy development from the perspective of sustainability. Major companies are building recycling centres as part of the 'circular economy'. Both mechanical and chemical recycling are being deployed. In mechanical recycling, the waste to be recycled is mechanically segregated, whereas in chemical recycling, chemicals are used to dissolve the required materials leaving behind the remaining unwanted waste. For plastic waste to be recycled, it typically needs to be segregated into plastics of a particular type.

Tyre recycling centres are particularly notorious for fires. During the recycling process the tyre is shredded, providing ideal conditions for burning.

Much waste which cannot be segregated for recycling goes to landfill. Any toxic components of the waste will tend to leach out over time and potentially contaminate rivers. Methane can accumulate, migrate along rock strata and be emitted in residential areas where explosions can and have occurred.

Incinerators

Incinerators destroy waste by burning it. Carbon dioxide and water vapour are discharged from the incinerator stack. There is always an unburnt portion of the waste, the ash. So incinerators are often considered as reducing the volume of the solid waste rather than eliminating it. There are environmental concerns about incinerators. Incinerators are increasingly developing new technology with many precautions to prevent the discharge of toxic gases and particulates.

What is of concern in this book is the potential for an incinerator to malfunction and potentially explode.

The majority of incinerators are equipped with grate furnaces. This technology optimises the waste movement through combustion chambers, contributing to better efficiency and completeness of incineration. The waste is introduced into the incinerator through a throat at one end of the grate. From here, it moves through furnaces to an ash pit in the other end. Incineration usually takes place at temperatures from 750 to 1000 degrees Celsius. The generated heat is subsequently transformed into steam, used for heating or electricity production.

There are two main causes of catastrophic potential which might result from misoperation of the incinerator:

1. Fire or explosion
2. Toxic gas release

An example is given in an IMCA Safety Flash of a 'near miss' when a fire occurred in an incinerator that burnt sludge at a temperature between 810°C and 850°C. A flame failure occurred, but the instrumentation to detect was not set at the fight alarm set point. Therefore, the interlock which should have prevented further fuel entering the incinerator did not operate. 'Back-up' of fuel occurred within the furnace which exploded when an automatic diesel burner reignited the mixture. This resulted in damage to the incinerator and release of smoke. The burning could not be immediately stopped (IMCA Safety Flash 19/19).

The intention of incineration is to obtain complete combustion of the materials. However, complete combustion is not always achieved. Incomplete combustion risks the emission of toxic materials such as dioxins and furan. Furan is a possible human carcinogen. Dioxins are persistent environmental pollutants, and they accumulate in food chains. Dioxins are highly toxic and can cause reproductive and developmental problems, damage the immune system, interfere with hormones, and cause cancer.

It should also be noted that dioxins can be inadvertently produced from incomplete combustion of many materials, including wood, especially damp wood – hence the ban in the U.K. on using wood in domestic wood burners unless it has been kiln-dried (World Health Organisation: Dioxins and their effect on human health).

Incident Examples – Power Generation (Fossil Fuels)

Kleen Energy Natural Gas Explosion (2010) – Explosion

In 2010 there was an explosion at a power plant in Connecticut, U.S.A, which killed six people and injured at least fifty. This natural gas-fired plant was newly constructed and was being commissioned. It was common practice at the time to clean out the pipework of debris using a 'natural gas blow'. In this operation, flammable natural gas is blown through the pipe and vented to the atmosphere. The U.S. Chemical Safety and Hazard Investigation Board investigated this incident and pointed to other incidents associated with the use of the 'natural gas blow'. They pointed out the ease with which the natural gas might reach an ignition source or be self-igniting as debris ejected might strike other material, causing a spark. Replacement of the natural gas blow with air blow was recommended (U.S. Chemical Safety and Hazard Investigation Board 'Kleen Energy Natural Gas Explosion', 2010).

Incident Examples – Hydroelectric Power

Sayano-Shushenskaya Power Station (2009), Russia – 75 Fatalities

This disaster was caused by a turbine failure at a hydroelectric power station in Russia. Seventy-five people lost their lives.

Vibration measurement is a valuable method of detecting progressive degradation of rotors in rotating equipment (see Chapter 22 on Equipment Maintenance). The turbine in question had experienced an elevated level of vibration, although apparently within specification.

The rotor of the turbine itself was 920 tonnes. When the turbine self-destructed the machinery hall was flooded. Many of the dead were plant personnel who had gathered in the machinery hall to help investigate and manage the misoperating turbine (Wikipedia: Sayano-Shushenskaya Power Station).

Laos Dam Collapse, (2018) Laos – Upwards of 45 Fatalities

Part of a hydroelectric dam under construction collapsed, leading to widespread destruction and homelessness. Forty people were confirmed dead. At least 98 more were missing and 6,600 people displaced.

The collapse happened after heavy rains resulted in the overtopping of the dam (Wikipedia: List of industrial disasters).

Incident Examples – Sewers

Guadalajara (1992) – Approximately 250 Fatalities

In 1992 a series of ten explosions took place in Guadalajara, Mexico. The number of people who died has been estimated at around 250, possibly more. It is understood that new water pipes made of zinc-coated iron were built too close to an existing steel gasoline pipeline. An electrolytic reaction commenced between the two dissimilar metals, eventually causing the gasoline pipe to corrode. The leaking gasoline entered the sewers. There was also an issue with the sewer pipe design, which had inverted siphons. These are understood to have held up the lighter gasoline which had leaked into the sewer. The explosions occurred when the gasoline vapours drifted out of the sewer vents and met ignition sources, probably from vehicles in the streets above.

Incident Examples – Waste Disposal

Currenta (2021), Germany – Six Fatalities

Six fatalities occurred after an explosion at a chemical waste site. This chemical waste storage and treatment site serves an industrial park comprising several different chemical manufacturing processes under different company ownership. Such chemical parks are quite common in Europe and the U.S.A., having initially grown from one large company, but with different parts of the operation being subsequently divested.

Whilst the incident is still under investigation at the time of writing this book, one presumes that incompatible chemicals were inadvertently mixed in the same tank.

Midland Resource Recovery, (2017), West Virginia – Two Fatalities

Two people died following an explosion, and a further died from a second explosion approximately one month later. The cause of the incidents was reactive, unstable chemicals that exploded when attempts were made to drain the unexpected liquid from scrap equipment for recycling (U.S. Chemical Safety and Hazards Investigation Board – Midland Resource Recovery, West Virginia, 2017).

Veolia (2009), Ohio

Highly flammable vapour released from a waste recycling process ignited and violently exploded.

Two personnel were severely injured. Multiple explosions afterwards significantly damaged the site as well as neighbouring residences and businesses. The incident occurred in the solvent recycling part of the operations. The process involved a flammable solvent that reacts readily with oxygen, and which can form unstable peroxides. To prevent this, an inhibitor is normally added, but it seems likely that insufficient was being added. The vent systems at the plant were not designed to cope with the vent load which resulted (U.S. Chemical Safety and Hazard Investigation Board – Case Study – Explosion and Fire in West Carrollton, Ohio, 2010).

Incident Examples – Incinerators

Tuas Incineration Plant (2021), Singapore – One Fatality

One man was killed, and two others were injured when an explosion occurred at an incineration plant. In this case, the issue may have been with electrical switchroom work rather than the incineration process itself.

CONCLUSION

Fossil Fuel Plants – Potential for Future Catastrophes

With the potential low, with the main potential on furnace explosions. I consider the potential for future catastrophes as follows:

Not possible	Unlikely	Possible (50/50)	Likely, no more than one	Very likely/ maybe several
☐	☒	☐	☐	☐

Water, Wastewater, and Effluent

The catastrophic potential here is very low, and therefore rate the potential for future catastrophes as follows:

Not possible	Unlikely	Possible (50/50)	Likely, no more than one	Very likely/ maybe several
☐	☒	☐	☐	☐

Hydroelectric

The catastrophic potential here is significant at the dams. More dams are being built, whilst there are many aging dams. There do not appear to be any specific initiatives likely to correct the vulnerability. I consider the potential for future catastrophes as follows:

Not possible	Unlikely	Possible (50/50)	Likely, no more than one	Very likely/ maybe several
☐	☐	☐	☒	☐

Waste Disposal

With waste recycling there is an increased risk of mixing the wrong materials. Industrial parks comprising different organisations sending waste to a common location is a specific increasing risk. I find the potential for future catastrophes as follows:

Not possible	Unlikely	Possible (50/50)	Likely, no more than one	Very likely/ maybe several
☐	☐	☐	☒	☐

Incinerators

The catastrophic potential here is limited, and incinerators are being designed in improved ways with increased automation. I suggest the potential for future catastrophes as follows:

Not possible	Unlikely	Possible (50/50)	Likely, no more than one	Very likely/ maybe several
☐	☒	☐	☐	☐

Chapter 12

Agriculture, Forestry and Wood Processing, Food and Drink

OVERVIEW OF THIS DIVERSE SECTOR

This sector is not immediately thought of as potentially responsible for catastrophes.

It is indeed a sector where state regulations and management practices may be weaker than in the other sectors discussed in this section of this book.

For example, in 2013, at least 100 people were killed when a blaze tore through a locked poultry slaughterhouse in China. Here there is nothing specific about the hazards of agricultural activity simply. It seems that poor fire precautions contributed to the gravity of the incident (Wikipedia, Jilin Baoyuanfeng poultry plant fire).

In the field of Occupational Safety, the agricultural sector has the highest incidents rate of fatal accidents per 100,000 workers (HSE: Fatal injuries in agriculture, forestry and fishing in Great Britain 2020/21). However, at first sight, these types of industries do not seem to have the potential for creating a significant catastrophe involving multiple fatalities.

Let's set aside the potential for failure of agriculture due to disease or natural catastrophes to cause many fatalities due to lack of nutrition. This kind of catastrophe is beyond the scope of this book.

There are, however, materials used in these industries that could potentially give rise to a catastrophe because of their hazardous nature. Anhydrous ammonia is stored at farms, especially larger ones, for fertiliser use. It is also applied directly to land (at least in North America), which requires that the anhydrous ammonia be transferred from the storage tank into 'nurse' tanks towed behind a tractor. Ammonia is applied from rotating wheels into the ground. The material is stored cold and under pressure. If this toxic material escapes, the consequences could be severe. There have been incidents on farms resulting from the rupture or overturn of the 'nurse' tanks, although given that these occur in the open air consequences have been limited. There have been many incidents

DOI: 10.1201/9781003360759-14

with anhydrous ammonia in storage and transportation. Ammonium nitrate is stored and used as a fertiliser in agriculture. If improperly stored, it can be explosive. See Chapter 13 on Process Industry Infrastructure – West, Texas, and Beirut catastrophes. At West, both Ammonium Nitrate and Anhydrous Ammonia were involved.

In poultry processing, highly toxic chorine may be stored and used for treating poultry to kill harmful bacteria. Liquid nitrogen, an asphyxiant, has a variety of uses in food processing including the freezing of products.

Perhaps the most significant catastrophic potential with agriculture, food, and beverages is the distribution of contaminated products, which cause widespread illness and possibly fatalities. Wikipedia lists over 100 notable food contamination incidents (Wikipedia: List of food contamination incidents).

Forestry itself may not have catastrophic potential, other than by wildfire (see Chapter 39), but wood processing certainly does due to the creation of large volumes of flammable wood dust.

In beverages, we have ethyl alcohol, a highly flammable material which can form explosive mixtures with air.

Illustrative Hypothetical Case – Whisky Distillery

So, let's look at our hypothetical case of the Whisky Distillery. The purpose here is to demonstrate that non-technical people can carry out rough calculations to assess if there is a cause for concern or not.

We have an on-site warehouse. How concerned should we be about ventilation and the potential to accumulate an explosive mixture? In drawing A4 (see Appendix A), this warehouse is 68 m × 20 m and is 15 metres tall. A typical barrel can hold approximately 200 litres and is 21 inches (53.3 cm) in diameter and 36 inches (91.4 cm) high.

Allowing for some space for alleyways in the warehouse, the reader will be able to calculate that approximately 36,000 barrels can be stored. (120 barrels along the length, 30 barrels along the width and racks 10 barrels high.)

This is equivalent to 7,200 m^3 of whisky. The approximate loss of liquor during ageing (the angels share), evaporation through the pores of the oak cask, is around 2% per year (see What is the Angel's Share https://www.distillerytrail.com/blog/what-is-the-angels-share/'. This is for a cool climate such as in Scotland. Distillers in hotter climates like Kentucky need to allow for considerably more. 2% of 7200 m^3, the total stock, is around 165 m^3 of liquid per year or 360 kgs/day.

The total volume of the warehouse is around 20,000 m³. We need to subtract the volume of the barrels – around 7,200 m³, ignoring the small additional volume of the oak barrel and other equipment such as storage rack in the warehouse. This gives a volume of air of around 13,200 m³.

The Lower Explosive Limit (LEL) of ethanol is 3.3% v/v. (this can be obtained from any Material Safety Data Sheet on ethyl alcohol). To get to the LEL we need 435 m³ of ethanol as vapour, equivalent very approximately to 533 kgs of ethanol (assumes density of ethanol vapour is around that of air). If there is no ventilation and we start with no ethanol in the atmosphere, it would take around between 1 and 2 days for the atmosphere inside the warehouse to reach the LEL.

This is not a desirable situation, clearly a ventilation system is needed!

It would also be helpful to consider the consequences of spilling the contents of a barrel. For simplicity, let's assume the barrel's contents are 100% ethanol. For this assessment, we can make an estimate using the free software programme ALOHA introduced in Chapter 4 and explained in some detail in Chapter 28. As input data, I used 200 litres of ethanol. I conservatively used just one air change per hour – similar to having no ventilation with the warehouse doors open on a calm day. Curious readers may want to try this themselves.

I found that the plume of ethanol vapour extends up to 160 metres (525 feet) from the source of the spillage. With ALOHA, it is easy to go back and see how sensitive the risk is to wind speeds, temperatures etc. It can also produce attractive maps of the danger zones showing the extend of the plume both downwind and laterally from the spill (Figure 12.1).

A similar analysis can be done to assess with regards to toxicity from such a spillage.

In this case, ALOHA gives us the distances from the spill below which the ERPG-2 limit is exceeded. This limit is the maximum airborne concentration below which nearly all individuals could be exposed for up to 1 hour without experiencing or developing irreversible or other serious health effects or symptoms that could impair an individual's ability to take protective action.

This analysis further confirms the need for ventilation where barrels of whisky are being handled to reduce fire hazards and toxic hazards should a barrel be punctured.

Incident Examples – Alcoholic Beverages

MGPI (2016) – Hospitalisation Including Members of the Public

In October 2016 an incident occurred at a plant producing distilled products and food ingredients, which included the production of alcohol by

meters

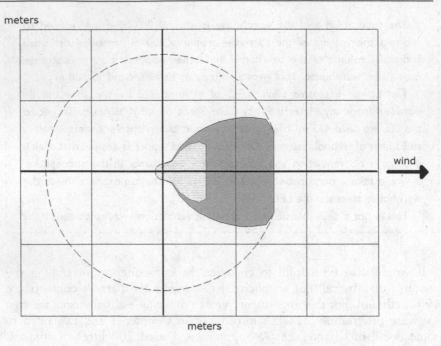

meters

Figure 12.1 ALOHA Output on Flammable Zones for Whisky Barrel Spill.

distillation. One employee and five members of the public required hospitalisation, and 140 people sought medical attention due to the toxic cloud produced by an inadvertent reaction. The incident has many lessons concerning chemical unloading and accidental mixing.

On the day of the incident, a road tanker arrived with a scheduled delivery of 30% sulphuric acid. However, due to a communication error, the sulphuric acid was accidentally discharged into a tank of sodium hypochlorite. The two unloading lines were only 18 inches apart. Normally locked, it seems that the sulphuric acid line was properly unlocked by the operator to allow unloading, but that the sodium hypochlorite line had inadvertently also been left unlocked.

This mixture of incompatible materials resulted in a reaction that promoted the release of a cloud containing chlorine gas and other compounds. The chlorine gas emission was already so dense by the time the truck driver detected the smell, that the truck driver could not get to the shut off valve either on the truck or at the unload point. Access to emergency equipment for the control room operators was inhibited because the equipment was locked away. The mixing was eventually stopped by Emergency Responders wearing appropriate protective gear (US Chemical Safety Board, Key Lessons for Preventing Inadvertent Mixing During Chemical Unloading Operations – MGPI Processing, 2018).

Incident Examples – Wood Processing

Bosley Mill, UK (2015) – Four Fatalities

A wood treatment plant went on fire and exploded catastrophically, destroying the factory and killing four workers. Three separate explosions are understood to have occurred. It took several days for the fires to be extinguished.

Issues with wood dust were apparently a long-standing issue. Maintenance at the plant was allegedly deficient. A corporate manslaughter case was taken against the company managing director, who was subsequently acquitted.

Apparently, a prior report made by a risk analyst, for insurance purposes, commented that "While normal business operations do not result in significant dust explosions, there is a risk that a small explosion, or other disturbance, may cause settled dusts to become airborne, perhaps in sufficient concentration to provide an explosive mixture" (Dust Safety Science: A Dark Day in Cheshire: Bosley Mill Explosion).

We will return to this incident in Chapter 33 when the issue of liability and legal penalties is discussed.

Whisky and Rum Distilleries

Various serious explosions and fires have occurred at whisky and rum distilleries, mainly at the warehouses where the product matures.

For example, an Australian distillery was struck by lightning in 1936, causing widespread damage, but fortunately, on this occasion, no fatalities. In 1996, in the U.S.A., 90,000 barrels of whisky were destroyed when fire raged through seven warehouses. In a similar incident, 17,000 barrels of whisky were destroyed in 2000 when a fire occurred in a maturation warehouse. Severe pollution occurred in the Kentucky River, killing an estimated 228,000 fish.

In 1960 a whisky warehouse in Glasgow, U.K., exploded, killing 19 firemen. Around 450 firefighters battled the fire, which took an entire week to extinguish (The Spirits Business: The world's 10 worst distillery disasters).

Incident Examples – Food Processing

Foundation Food (2021) – Six Fatalities

Liquid nitrogen is also used in poultry processing plants. In 2021, six people died from asphyxiation after a release of liquid nitrogen from a cryogenic poultry process line. The investigation was ongoing at the time of writing this book. The focus of the investigation is on maintenance and on the conversion of the line from freezing using ammonia to a different process using nitrogen (U.S. Chemical Safety Investigation Board: Update on Poultry Plant Incident, 2021). The nitrogen process sounds inherently safer, but clearly there were omissions in planning the transition.

Georgia Sugar Refinery, U.S.A. (2008) – 13 Fatalities

Fourteen people were killed and 40 injured following a dust explosion at a sugar refinery.

The explosion occurred in the centre of the refinery, where bagging and storage facilities fed completed product by a network of elevators and conveyor belts. Many of the buildings were six to eight stories high with narrow gaps in between. Accumulated sugar had occurred in many parts of the sugar refinery. The investigation found that a below-ground sugar conveyor had been modified. Whereas the conveyor and tunnel as de-signed were ventilated, an enclosure was built around the conveyor to reduce the potential for product contamination. This enclosure was not ventilated and likely accumulated a mixture of sugar dust and air inside of the explosive limits. Once this met an ignition source, such as a hot bearing, an explosion occurred, leading to secondary explosions in parts of the refinery where sugar dust had accumulated. The refinery was es-sentially destroyed (U.S. Chemical Safety Investigation Board: Imperial Sugar Dust Explosion and Fire).

Relevant Safety Management Techniques which could have helped avoid this catastrophe (with references to section 3) (Table 12.1).

Potato Factory, U.K. (2018)

An incident occurred at a potato factory in which workers were exposed to sulphur dioxide gas causing significant injury. I have included this event to demonstrate the potential for catastrophe within the food industry. The factory had had a new potato processing machine installed. The preserva-tion process involves exposing the potatoes to a preservative so that, when cut, they do not go brown. The preservative gives off a small amount of sulphur dioxide when used. The factory accepted the new line, tested it with

Table 12.1 Safety Management Failures (Georgia Sugar Refinery)

Aspect of Safety Management	Evidence of	See Chapter
Governance of Safety	Internal audit of factory condition Absence of management of change when building the conveyor belt enclosure	31
Operators	Housekeeping issues	19
Consequence Analysis	No appreciation of the potential consequences from the dust explosion and secondary dust explosions	28

water, and started to commission the preservative system. This is when the worker exposure occurred. The company was fined a considerable sum for inadequate risk assessment (IOSH Magazine Jan/Feb 2022).

CONCLUSION

Agriculture, Forestry, Wood Processing, Food, and Beverages – Potential for Future Catastrophes

The potential is low, but there are many small operations in both wood processing and food preparation. There seems to be no reason to be confident that incidents like Bosley Mill and Foundation Food will not recur:

Not possible	Unlikely	Possible (50/50)	Likely, no more than one	Very likely/ maybe several
☐	☐	☐	☒	☐

Chapter 13

Process Industry Infrastructure

RISKS ASSOCIATED WITH WAREHOUSES, STORAGES, TERMINALS, PIPELINES

Some catastrophic incidents have occurred, not at manufacturing plants or locations of obviously high risk, but in parts of the process industry infrastructure. By infrastructure, I am referring to:

- Warehouses
- Storages – tank of liquid or gaseous hazardous material
- Terminals
- Pipelines

Compared with the other examples of high-risk industries covered in this book, the technology is straightforward but perhaps not as free of risk as it might first appear. No chemical reactions are occurring and very little in terms of materials being heated, boiling, condensing, etc. So, from where does the hazard arise?

Warehouses store materials sometimes for long periods. The materials may be liquid stored in steel or plastic barrels. If the materials are solid, they may be stored in bags of various sizes, or in silos. The hazardous materials within can start to degrade. Containers may become punctured by, for example, the forks of a forklift truck. Unlike the process industries that produced the material, there may be limited fire protection in terms of firewater hydrants, deluge and foam systems, and dedicated emergency response teams. 'Ownership' of the materials inside can become lost, as we shall see in the examples below.

Tank farms with liquid or gaseous storages are typically associated with terminals. But they may also be associated with manufacturing facilities or the facilities that consume the product. For the most part, airports have tank farms to store the fuel for the aeroplanes. There are tank farms at ports to fuel the ships and railway hubs to fuel the locomotives. Often these storages are remote from the facility with which they are associated.

DOI: 10.1201/9781003360759-15

The CSB identified 22 major tank incidents over the last 50 years (U.S. Chemical Safety and Hazard Investigation Report: Caribbean Petroleum Tank Terminal Explosion and Multiple Tank Fires).

Tank farms are a familiar sight in the landscape of industrialised nations. They typically contain large quantities of hazardous and flammable hydrocarbons, including crude oil, gasoline, diesel, and fuel oil. Fires can occur within the vapour space of the tank. Some materials, especially crude oil, are held in floating roof tanks, preventing a vapour space from forming. Such tanks may develop fires at the rim seal, which seals the floating roof from the wall of the tank. Corrosion can occur within and outside the storage tank, especially where moisture collects, resulting in leakage. Filling and emptying operations may result in overfill and release of flammable materials from the vent of the tank. Tank farms may be isolated and unmanned, being remotely filled and discharged by pipeline from a control room which may be hundreds of miles away from the storage area.

Tank farms may also have the familiar pressurised spheres containing highly flammable liquified petroleum gases (LPG), including propane and butane. Any release will quickly evaporate, turning the liquid into a flammable gas, which will drift in the wind – until it finds a source of ignition!

Terminals usually include storage tanks but additionally have means for importing and exporting materials. These means might consist of:

- Road tanker loading and unloading facilities.
- Rail tank wagon loading and unloading facilities with railway spurs.
- Jetties for the unloading and loading of barges and ships.
- Pipelines and pumping stations for the importing or exporting of materials.

Terminals have the additional hazards associated with filling and emptying the road tank cars, rail tank cars, and barges and ships. A particular point of risk of leakage is at the connections, hoses and 'hard arms'. A 'hard arm' is comprised of articulated pipework with sealed joints. Hoses have a risk of rupture. 'Hard arms' reduce this risk, although there is still potential for leakage from the sealed joints.

Pipelines may carry either liquid or gaseous materials and will typically be above ground at some points and below ground at others. Pipelines extend over thousands of miles in open country and, in some cases, through built-up areas. They may be exposed to direct damage from persons carrying out excavations without taking due precautions to identify if pipework is present, or by impact from vehicles and farm machinery. Leakage can also occur due to corrosion. This is a particular problem where the pipeline is buried. A method whereby an electrical charge is constantly applied to the pipe is often used to reduce corrosion. This method is known as cathodic protection. However, it needs to be frequently checked and tested for continuity.

Static electricity can be the source of ignition. In 1982, in Venezuela, a group of tanks blew up whilst unloading crude oil to storage tanks. The fire lasted for four days. One hundred forty-five workers died, and more than 500 people were injured. The investigation found that this accident was caused by static spark discharge resulting from a high oil flow rate. The faster liquid flow velocity is, the easier static electricity is produced. The oil flow must keep within a specific speed limit! (IOP Science: 7th International Conference on Applied Electrostatics. A case study of electrostatic accidents in the process of oil-gas storage and transportation.)

Some gas plants have additional facilities to liquefy the natural gas (liquified natural gas is known as LNG). This liquefaction requires temperatures of approximately -160°C. The LNG can be transported by ships and even by road. The development of LNG has been rapid in the decade starting 2010. It is stored and transported at close to atmospheric pressure in well-insulated tanks. Unless refrigerated, these tanks are constantly venting some gas. There is, therefore, a considerable catastrophic potential. See also Chapter 9 on Transportation. So far, there have been few serious LNG incidents, perhaps because of good procedures and engineering, perhaps because it is a relatively new industry. A specific challenge for LNG storage and shipment is the prevention of a phenomenon called 'rollover'. LNG contains some contaminants. Whilst primarily methane, there will also be traces of ethane and higher molecular weight hydrocarbons. In storage, some LNG vaporises off, leading to layers of liquid with a slightly higher density due to the slight concentration of the contaminants. Stratification of layers of LNG in storage can also occur when LNG from a different source or with different temperatures or contaminants are loaded. If the denser layers are at the top, a sudden movement can occur as the higher density layer descends, creating mixing and vapourisation of the lower superheated layers. Overpressure of the tank can occur. This overpressure is normally relieved through a pressure relief valve passing to the atmosphere, potentially causing a significant vapour release.

Incident Examples – LNG

Cleveland (1944)

One hundred twenty-eight people died following a vapour cloud explosion and fire. It is understood that the tank used was old and brittle and that the tank did not have a retaining wall.

Algeria (2004)

An explosion at an LNG liquefaction facility killed 27 people and injured 56. Three LNG manufacturing plants were destroyed. The total loss was USD 900 million. It is possible that the initial cause of the explosion was not the LNG, but the hydrocarbons which formed part of the refrigeration system.

Another theory is that the LNG vapour entered a boiler through an inlet fan creating an explosive mixture and that the heat load from the explosion reached the leak source, causing a second explosion (Fire Safety Journal 61 (2013) 324–336).

Incident Examples – Warehouses

Beirut Ammonium Nitrate Warehouse Explosion (2020) – Over 200 Fatalities

In August 2020, a massive explosion occurred in Beirut, Lebanon. There were over 200 deaths. Many thousands of people were left homeless. The source of the explosion was a warehouse at the port.

Whilst formal investigations continue, it is understood that the warehouse stored ammonium nitrate. Fireworks were also understood to be stored. The ammonium nitrate came from a ship taking the material to Africa for use in the mines. En route, the ship stopped in Beirut for maintenance. Then the shipping company became insolvent. The material was confiscated and stored in the warehouse in Beirut. Ownership and responsibility for the material were not clear. Under such circumstances, it seems likely that the material degraded and became dirty.

Pure compact ammonium nitrate is not spontaneously explosive, but when it becomes dirty, for example, with leaked oil, it can form an explosive substance, and many disasters have resulted.

Indeed, the particular type of ammonium nitrate in storage in Beirut was destined for mining operations in Africa. This grade is manufactured with deliberately high porosity. The ammonium nitrate is soaked in an appropriate oil at the mine site to create an explosive material.

Warehouses typically contain a variety of materials, sometimes belonging to different organisations, and presenting very different hazards.

There can be little ownership for the material inside, the material may not be regularly inspected, and the material stored can degrade.

(Wikipedia: 2020 Beirut Explosion)

Referring to the chapters which follow' on Catastrophe Prevention, we see above failings in the following as aspects of safety management (Table 13.1).

West Fertiliser Explosion and Fire (2013) – 15 Fatalities Including Public

This disaster occurred at a fertiliser storage and distribution facility in West, Texas. Twelve volunteer firefighters and three members of the public died in the event. A school and a nursing home in the vicinity were severely damaged. Again, ammonium nitrate was involved and probably degraded in storage due to contamination or potentially overheating. A fire preceded the explosion. The local fire brigade was on scene in the 20 minutes between the fire being reported and the explosion.

Table 13.1 Safety Management Failures (Beirut Warehouse)

Aspect of Safety Management	Evidence of	See Chapter
Operating Procedures	Segregation of material and housekeeping	19
Human Factors	Confused ownership	20
Laws, Regulations, Standards	Government responsibilities unclear	32
Governance of Safety	Management of change regarding the storage of the material	31
Legacy Issues	Old material – not anyone's problem?	34

It seems that the local responders were not well informed about the dangers of explosion and, in their determination to fight the fire, became victims as the material exploded.

At the time of the incident, in the early evening, there were no personnel at the site. Residents were the first to observe the smoke and alert the fire department, by which time it was likely that a severe fire was in progress.

The incident highlights some particular hazards associated with warehouses:

- Warehouses are typically unmanned outside of daytime working, so no one on-site can detect and give an early response to any developing issue.
- Emergency responders may not know, or may not have time to find out, the dangers of the multiplicity of materials inside a warehouse.
- Warehouses and residential and community facilities are often physically close together.

(U.S. Chemical Safety and Hazard Investigation Board. Investigation Report – West Fertilizer Company Fire and Explosion (2013)).

Warehouses are typically owned by smaller companies that distribute goods, and who do not have 'deep pockets' in the event of an incident. It is also worth noting that the company which owned the West facility declared bankruptcy shortly after the incident. Following both the Beirut explosion and the West, Texas explosion, no one responsible could be brought to account. No one was available for the massive damages caused, other than some damages covered by insurance.

Incident Examples – Storages – Tank Farms of Liquid or Gaseous Hazardous Material

See Incident Examples – Terminals, below, for storage tank incidents relating to hydrocarbons.

The following environmental pollution example is selected to give a different perspective.

Freedom Industries 2014 – Toxic Release

Hundreds of thousands of people were left without clean drinking water following a storage tank leak.

The extent of the leak was initially underestimated. As the crisis evolved, residents in the area where drinking water may be contaminated were given unclear and conflicting announcements because of the changing information from the company owning the tanks and government agencies. This increased public uncertainty about the safety of the drinking water. The leak was caused by holes that had developed in the base of one of the tanks. The holes were caused by gradual 'pitting' corrosion that degraded the thickness of the tank floor from the interior.

Tanks are usually provided by secondary containment, typically a bund surrounding the tank, to catch any leakage. Such secondary containment was provided but apparently not in a good state of repair.

This incident highlights some particular hazards associated with storages:

- Storage tank internal inspections are hard to perform. The tank must be isolated, emptied, and cleaned before specialists can enter the tank to carry out inspection. The time between inspections can extend to several years. These inspections include measuring the thickness of the remaining metal. Thickness measurements themselves may not spot small, localised areas of corrosion, especially 'pitting' corrosion which is not uncommon in such storage tanks.
- Storage tanks encountered in industry are often quite old, dating back to when the site was originally constructed. Fifty-year-old tanks are not uncommon.
- Tanks should be installed on concrete with a surrounding bund wall. If directly on the ground with bunds made from mounds of fill, they should at least have an impermeable membrane to contain the fluid in case of a release. The state of maintenance of secondary containment, on my observations, is often not good.

(U.S. Chemical Safety and Hazard Investigation Board. Investigation Report – Freedom Industries Chemical Release (2017)).

Referring to the chapters which follow on Catastrophe Prevention we see above failings in the following as aspects of safety management (Table 13.2).

Incident Examples – Terminals

Buncefield Oil Storage Facility (2005) – Explosions and Fire

Multiple explosions and fire caused widespread damage to the terminal and resulted in blast damage to surrounding areas. It was very fortunate that

Table 13.2 Safety Management Failures (Freedom Industries)

Aspect of Safety Management	Evidence of	See Chapter
Equipment Maintenance and Inspection	Condition of tank and bund	22
Emergency Response	Confused, especially in liaison with local authorities	23
Laws, Regulations, Standards	ineffective in relation both to Inspection and Emergency Response	32

the incident occurred on a Sunday morning. There were few people either at the terminal or in the area. In the vicinity of the site, there is an industrial estate and a school.

The incident was primarily caused by the overfilling of a tank which caused a plume of gasoline vapour to drift over the site and neighbouring premises until the cloud found a source of ignition. The resulting explosion damaged adjacent tanks, which in turn caught fire.

Secondary containment in the form of bunds failed to contain the mixture of leaking hydrocarbons plus the firefighting water and foam applied, primarily because of leaking joints in the concrete bunds.

Some such terminals have tertiary containment whereby spillage from the site which escapes from the secondary containment is directed by kerbing is collected in ponds. At Buncefield, the only tertiary containment was the site's drainage system designed for rainwater and minor spills which would flow via interceptors (to catch small amounts of floating hydrocarbon) and then to the site's effluent treatment plant.

Considerable pollution resulted both through hydrocarbons and firefighting foam entering neighbouring streams and soaking away into the groundwater.

The cause of the overflow that led to the incident was a failed level instrument used to gauge the tanks' contents, together with a failed independent high-level switch. The level gauge was recognised to have a recurring problem and had stuck repeatedly in months preceding the incident. The independent high-level switch had apparently not been installed with a padlock to keep the switch in the raise position, as was required for normal service. Removal of the padlock was for testing only.

The overfilling occurred because the control room operator was only seeing a partially full tank due to the instrument failure and was not alerted when the tank reached a high level since the independent level switch was not operating.

Referring to the chapters which follow on Catastrophe Prevention we see above failings in the following as aspects of safety management (Table 13.3).

Table 13.3 Safety Management Failures (Buncefield)

Aspect of Safety Management	Evidence of	See Chapter
Operators	Did not recognise overfill	19
Equipment Maintenance and Inspection	Level instruments unreliable and poorly maintained	22
Consequence Analysis	Potential magnitude of consequences not identified	28
Emergency Response and Readiness	Scenarios did not take into account the amount of fire water, and where it would drain to	23

(HSE: COMAH (2011). Buncefield: Why did it happen?)

CAPECO Puerto Rico (2009)

In a similar incident, a tank of gasoline overflowed. The gasoline formed an aerosol-like spray, which ignited when the vapour cloud reached the Wastewater Treatment Plant. Seventeen of the plants 48 petroleum storage tanks were destroyed, along with significant damage to neighbourhoods and businesses in the area. Petroleum products leaked into the soil, nearby wetlands, and waterways. The overfilling occurred because of the failure of a level gauge. Despite the Buncefield incident some four years earlier, the tanks at CAPECO were not equipped with independent high-level switches.

Secondary containment was provided, with dike valves that could drain the dikes after rainstorms. On the night of the incident, some of the dike valves had been left open. These valves may have been hard to see in the dark.

These incidents highlight some particular hazards associated with terminals:

- The importance of reliable tank gauging with regular instrument testing and maintenance.
- The benefit of a proven independent high-level switch preferably with automatic shutdown of the filling valves or pump. Note that this is not always practical due to sometimes complex valve routing in a terminal and because of concerns with regard to hydraulic hammer.
- Bund walls and kerbing are sometimes not adequately maintained. Valves in dikes that are normally closed to contain any spill need to be routinely audited to ensure they are left in the correct position.

(U.S. Chemical Safety and Hazard Investigation Report: Caribbean Petroleum Tank Terminal Explosion and Multiple Tank Fires).

Incident Examples – Pipelines

Ghislenghien, Belgium (2004)

A major natural gas pipeline exploded, killing 24 people and leaving 122 wounded. Several of the dead and injured were members of the emergency services. Some weeks before the incident, a mechanical excavator impacted the pipeline. The fire brigade was at the scene on the day of the incident since gas had started to leak (Wikipedia: Catastrophe_de_Ghislenghien).

Qingdao, China (2013)

An oil pipeline exploded, killing 62 people due to a pipeline rupture. The pipeline ran close to or through the city. Crude oil leaked into the city's storm drains. After the leak, there were several explosions and fires.

Oil pipelines, especially those close to major population centres, run the risk of tampering to collect leaked oil.

These incidents highlight some particular hazards associated with pipelines:

- Pipelines are subject to mechanical damage from construction activities. The pipe may be weakened but not immediately start to leak. Weeks can separate the incident of damage to a leak occurring.
- There is significant exposure to risk to inhabitants of property along the line of the pipeline, and this exposure becomes particularly concerning when the pipeline passes through an area of dense habitation.

(Wikipedia: Qingdao oil pipeline explosion).

Prudhoe Bay (2006)

Over approximately five days, some 21,2000 U.S. gallons of crude oil spilt from a 0.25-inch hole in a 34-inch diameter pipeline on the Alaskan north slope. As a cost-saving measure, the injections of a corrosion inhibitor had been discontinued. Inspection of the pipeline using a 'smart pig' had not been carried out. 'Smart pigs' are instruments carried within a foam or rubber body that fits inside the pipeline. They are pushed along the pipeline under slight pressure and carry out, and record measurements along the entire length travelled. Sixteen miles of the Alaskan pipeline had to be replaced.

(Perrow (2007) and Wikipedia: Prudhoe Bay Oil Spill).

CONCLUSIONS

Process Industry Infrastructure

The potential is high in all sectors with recent disasters of sizeable proportions. Since Buncefield a lot has been done with regards to independent hi – hi level alarms. Apart from that I don't see much practical activity to reduce the risk.

Not possible	Unlikely	Possible (50/50)	Likely, no more than one	Very likely/ maybe several
☐	☐	☐	☒	☐

Chapter 14

Metallurgical, Fabrication, and Product Assembly Industries

RISKS ASSOCIATED WITH THIS DIVERSE SECTOR

Steelworks and other metallurgical industries are not usually thought of in the context of catastrophes. But we are looking for catastrophic potential, not necessarily proven through past incidents.

The industries that convert the metal ore into raw metals and alloys include steelworks and makers of specialist metals. These processes involve very high temperatures and heavy equipment. Occupational safety injuries are common, although there are good signs that these are being reduced. Explosions also occur. One of the most common causes of an explosion occurs when moisture gets mixed with molten metal at high temperatures. Any water present expands 1700 times when it turns into steam, causing an explosion which can seriously injure workers. Incidents have also occurred with the dumping of slag and breakages of the massive ladles used to carry and pour out the molten metal. These events can cause traumatic injury to the workers in the vicinity.

In 2019 a dramatic explosion occurred at the Port Talbot steelwork in the U.K. (see Incident Examples below). Local residents described it as a massive explosion. Two employees suffered minor injuries. An internet search over the past few years indicates a significant number of explosions at steelworks around the world, which fortunately did not lead to fatalities.

Many industries, such as the automobile industry, which convert metal into mechanical components have minimal catastrophic potential. Potential serious consequences tend to be limited to the solvents and paints that these industries might use to create the finished product.

However, of more concern are the metallurgical works that manufacture their own coke. In a coker, the coking coal is loaded into cells which are then heated to around 1000°C. Liquid and gaseous hydrocarbons are emitted, which are collected and form useful by-products such as phenols. Once the hydrocarbons have been distilled off, the cells are cooled and the

DOI: 10.1201/9781003360759-16

purified coke discharged for use in the metallurgical furnaces. Incidents can occur related to the flammable chemical by-products. These types of incidents are akin to those we found in Chapters 6 and 7 on Oil and Gas and the Chemical Industry.

Notes from Trevor's Files: Kazakhstan Coker Operator

When working on safety improvements at a coal and steel operation in Kazakhstan, I climbed to the top of a long coker comprising many coking cells. Here, the coking coal is loaded into the cells using a crane in regular motion across the top of the Coker. Some of the cells emit flames when they are opened, especially if not allowed time to cool down fully. Some such fires were occurring at the time. There was much dust and fumes, together with a smell of phenolics and aromatics, known human carcinogens. I approached the operator assigned to keeping the top of the coker 'clean', an operation primarily involving clearing coal from the rail tracks on which the mobile crane operated. He was working along the open edge of the Coker which has no edge protection protecting from a fall which in the poor visibility seemed a significant risk. Providing such edge protection would be very difficult to achieve given the layout of the Coker. I wanted to find out something about the operator's attitude to safety, so I asked him, via a translator, if:

- He had any safety concerns.
- If any safety improvements could be made.

He boldly answered 'no' to both questions – everything is very safe here!

Shipbuilding and ship repair are also worth mentioning here as examples of heavy industry. Whilst there is potential for many occupational injuries, few risks would give rise to catastrophes. However, ships are typically put into a dry dock or constructed in drydock, and there have been several incidents where the drydock gates have failed. One of the most recent is the 2002 tragedy at Dubai drydock where 21 workers drowned. The dock had five vessels undergoing work, and there were a total of 241 workers in the dock at the time (Dieselduck).

Other fabrication and product assembly industries such as automotive pose little threat of catastrophes involving multiple fatalities, etc., although they have plenty of potential for occupational injuries.

Incident Examples – Steelworks

Port Talbot (2001) – Three Fatalities

Three workers were killed and 12 others seriously injured by a furnace explosion.

An over-pressure occurred within the furnace because of water mixing together with a considerable amount of hot materials, including molten iron. A violent release of energy occurred, sufficient to vertically raise the entire 5,000 tonnes or so of the furnace and contents by around 0.75 m (2.5 ft). This reaction resulted in the loss of containment at a lap joint and discharge of large quantities of hot gas and debris into the furnace cast house floor area. It is understood that this particular blast furnace was of an older design and had this lap joint had been eliminated or substantially improved in future designs. This particular blast furnace was planned to be taken out of service.

Events and failures leading up to the incident included:

- Failure of electrical cooling pumps due to electrical fault at the power plant – insufficient alternative backup available.
- Continued operation of the blast furnace on full blast with only 55% of cooling capacity.
- Failure of some coolers due to previous damage and corrosion.
- Delays occurred in locating leaking coolers, and further coolers failed, eventually allowing perhaps eighty tonnes of water into the furnace.
- The furnace cooled too much, and attempts were made to 'recover' the furnace with oxygen lances. The water inside the furnace came into contact with the hot molten metal and slag, producing vast amounts of steam. The vessel failed at the lap joint releasing the contents of the furnace.

Referring to the chapters which follow on Catastrophe Prevention we see above failings in the following as aspects of safety management (Table 14.1).

Table 14.1 Safety Management Failures (Port Talbot)

Aspect of Safety Management	Evidence of	See Chapter
Equipment Maintenance and Inspection	Condition of Coolers and Bolts	22
Risk Assessment	Study of the effects of electricity failure	25
Legacy Issues	Lap joint of a design no longer used	34

CONCLUSIONS

Metallurgical, Fabrication, and Product Assembly Industries

The potential is moderate in facilities which handle molten metal.

Not possible	Unlikely	Possible (50/50)	Likely, no more than one	Very likely/ maybe several
☐	☐	☒	☐	☐

Civil Infrastructure, Small Industrial Estates, Retail, Recreational, Residential

OUTLINE OF RISKS

This sector is not usually associated with catastrophes. But we are looking for catastrophic potential, not necessarily proven through past incidents. Then there is the Grenfell Tower disaster.

Hospitals, Laboratories, Industrial Estates, and Retail establishments often handle toxic or flammable materials or biological hazards such as bacteria and viruses that can cause catastrophes.

Agreed, this chapter is something of a 'hotchpotch' of 'other' types of hazardous facilities. Let's work through them in order.

Hospitals

Hospitals help people overcome injury and sickness. At first sight little potential for catastrophe here. Unfortunately, hospitals gather sick people together, mainly in the same building. Typically, a hospital will have segregated air flows for various parts of the building with filtration systems. Nevertheless, a hospital is a source of biological hazards. Germs and viruses can infect entire hospitals. Norovirus outbreaks are common in semi-enclosed environments such as hospitals, nursing homes, schools, and cruise ships and can also occur in restaurants and hotels. Such events come within the scope of a catastrophe, especially since infection by norovirus or other diseases can be fatal to the many patients who are already health impaired.

Laboratories

Laboratories may be establishments with several different purposes:

- Carry out tests to, for example, determine the presence of a virus or to assess a blood sample and identify any health-related issues. Tests might be carried out on the water to identify any chemical causing pollution, etc.

DOI: 10.1201/9781003360759-17

- Carry out experiments to extend our scientific knowledge and identify new inventions.
- Small-scale manufacturing, especially of pharmaceutical or household chemical products.

Experimental laboratories contain particular hazards since, by definition, the experiments are new developments about which not a lot is known. Whatever the type of laboratory, they may handle a variety of hazardous materials or produce hazardous materials either deliberately or accidentally. Laboratories are often secretive about their business. It may not obvious what goes on or the potential hazards. The inventories of hazardous materials at a laboratory are relatively small compared with many other industrial establishments discussed in this book. Still, sometimes it only takes a small amount of highly toxic chemical or biohazard to create a catastrophe. It remains one possibility that it was the accidental release of a coronavirus from a laboratory that caused the COVID-19 pandemic.

Note: for the Los Alamos Scientific Laboratory incident 1958, see Chapter 16 on Explosives Manufacture and Military Installations

Small Scale Industrial Estates

Industrial estates often contain small to medium-size manufacturing and extensive warehousing. There are a wide variety of materials in use, typically in small quantities from drums, sacks, and totes. There is potential for mixing the wrong materials, for puncturing drums of hazardous materials with the toes of a forklift truck, for drums to fall off lorries, etc. The nature of the business or the hazards within may not be evident from the outside. Due to the sheer number of small units, it is unlikely that the enterprise will receive much attention from the regulator or local authority unless something goes wrong.

Retail

It seems strange to include retail. We go shopping not expecting to be exposed to hazards. Indeed, the potential is low and confined mainly to occupational health and safety incidents.

Gasoline stations are a particular risk. Fires at the car being refuelled do occur, typically ignited by static electricity from the person refuelling.

Static electricity is one of the main causes of fires at gas stations.

Using cell phones at gas stations can give way to activities that can statically charge customers as they refuel their gas tanks. For example, a customer may receive a call and re-enter the vehicle to answer it. By touching or rubbing the seat cover, their bodies may well become statically charged. Whilst still on the phone, the customer may step out of the vehicle and directly touch the nozzle or the tank cap creating a static discharge.

That's why it's advised not to re-enter a vehicle while you are still refuelling. If you must, then make sure you touch a metal part of your vehicle to completely discharge your body before you touch the filling nozzle. Since a customer is less likely to remember these instructions when using a cell phone at a gas station, it is advisable not to use cell phones. This is why gasoline stations have warning signs pasted onto their walls, instructing customers not to use cell phones while at the facility (Wkblog: Fire at Gas Stations).

A gasoline station fire on its own is unlikely to be catastrophic. But what if it spreads to the road tanker unloading petrol, or neighbourhood premises?

The NFPA, (NFPA 2020 Fires in or at service of gas stations), reports an annual average of 2340 vehicle fires every year in the U.S.A. with typically one fatality and eight injuries every year.

Recreational

Recreational activities are not generally thought of as potentially catastrophic. However, multiple illnesses have resulted from poor control over ventilation in recreational facilities, e.g. cruise liners through contacting biological agents such as norovirus. Swimming pools and spas have occasionally caused illnesses when improperly managed. Most swimming pools are kept clean by adding chemicals that contain chlorine as the active antibacterial agent. Household cleaning agents also commonly contain chlorine, such as sodium hypochlorite. However, if mixed with inappropriate chemicals, particularly if the water becomes acidic, chlorine can be evolved. It is hard to see this as a potential catastrophe, but it is worth remarking on nevertheless. In August 2021, 24 people were taken to hospital having inhaled fumes at a hotel spa in the U.K. These fumes are understood to have been chlorine gas. A fatality occurred in November 2019 at a fast-food restaurant in the U.S.A. An employee had accidentally spilt some cleaning fluid onto the floor. Later another employee started to clean the same floor with another cleaner. One cleaning fluid was acidic, the other contained hypochlorite. Hypochlorite releases chlorine when mixed with acid.

Notes from Trevor's Files: Local Gym

Whilst preparing for a swim at my local gym, the duty manager happened to be opening the door to a small equipment room, which had some hazard warning labels on the door. I could see some gas cylinders inside. I just had to ask the following question. "Sorry for being nosy, but may I ask what is in this room". "Sure", he replied, "some equipment for the steam room and sauna". I could see two sizeable blue gas cylinders and a smaller black cylinder inside this small room. "What do you have in those cylinders", I asked. "I Don't know", was the reply. "Do you think it would be wise to find out in case something would happen?" I asked inquisitively. "I suppose so, but we have all the COSHH sheets". COSHH means "Control of Substances Hazardous to Health", and it is a regulatory requirement to have such sheets in the U.K. I still don't know if he understood my point. If you don't know what the substances are, COSHH sheets are not of much use.

You might think that using the standard colour coding of gas cylinders, would be enough to identify the contents. Note that I was not allowed to get close enough to read the label in the room. The colour coding of gas cylinders is quite confusing. There is a standard BS EN 1089, which identifies gas cylinders by the pattern on the collar of the cylinder, but I found this did not identify the cylinders I had observed. This standard excludes LPGs, and it seems quite likely that the blue cylinders were propane or butane for heating the sauna. LPG cylinders could be red or blue!

At the time of writing this chapter, I am reminded of the deaths of six children in Tasmania when a bouncy castle was lifted into the air. Other bouncy castle deaths have occurred. Risk assessment in recreational activity is frowned upon as a burdensome 'paperwork' task, but very necessary.

I am also reminded of the 1993 disaster in Lyme Bay where four teenagers on a sea kayaking trip died. The incident led to legislation in the U.K. to regulate activity centres working with young people. As a former cub scout leader, I must confess I found all the additional restrictions burdensome.

Notes from Trevor's Files: Church Bazaar

I recall a personal experience of being involved in a church bazaar. Everything seemed to be going well. I helped at the barbecue, and we even had a fire extinguisher ready. Then somebody pointed out that there was no way a fire engine or ambulance could access the bazaar nor the adjacent residential properties due to the streetside parking. We had not carried out a proper risk assessment.

Hotels

Hotels don't entirely fall into the category of catastrophe that has been explored elsewhere in this book, but we cannot leave it without a mention.

Notes from Trevor's Files

I often needed a hotel stay when travelling to work offshore out of Aberdeen. On one particular visit, the only hotel available was a grand old hotel used more by tourists than 'oilies'. In the middle of the night, the fire alarm sounded. Of course, having been well trained in emergency response, I immediately left my room, having thrown on a coat over my 'jammy joes' to protect against the cold autumn night and descended the fire escape. The muster point was across the road, where I waited whilst the hotel headcount checker did his work. The fire brigade arrived and commenced preparations, although there was no evidence of fire. About 30 minutes after the fire alarm sounded, a group of elderly ladies appeared through the grand entrance, fully dressed with their suitcases packed and in hand. The hotel was of an old design with wooden staircases and panelling. I hate to think of the consequences if the fire had been significant.

During the time when I was writing this book, I visited several hotels during my travels; at some stage in each hotel, the fire alarm sounded. I will relay just one of these stories when my wife and I were on a short vacation to celebrate New Year's Eve. On the evening of our arrival, the fire alarm sounded. My wife and I evacuated promptly, as did at least some of our neighbours. Of course, the alarm was quickly cancelled, with no explanation, and we returned to our rooms — since everyone else had done so and it was very cold and raining outside. On the morning of New Year's Day at 06:30, the fire alarm sounded once more. Not a soul moved, including, I am ashamed to admit, myself. We all guessed that someone starting breakfast had burnt the toast. It was also straightforward to escape from our room through the ground floor patio door, or at least that is our excuse!

There have been some dreadful hotel fires. Often the consequences of a hotel fire are aggravated by residents who have not consulted the escape procedures, and there is confusion about what to do. Disorientation by smoke is a severe risk. 'False alarms' in hotels are so common that they are widely ignored. Language difficulties are often apparent in instructions involving emergency escape. Emergency escapes may be blocked or otherwise unusable.

Note from Trevor's Files

Whilst staying at a hotel close to a steel plant in Kazakhstan, I did my regular routine of checking the emergency escapes making sure I knew where they were and also that they were clear. What I found was that every single emergency escape was locked closed. When we addressed the issue with the hotel management, the reason explained was that if they unlocked the emergency doors, we were in danger from thieves. As it happened, the steelworks also managed the hotel.

Hotels are also open to deliberate acts of sabotage, arson, and terrorism.

Residential

Residential catastrophes also do not fit easily within the context of the types of catastrophes being discussed in this book. However, because of the enquiry's findings into the Grenfell tower disaster, there is no question that this was an avoidable man-made disaster.

Prior to this tragedy, I had been somewhat amused by the several tower block fires that seemed to occur in Dubai, a city I often visited regularly. Fortunately, these fires did not result in serious injury. A colleague of mine conjectured that the fires were due to corrupt building practices, where cheap and less fire-resistant materials were used instead of those intended or required by local building codes. I had the utmost faith in British building codes, and their application, until Grenfell Tower.

At Grenfell Tower in London a fire gutted the tower block killing 72 people. I am probably not alone in thinking that standards were better in the U.K. A refurbishment had installed cladding on the tower which did not adequately resist the spread of fire. Indeed, the public enquiry suggests that the cladding actively promoted the spread of the fire. It is also noteworthy that the tower was owned by the local authority and was managed by a tenants association, with eight elected tenants, four councillors and three independents. The tenants association was not a 'for profit' organisation. The idea behind the tenants association was to give residents power over how their buildings were managed. It was the tenants association which initiated the refurbishment which installed the cladding.

Of course, people are rightly enraged that this happened. Just to put this into perspective, if you look at the average two-storey house, a substantial part of the construction is wood. Some houses even have timber frames and wooden cladding. Wood as we know is flammable. This catastrophic danger rises as the height of the residential property increases. Wood is also relatively cheap and can be readily cut and shaped to form the desired pieces for the construction of the house including the stairway, etc.

For more on Grenfell Tower, see Incident examples below and Kernick (2022): Catastrophe and Systemic Change – Learnings for the Grenfell Tower Fire and Other Disasters.

Incident Examples – Laboratories

Bio Lab Chemical Release (2020) – Thermal Decomposition

In September 2020, a thermal decomposition occurred at a laboratory facility in the U.S.A. Chlorine-containing materials were involved. The facility is a manufacturer of pool and spa treatment products. At the time of writing the book, the investigation was ongoing. (US Chemical Safety Board: Bio Lab Chemical Release).

Marburg Virus Disease (1967)

This event followed an escape of the virus from a laboratory using imported monkeys. The monkeys were being used to develop cures for tetanus and diphtheria. Seven people died. It is understood that a contribution to the escape was the failure to wear the required Personal Protective Equipment.

(Wikipedia: Marburg virus disease).

Incident Examples – Industrial Estates

Plastics Manufacturing on Industrial Estate (2021)

A fire and subsequent explosions occurred at a plastics facility at an industrial estate in the U.K.

Polyurethanes were involved. There was one fatality. Decomposition of polyurethanes can generate cyanide gas and several other potentially toxic materials (The Chemical Engineer October 2021).

In the same month, a major fire occurred at another factory in an industrial estate in the same region of the U.K. Pollution incidents occurred, and 14 schools in the area were temporarily closed.

DPC (2002) – Toxic Gas Release

This facility receives chlorine, a highly toxic material, in 90-ton tank cars and repackages it into 150-pound cylinders and 1-ton containers. It employed just 12 full-time personnel in a one-shift Monday-Friday Operation. Viewing the site from Google Earth Street View, it is easy to see that this is a small operation in a rural location with an equally small industrial estate on the other side of the highway. Anyone passing by would not be aware of any specific hazard. The site has a few tanks. There is a simple gate and fence with minimal intruder protection. A mobile home park is located close to the facility, with the nearest mobile home some 150 yards from the chlorine handling operations. Further to the north, a hospital, a school, and a municipal airport are located. The incident was investigated by the CSB (US Chemical Safety Investigation Board, 2003, Chlorine Release, DPC Enterprises).

The chlorine tank cars arrive by railroad and are connected to the site's repackaging process by a hose. At the end of the working day, the operator closes a manual valve on the top of the rail tank car. Residual chlorine in the piping system is directed toward a small bleach production process. A vacuum is pulled, and the system is left under negative pressure. An Emergency Shutdown Button is pressed to close all the Emergency Shutdown Valves as a final step. The rail tank cars had two valves for liquid chlorine discharge and a 'pad air' valve to pressurise the tank during discharge.

This incident investigation focuses on the failure of one of the chlorine unloading hoses. Imagine a hose 11 feet in length and 1 inch in diameter. The hose specifications as required by the owning company were for the hose to have a PTFE inner liner, a Hastelloy C-276 structural reinforcement braid layer, overlain by a high-density polyethylene spiral guard for abrasion protection. I have used such hoses myself during my life in the chemical industry.

The site had chlorine detectors and air-actuated ball valves (the ESD valves) at the unloading stations. The chlorine detectors were set up to alarm at 5 ppm chlorine and automatically close the ESD valves at 10 ppm. Each rail tank car also had five ESD valves.

During this incident, 48,000 pounds (21,800 kgs) of chlorine from a tank car originally containing around 180,000 pounds were released when an unloading hose failed. The chlorine alarm sounded, and an employee pushed the ESD button. However, several crucial ESD valves failed to close.

According to the CSB report, dispersion modelling showed concentrations of 3 ppm or above could have extended approximately 3.7 miles from the point of release.

Emergency response HAZMAT personnel plus an employee trained in HAZMAT work eventually accessed the area of the release. Two personnel in very poor visibility closed the liquid valves that supplied the ruptured hose, although that on its own did not stop the release, and another valve on the rail tank car had to be closed.

An accumulation of chlorine hydrate was observed close to the tank and measures were taken to remediate this using lime. This was only partially successful, and this material released relatively minor amounts of chlorine for the next two days.

Laboratory testing of the ruptured hose versus the intact hose revealed that the ruptured hose's braid layer was 316L stainless steel. In contrast, the intact hoses were correct - of Hastelloy C-276.

It is concluded that the cause of the rupture was corrosion of the 316L stainless steel.

The ESD system control system was found to be working correctly. However, the torque necessary to close the valves far exceeded that available from the valve actuators, probably due to corrosion of the valves and valve balls.

The two different hoses appear identical, and there were no adequate mechanisms such as colour-coding or stamping to distinguish between them. The investigation included examining the procedures of the hose supplier. It was found that there was potential for a mix-up during the assembly of the two different types of hose braid. The suppliers did not carry out Positive Materials Identification using an analyser, neither did the user (see Chapter 21 on Equipment Suitability and Reliability – Materials of Construction and Corrosion). The paperwork associated with the hose shipment identified the hose braid as Hastelloy, not as 361L.

The CSB also found issues with the mechanical integrity programme at the facility. It did have one as required by the U.S. authorities, and this was seen as adequate for dry chlorine. However, it was judged that there was inadequate supervision of inspection and test personnel and insufficient training of employees in the catastrophic potential of corrosion-induced failure. Such training would have emphasised the importance of keeping the system free of moisture and auditing the positive verification of ESD valve closures.

The facility was required to have an Emergency Response Plan under the Emergency Planning and Community Right to Know (EPCRA), EPA RMP, and OSHA PSM standards. However, the CSB found that the facilities emergency response plan lacked clear guidelines and mechanisms for community notification. Responsibilities for facility emergency response personnel were inadequately defined, there was a lack of training, emergency response equipment was in an inaccessible location, etc.

Fortunately, there were no fatalities or long-term injuries, although 66 people sought medical attention.

I have spent some time on this particular incident because it demonstrates some of the hazards that may not be immediately apparent in small, unassuming facilities located on industrial estates. Such small facilities rarely have access to the expertise typically available in larger organisations and may not have the background and training to conduct a thorough risk assessment. The CSB report does not indicate if a risk assessment in the form of a HAZOP, or similar was carried out on the site operations. I tend to think that a HAZOP might not have identified this particular failure mode. HAZOP teams do not necessarily recognise risks which can emerge from mechanical integrity failures and the guide words used in a HAZOP do not necessarily lead them to consider this. For description of HAZOP see Chapter 26.

The following Bow-Tie diagrams summarise the incident (Figures 15.1 and 15.2).

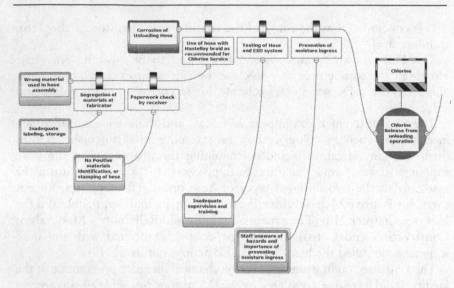

Figure 15.1 Chlorine Release Incident – Left-Hand Side of Bow Tie.

Source: Created using BowTieP Software © CGE Risk Management Solutions B.V. 2021.
BowTieXP is a registered trademark of CGE Risk Management Solutions B.V.

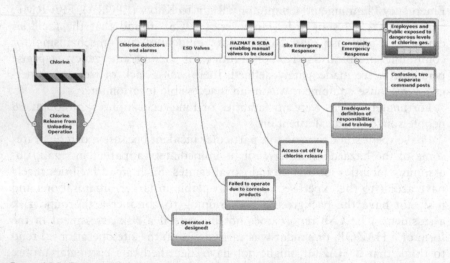

Figure 15.2 Chlorine Release Incident – Right-Hand Side of Bow Tie.

Source: Created using BowTieXP Software © CGE Risk Management Solutions B.V. 2021.
BowTieXP is a registered trademark of CGE Risk Management Solutions B.V.

Table 15.1 Safety Management Failures (DPC)

Aspect of Safety Management	Evidence of	See Chapter
Equipment Maintenance and Inspection	Hose Inspection and Quality Control	22
Equipment Suitability	Hoses	21
Risk Assessment	No mention in investigation	25
Emergency Response	Inadequate, not rehearsed	23
Laws, Regulations, Standards	Not adequate for the control and the installation and community emergency response	32

Referring to the chapters which follow on Catastrophe Prevention, we see above failings in the following as aspects of safety management (Table 15.1).

Incident Examples – Residential

Grenfell Tower (2017), London

A fire gutted the tower block killing 72 people. At the time of writing this book the inquiry hearings have not been fully concluded.

We know from part 1 of the inquiry that the fire was started by a domestic appliance in one of the flats lower down the tower block (Grenfell Tower Inquiry: Phase 1 Report Overview, October 2019).

Fire occasionally occurs in domestic appliances – I burnt out my tumble drier when I was posted to Kazakhstan due to inadequate clean-out and, on another occasion, nearly let a frying pan go on fire. A tower block design and procedures need to allow that fires will occasionally occur. The fire at Grenfell found its way to cladding panels on the outside of the tower block and propagated up the outside of the building. The inquiry concludes that the reason the flames spread so rapidly up, down, and around the building was the presence of the aluminium composite material (ACM) rain-screen panels with polyethylene cores, which acted as a source of fuel. The suggestion being pursued in phase 2 of the inquiry is that 'the building failed to comply with the building regulations in that the external cladding did not adequately resist the spread of the fire'. The cladding had been retrofitted to the building to improve weatherproofing. Phase 2 will clearly examine the management of change issues involved.

There were further issues with fire safety in the building, although they are overshadowed by the flammable cladding.

The fire brigade struggled to handle the fire. The otherwise experienced incident commanders and senior officers attending the fire had received no training in the particular dangers associated with combustible cladding, even though some senior officers were aware of similar fires that had occurred in other countries, and of the fact that construction materials and methods of construction were being used in high-rise building facades with a limited understanding of their behaviour and performance in a fire. Apparently, the fire precautions adopted the 'stay put policy' and it is suggested that this was not appropriate for this particular building. There was undue delay before 'stay put' was changed to 'evacuate'. (Grenfell Tower Public Inquiry Update: Meeting 18 December 2019 – Item 19/59).

There are many participants in such a refurbishment:

- Council
- Tenants association
- Building control organisation
- Main refurbishment contractor
- Refurbishment contractor
- Architects
- Fire engineer
- Manufacturer of the cladding panels
- Manufacturer of the insulation

As the reader can imagine neither no one party accepts responsibility for what happened.

CONCLUSIONS

Hospitals

The potential is moderate, given some risk arising from biological agents.

Not possible	Unlikely	Possible (50/50)	Likely, no more than one	Very likely/ maybe several
☐	☐	☒	☐	☐

Laboratories

The amounts of materials handled are small, but the potential is moderate, especially given the potential lack of control and risk assessment in an experimental facility.

Not possible	Unlikely	Possible (50/50)	Likely, no more than one	Very likely/ maybe several
☐	☐	☒	☐	☐

Industrial

The amounts of materials handled are low by comparison to chemical plant, but the potential is moderate, especially given the potential lack of control and risk assessment in small facilities facility.

Not possible	Unlikely	Possible (50/50)	Likely, no more than one	Very likely/ maybe several
☐	☐	☒	☐	☐

Retail

Amounts of hazardous material are small.

Not possible	Unlikely	Possible (50/50)	Likely, no more than one	Very likely/ maybe several
☐	☒	☐	☐	☐

Recreational

Multiple fatality rare, but I am not convinced that proper thoughtful risk assessments are being done.

Not possible	Unlikely	Possible (50/50)	Likely, no more than one	Very likely/ maybe several
☐	☐	☐	☒	☐

Hotels

High probability of multiple fatality. Means of prevention are clear, but not taken seriously.

Not possible	Unlikely	Possible (50/50)	Likely, no more than one	Very likely/ maybe several
☐	☐	☐	☐	☒

Residential

In tower blocks, there is a high probability of multiple fatality rare, and I am not convinced that proper building codes are being applied.

Not possible	Unlikely	Possible (50/50)	Likely, no more than one	Very likely/ maybe several
☐	☐	☐	☒	☐

Chapter 16

Explosives Manufacture, Military, and Defence Establishments

ASSESSING THE RISK

Establishments that manufacture explosives are, by definition, hazardous facilities often containing inventories of explosives, sources of radiation and radioactivity, and weapons.

Commercial explosive manufacturing making products for use in, for example, mining and civil engineering are relatively easy to identify and are typically registered under Seveso. Fireworks manufacturing for recreational purposes may be less well controlled.

We will not consider incidents that occurred in conflict or consider military aircraft and ships (risks here are largely as in Chapter 9 on Transportation).

Of more interest here are locations where munitions and explosives are made and weaponised.

Military facilities are often secretive. They are excluded from the Seveso regulations because of their sensitivity, although 14 were registered in the UK COMAH list (equivalent to Seveso) in 2018.

Establishments that produce regular explosives (non-nuclear). Given the clear potential for explosion. Nitro-glycerine, TNT, and more contemporary explosives such as RXD and PENTA all involve mixing relatively benign materials with concentrated acids. An explosive solid results which can be detonated by impact or employing a detonator.

Nuclear facilities provide a more severe catastrophic potential, not so much through explosions but through the release of radiation. But it is also possible that the processing of nuclear materials has a civil use, for example, in nuclear power generation. In this book, we are treating facilities handling nuclear materials, other than for power generation, as military establishments.

AWE Aldermaston Example

As an example of how the layman can make a reasonably extent of assessing risk, even for a secretive defence establishment, I selected the example of the U.K. Atomic Weapons Establishment at Aldermaston.

DOI: 10.1201/9781003360759-18

Incidental Case Study: AWE Aldermaston

During a pleasant drive in the British countryside, I came across a large sprawling factory quite unexpectedly. It was the UK Atomic Weapons Establishment at Aldermaston. I take this as an example of the kind of facility you can find in your own country. Not being a nuclear engineer, I had little basis for assessing the facility's risk, unlike many of the other facilities discussed in this book, where my background experiences as a chemical engineer have contributed to my understanding.

Taking this as a case study, what can be found out by information in the public domain? It should be pointed out that I undertook this research with an open mind. I have no particular bias pro-nuclear or anti-nuclear, neither do I have any fixed views on nuclear disarmament. I approached this research to see if, based on public domain information, I could form an opinion on the risk of this site.

There may be some readers who think I could be putting the country's safety at risk by engaging in this research. I am confident that anything I learnt during my brief research of public domain information available on the internet to all, is already well known by any potential enemies of the U.K.!

AWE ALDERMASTON – GENERAL DESCRIPTION

From the Atomic Weapons Establishment website, we can find out that the Aldermaston facility was formerly an RAF airfield and became the site of the UK's Atomic Programme in 1950. The first fissile material building at the site became operational in 1957. In 1998 the plutonium facility at Aldermaston became fully operational. The facility is ISO 9001 certified (Quality) and ISO 14001 certified (environment) and is licensed to operate by the Office for Nuclear Regulation. There has been substantial investment in laser facilities in recent years. In 2011 building work commenced on the replacement of the nuclear materials processing facility. The AWE employs approximately 4500 personnel, and the site covers 750 acres. (Wikipedia: Atomic Weapons Establishment). Aldermaston is responsible for making Britain's Trident nuclear warheads. It also stores nuclear waste from Royal Navy submarines. The website indicates that there was a fire in 2010, more on that below.

Further information can be gleaned from a variety of sources including:

- The AWE Website
- The AWE Aldermaston Site Development Strategy Plan (An AWE Public Information Leaflet)

- Global Security: AWE Aldermaston – United Kingdom Nuclear Forces
- Nuclear Weapons Archive: Britain's nuclear weapons – British Nuclear Testing.
- Google Earth Images
- 2019 Aldermaston Consequences Report available from West Berkshire Council Website.
- REPPIR – Radiation Emergency Preparedness and Public Information Regulations: What you should do if there is a radiation emergency at the AWE Aldermaston or Burghfield Sites available from the Basingstoke Council Website.
- "An evaluation of the degree to which potential accidents with off-site radiological consequences occurring at AWE Aldermaston are a material consideration in defining the overall future pattern of development of Tadley" – available from the Basingstoke Council Website.

From this data it is possible to infer a rough layout of the site as shown in Figure 16.1

From aerial photographs, the runways of the original aerodrome are clearly visible and shown in dashed lines.

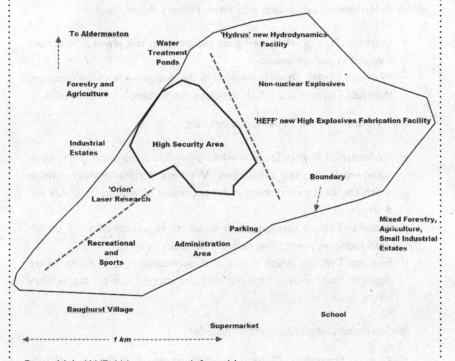

Figure 16.1 AWE Aldermaston – Inferred Layout.

The inner high-security area is understood to be where the vast majority of processing of nuclear materials is carried out including the glove boxes where the radioactive materials are formed into components for assembly into warheads.

History

The plutonium component manufacturing buildings were opened in the early to late 1950s. It is understood that these became badly contaminated in 1978 and were closed. The same area was reopened in 1982 to manufacture an improved design of warhead, known as the Chevaline. Operation of this facility continued long after its initially intended closing date, and later manufactured the first Trident warheads. A complex began construction in 1983 and, after five years of delays, went into operation in 1991. This complex has 300 glove-box production units for Trident plutonium component production.

Non-nuclear explosives manufacture, and storage are located to the east of the site and occupy buildings and bunkers which are well spread out – this area is some 40% of the total site area. A new high explosive building was constructed here in 2015 (shown in Figure 16.1 as HEFF)

Organisation

AWE Aldermaston is organised into three primary departments:

- Warhead Physics which performs research on the physical processes involved in nuclear weapons.
- Warhead Design, which is involved in the design work of the weapons.
- Materials Department, which develops the materials and processes.

The Warhead Physics Department includes:

- Mathematical Physics Division which conducts theoretical work, computer modelling, and simulations. Warhead Hydrodynamics division which conducts experiments in the processes of weapon assembly and disassembly.
- Radiation Physics Division which conducts experimental work on nuclear radiation physics and radiation hydrodynamics.
- Foulness Division, which is not at Aldermaston, but in Essex where explosive experiments (non-nuclear) are carried out on the artillery testing area.

The Warhead Design Department includes:

- Weapon Engineering Division, the detailed design of the physical construction of the weapon.

- Weapon Diagnostics Division, which is involved in system testing and nuclear hardening. Nuclear hardening is the process of making equipment and electronics less vulnerable to nuclear attack.
- Electronic Systems Division which develops fusing and arming systems.

The Materials Department includes:

- Chemistry and Explosives Division
- Chemical Technology Division
- Metallurgy Division.

One can see that the Aldermaston site is not confined to manufacturing activities. There is much scientific, engineering, and research work.

The site is growing in expertise in non-nuclear activities. This includes:

- radiation protection (otherwise known as health physics)
- waste management
- decommissioning
- high-energy systems (including explosives) and high-speed phenomena are modelled and analysed using powerful theoretical and experimental tools.

Non-Nuclear Safety Incidents

In the various sources indicated above, it is understood that there have been serious safety incidents at the site not involving nuclear materials:

In July 2006, two fires occurred due to the combustion of specialist metals. It is important to emphasise that these fires were not in the areas handling nuclear materials and there was no danger of radioactive release as a result of these incidents. However, the fires led to significant delays in AWE's nuclear weapons decommissioning program.

Another fire broke out in August 2010 when a solvent used in explosives manufacturing burst into flames. The fire occurred in a concrete bunker that housed non-nuclear explosives and machinery. A precautionary cordon was placed around the facility, and houses in nearby Red Lane (the road to the East of the site in Figure 16.1) were evacuated. The blaze injured one man.

Local newspapers tend to take a more sensationalised approach (see internet versions of the Basingstoke Observer: Average of four fire calls a week at AWE, Safety fears under the spotlight again at AWE, BBC News: AWE Aldermaston nuclear waste deadline expires, Berkshire Live: AWE Aldermaston reported 82 safety incidents in three years).

From 1 April 2000 to 5 August 2011, it is reported that 158 fires broke out at AWE sites, with the fire brigade being called out to deal with alarms on average four times a week over this period, most relating to minor incidents such as fires in bins and kitchens.

The newspapers, making enquiries under the freedom of information act, further revealed that just days after the first 2010 fire, which broke out when a solvent used in making explosives burst into flames, two further fires were recorded – on 5 August 2010 and 16 August, 2010. There were also four chlorine leaks and one nitric acid spill.

AWE Aldermaston reported 82 incidents over three years - an average of more than two a month - and 137 between 1 April 2001 and 31 March 2015, the largest number of incidents of any nuclear weapons site U.K. It is understood that 22 of the 82 incidents were cases where plant operating rules, limits, or conditions were breached. Three cases involved personnel receiving an intake or suspected intake of radioactive material, and two were fires.

Office of Nuclear Regulation Opinion

What does the Office for Nuclear Regulation (ONR) have to say? In 2018 the ONR reported that Aldermaston required 'enhanced attention'. The site was criticised for delays in improving safety, its reliance on ageing production facilities, and postponements to new building projects. The AWE was first placed in 'special measures' in 2013 (The Office for Nuclear Regulation (ONR): Improvements Required at AWE July 2019). ONR publications contain public sector information published by the Office for Nuclear Regulation and licensed under the Open Government Licence v3.0. The ONR as a government body is obliged to be open about its activities and conclusions. 'Special measures' means that the site is subject to enhanced levels of regulatory attention. This categorisation followed the discovery of corrosion in structural steelwork, which resulted in the closure of the top-secret A45 building, which manufactured enriched uranium components for nuclear warheads and fuel for nuclear submarines.

There are the audits by the ONR itself. The last four years of audits can be viewed in the Office for Nuclear Regulation 'Intervention Records' on the ONR's website, best read together with the 'Compliance Inspection – Technical Inspection Guides'. A technical inspection guide covers many topics to be audited; readers can view this in considerable detail. Summarising 14 audits from 2020 and 2021, 13 audits were rated 'Green' - Good, and 2 audits were rated Amber – Improvement Needed.

Off-Site Planning Zones and Local Authority Liaison

In accordance with the REPPIR regulations (The Radiation (Emergency Preparedness and Public Information) Regulations 2019), the Local Authority is recommended to have an off-site emergency plan.

Based on a risk assessment the boundary determined extends radially 1,530 metres from the centre of the Aldermaston site. This is the minimum distance to which urgent protective actions should be taken based on consequence assessments. The hazards of tritiated water, water where the hydrogen atoms are replaced with tritium – an isotope of hydrogen, are mentioned explicitly regarding the protection of babies from harm. The distance mentioned is based on Hazard Identification and Risk Evaluation process. The document "An evaluation of the degree to which potential accidents with off-site radiological consequences occurring at AWE Aldermaston are a material consideration in defining the overall future pattern of development of Tadley" is quite detailed and reassuring in demonstrating the risk to residents in the neighbourhood of close by Tadley is acceptable. It makes frequent reference to the HSE Document Reducing *Risks, Protecting People,* otherwise known as R2P2, which we have referred to previously in Chapter 2, with regards to tolerable risk.

In practice, the Off-Site Detailed Emergency Planning Zone (DEPZ) is set more conservatively in an irregular polygon at around 2–2.5 km from the centre of the Aldermaston site.

Within the Detailed Emergency Plan Zone, there is a residential home, a school, supermarkets and public roads, and forestry and farmland. Industrial estates and residential houses are to be found just 300 metres from the site's boundary.

The Outline Planning Zone extends to a radial distance of 15 km. This includes heavily populated areas, including parts of Basingstoke and Reading. Outline Planning Zones are areas around specific sites that require public protection plans. These plans define the way police, fire, and other emergency services should act in the event of an emergency initiated by an incident on the site.

Despite being of such importance to national security, and require a degree of secrecy, the site is nevertheless required to seek planning permission and maintain good relationships with the local council. Some local community liaison minutes are available in the public domain. For example, from the Minutes of the 101st AWE local Liaison Committee Meeting, 17 March 2021, we learn that the AWE was fined £ 660,000 following a charge brought by the Office for Nuclear Regulation (ONR). The charge related to an incident involving a flash-over of electricity when a contractor's tool made contact with a live conductor. It was described as a conventional health and safety matter with no nuclear risk. We also learned from the same meeting minutes that there were 13 Recordable Injuries in 12 months. There were two process safety events. One incident related to a solvent leak into a bund, and the second related to material being left in a foyer over the weekend.

Criticality Risks

There are no nuclear reactors at the site. The worst-case scenarios considered in the risk assessments determining the off-site emergency planning zone do not include critically events. It is understood that "An explosion resulting in a nuclear yield is not possible by virtue of the safety features in the design of the weapon". Such a criticality or 'nuclear yield' event is an accidental uncontrolled nuclear fission chain reaction, as takes place in a controlled way in the core of a nuclear reactor. The critical radius is around 10 cm for plutonium and 16 cm for enriched uranium. Above this radius, sufficient neutrons are emanating from the radioactive material to sustain a nuclear fission reaction within it. So, if too many plutonium, or enriched uranium, components would be brought into proximity, a criticality event could occur, resulting in, at minimum, a severe radiation dose to the operators present. Apparently, the volumes of radioactive material, and the administrative and process barriers in place at AWE are such that criticality incident at Aldermaston is not credible. In this context see the Los Alamos 1958 criticality incident described below, where a severe radiation dose resulted from a criticality event in a waste treatment facility.

Project Pegasus

Pegasus is a £634 million project to deliver enriched uranium storage and manufacturing capability which was first authorised in 2011.

A safety report prepared by the Office for Nuclear Regulation in 2013 to authorise construction of the facility's main enriched uranium store exposed a range of difficulties with design work for the project. Objections included criticisms that some documents were not sufficiently well developed and that some statements were not well substantiated. A re-submission was requested which resulted in a four-month delay to the project.

However, from a MOD letter dated March 2021 we learn that work has only just restarted on Project Pegasus. Work on the project was paused in 2014 or early 2015 due to 'mismanagement, delays and cost overruns'. The facility will be used to manufacture enriched uranium warhead components and nuclear submarine fuel pellets. The letter states that Pegasus suffered from 'poor contractor performance' and 'an overly complex technical solution'. This led to 'significant additional construction and safety case costs' and 'severe delays'.

Pegasus will produce solid, liquid, and gaseous waste. There is a particular hazard with nuclear waste. Plutonium solutions are susceptible to criticality events – the presence of water, which is a moderator, slows the neutrons and makes them more productive in hitting plutonium nuclei (see Los Alamos criticality event in the incident examples below).

Environmental Incident Involving Nuclear Materials

There have also been environmental incidents as reported by the Nuclear Information Service, along with the BBC (14 October 2013). Increased levels of tritium – a radioactive form of hydrogen – were found in the Aldermaston Stream. and in the North Ponds drainage system for the site. The subsequent investigation by the AWE found that the tritium increase had been caused by a ventilation fan being switched off during modification works to a radioactive waste store. Stopping the fan resulted in tritium discharged from the waste store accumulating on the roof, and then being washed by rainfall into surface water drains rather than being dispersed into the atmosphere as would normally be the case.

It is understood that the increased levels of tritium in the discharge did not represent an environmental hazard.

Conclusion

So, what's my conclusion? Although I feel some sympathy for this facility under close public scrutiny and that so much of the site's performance is published on the internet, such pressure is not unique to AWE Aldermaston. Many non-nuclear sites are also under close public scrutiny. I have certainly experienced many of the same pressures in chemical plant management.

The number of safety incidents, even if they are not a nuclear threat, suggests the site's safety culture is not well developed.

We also have many reports of delays and cost overruns to projects which do not speak well of a site capability.

I tend to agree, based on my limited experience of radiation matters, that the site does not present an unacceptable risk to the public. There will be many administrative and engineering barriers preventing the release of radioactive material in a sufficient quantity to be a cause of concern to the public.

However, risk, as we have seen in Chapter 4 and elsewhere comprises components of both consequence and probability. Given the critical nature of some reports above probability may be optimistically stated.

I see the site, based on the information available, as being a moderate risk, but not a bad one. In the insurance industry, we tend to grade insurance risks as good, moderate, poor.

Safety cases are not public domain documents, neither for chemical plants, never mind nuclear facilities. Therefore, I have limited tools to complete my consideration, especially regarding consequences of an incident. We will return to the issue of public domain safety cases in section 5 of this book.

The reader may be interested to carry out a similar internet search, in order to make their own assessment of a nuclear or other hazardous sites in which they are interested.

So, what would be the purpose of making such a detailed assessment of a facility based on publicly available information. It does take some time to put together.

Firstly, we all seem to have opinions about everything. I confess that I do. Often, as human beings, these opinions can be based on very little fact. They might be deep-seated in prejudices formed long ago. Making such as assessment allows us to at least have an informed opinion. Isn't an informed opinion so much better than an uninformed one? AWE has several things going against it when it comes to public opinion, including its key mission of atomic weapons and its association with nuclear operations.

There is a second more important objective of making such an assessment. I chose AWE just as an example. It could have been an oil refinery, a chemical plant, an airport.

Assembling a number of key facts about a hazardous facility enables the reader to ask really informed and potent questions. Asking questions of the facility manager, councillors, planners, etc. is a really good and ethical way of applying healthy pressure on an organisation to take measures which reduce risk.

If I lived in the vicinity of AWE Aldermaston, I might at least ask what measures have been taken to reduce the potential for a tritium release, given that we found out above that they had had one. I might also ask whether on a typical working day, if the level of background radiation increased and by how much. I would not be satisfied with vague reassurance about 'below levels which might cause a risk'.

Out of curiosity, I asked my grown-up children to get me a Geiger counter for Christmas. Whenever in the region of a nuclear facility I pass by with the Geiger counter clicking, and I am reassured to see that so far detected no increase in radiation at all.

Incident Examples – Fireworks

Donaldson Enterprises (2011), Hawaii – Five Fatalities

Fireworks, which had been seized because of illegal use, were being stored for disposal. The disposal method involved soaking the fireworks in diesel and then taking them to open ground where they were burnt. The open ground formed part of a shooting range. An accumulation of fireworks and firework explosives resulted in an explosion that killed five people. The US Chemical Safety and Hazard Investigation Board (CSB (2013) Investigation Report – Donaldson Enterprises, Inc Fireworks Disposal Explosion and Fire) concluded that there was insufficient regulation and risk assessment of the operation.

Incident Examples – Military and Defence Establishments

Los Alamos (1958) – One Fatality

On the day of the accident, the mixing tank was supposed to contain a 'lean' concentration of dissolved plutonium, meaning that there was little plutonium. The material was in a bath of highly corrosive nitric acid and a caustic aqueous emulsion. However, the concentration of plutonium in the mixing tank was nearly 200 times higher than the operator had anticipated due to at least two improper transfers of plutonium waste to the tank from undetermined sources. Additionally, the plutonium was distributed unevenly, with the upper layer of the solution containing exceptionally high concentrations of plutonium, dangerously close to criticality. When the operator switched on the mixer, a vortex began to form. The denser aqueous layer within the tank was forced outward and upward, forming a 'bowl'. The less dense, plutonium-rich layer was drawn toward the vessel's centre.

The dissolved plutonium in such a configuration reached criticality. Criticality occurs when there is sufficient radioactive material for neutrons to bombard the nuclei of the solution's plutonium atoms with the frequency necessary to cause these atoms to break apart and release another neutron in a sustained nuclear chain reaction. It is understood that the chain reaction lasted only around 200 microseconds but was sufficient to release a massive burst of neutron and gamma radiation. The operator later died as a consequence of the radiation exposure. Interestingly, the possibility of a critical excursion in the tank was considered virtually non-existent. The potential for a catastrophe of more significant proportions exists if the criticality had lasted longer and more people were present (Wikipedia: Cecil Kelley criticality accident).

Criticality accidents continue to occur, although the last one listed by Wikipedia was in 1999 when two workers died from severe radiation exposure in Japan. The workers were pouring uranyl nitrate solution into a

Table 16.1 Safety Management Failures (Los Alamos Criticality Incident)

Aspect of Safety Management	Evidence of	See Chapter
Operators and Operations Procedures	Plutonium containing material above prescribed concentration	19
Risk Assessment	Criticality was not believed possible	25

tank that had not been designed to hold a solution with this degree of uranium enrichment.

Referring to the chapters which follow on Catastrophe Prevention we see above failings in the following as aspects of safety management (Table 16.1).

CONCLUSION

Explosives Manufacture, Military, and Defence Establishments – Potential for Future Catastrophes

There are parallels with nuclear power generation when the facility is handling nuclear materials. For other general explosive manufactures, I tend to regard their precautions in high regard.

Not possible	Unlikely	Possible (50/50)	Likely, no more than one	Very likely/ maybe several
☐	☐	☐	☒	☐

Chapter 17

New and Emerging Technologies

So far, we have examined mainly industries and facilities where their hazards are known because of past incidents. However, we cannot wait until an incident occurs before learning about new catastrophic potential.

In this section, a number of new or emerging technologies will be discussed including, technologies arising from efforts in the field of sustainability and control of global warming, such as:

- Large-scale production storage, distribution, and use of hydrogen as a clean fuel.
- Large-scale production, storage, and distribution of ammonia as a non-carbon means of transporting energy.
- Large-scale storage of energy in batteries and battery banks.
- Electric cars.
- Solar farms.
- Carbon capture.
- Driverless cars and automated roads/motorways.
- Commercial space exploration.

A key concern to this author is that, given the pressure to move fast on global warming, not all new technologies will be tested well in advance. We will have to learn the hard way, yet again, through incidents and fatalities. Perhaps we are already seeing this issue in some of the technologies below, notably batteries.

Hydrogen

Hydrogen is a potential alternative to fossil fuels. Hydrogen has been produced for many years using steam methane reformers. This is 'grey' hydrogen since the process turns methane (CH_4) into Hydrogen (H_2) with the aid of steam (H_2O). By a simple mass balance, you can see that this results in the production of carbon monoxide (CO), a highly toxic gas.

DOI: 10.1201/9781003360759-19

$CH_4 + H_2O = CO + 3 H_2$. This requires high temperatures of around 900°C in a furnace and pressures of approximately 25 bar.

The reaction, therefore, cannot be stopped here, and a water-gas shift reaction using a catalyst is used to eliminate the carbon monoxide.

$$CO + H_2O = CO_2 + H_2$$

Therefore, grey hydrogen manufacturers produce a lot of carbon dioxide from both the furnace fuel exhaust and the above chemical reactions. Carbon dioxide is a greenhouse gas that contributes considerably to global warming.

From a carbon footprint perspective, there is an even worse form of hydrogen manufacture. This is known as 'brown' hydrogen, which comes from the gasification of coal.

'Blue' hydrogen is produced by the same well-known steam methane reforming process, but the carbon dioxide is separated out (carbon capture) and sequestered. Sequestration means that the gas is stored, out of the way. We will come back to this under Carbon Capture below.

'Green' hydrogen is manufactured entirely without fossil fuels by the electrolysis of plain old water. This sounds like a great alternative, but still requires a lot of electrical energy.

From a catastrophe prevention perspective, we need not be too concerned about the method of manufacture. But it seems that hydrogen will be more ubiquitous than before, and this has considerable catastrophic potential.

Hydrogen is extremely flammable and has very energy required for ignition. Hydrogen leaking from a pipe or small crack will burn without finding an ignition source. The energy of release is sufficient.

Therefore, hydrogen transportation and storage create a potential catastrophic potential if that hydrogen is accidentally released.

To date, significant hydrogen explosions have been rare. It is worth pointing out that the highly destructive explosions at Fukushima Daiichi (see Chapter 10) were due to hydrogen inadvertently created by the reaction between the fuel rods and water surrounding the rods for cooling. Under the extreme heat of the nuclear reactor which was out of control hydrogen was produced.

Ammonia

The distribution and storage of hydrogen is difficult. An alternative is Ammonia (NH_3). It can more efficiently store energy, which can be subsequently released by turning the ammonia back into hydrogen.

Compared with hydrogen, ammonia is easily liquified, and it can be stored in pressurised tanks at around 10 bar. This liquefaction of ammonia is well-known technology and is widely used in the chemical and fertiliser industries. Storage of ammonia is not, therefore, an emerging technology. The extent of ammonia storage is likely to expand dramatically, however, and therefore multiply the associated risks.

There is a problem, however, as ammonia is highly toxic, whereas hydrogen is not. So an ammonia storage tank has very considerable catastrophic potential. Whilst there have been many incidents with ammonia, and many have resulted in gas exposure, large releases, and injury, explosions and fatalities are rare. The use of ammonia is common in the food and food storage industry.

Batteries

There is now much development in battery technology. This book's readers probably have mobile phones and tablet computers that rely on batteries to keep their mobile devices operating when they are not plugged into an electrical supply.

Lithium batteries are made to deliver high output with minimal weight. These small batteries, typically lithium-ion batteries, have a known fire risk. Battery components are designed to be lightweight, and therefore have thin partitions between cells and a thin outer covering. The partitions or coating are relatively fragile, so they can be punctured. If this happens, a short circuit can occur. The resulting spark can ignite the highly reactive lithium. The battery can heat to the point of thermal runaway, where the increasing temperature causes reactions that further increase the heat evolved. The heat of the contents exerts pressure on the battery, potentially producing an explosion.

At first, the use of small batteries does not appear to have catastrophic potential, only the user is at risk. But consider mobile devices onboard aeroplanes either in the hold or in the cabin. There have been several incidents where lithium batteries have gone on fire during air transportation, potentially putting the plane and its passengers and crew at risk. The FAA reported in October 2021 that a total of 343 aviation-related thermal incidents involving lithium batteries.

More serious from a catastrophic potential is the building of larger and larger battery banks. Typically, these battery banks store solar energy when the sun is shining and supply the stored energy to homes or businesses both day and night.

Unrelated to the battery chemistry is the increased voltage of such systems, leading to an arc-flash risk. Since these batteries are used in domestic and small business settings there is the potential for unqualified people to interfere with the system and inadvertently expose themselves to injury.

Battery technology is improving, partly in response to incidents. For example:

- Improved cell quality, so that such decompositions of the battery are not initiated.
- Barriers between cells to limit or prevent cascading of overheating from one cell to another.
- Improve ventilation to ensure the dissipation of flammable gases which might accumulate within the containment building.
- Improved fire suppression.
- Training of first responders and fire brigades in managing battery fires.
- Electric cars.

Given the absence of gasoline in an electric car, one might think that the fire risk was negligible. But this is not the case. Incidents have occurred where on collision there has been a battery fire. However, since individuals are involved, we will not dwell on this further since the hazard has little potential to reach catastrophic proportions. I would further agree with the makers of electric cars that the risk is lower than for a gasoline-driven car.

Solar Panels and Solar Farms

Solar panels can and do go on fire. The incidence rate is reassuringly low, but the risk is there. These fires are primarily due to DC arc faults resulting from loose joints, corrosion of joints, damage to insulation, material ingress, etc. Poor installation is the primary cause, but defects in manufacture and assembly can also be involved.

This risk may not be comforting for those families who have installed solar panels on their roofs. However, I would emphasise that the incident rate is low and that in my research, I did not find any cases where the fire had resulted in harm to the families below.

Solar panel fires are understood to have the potential to emit toxic gases. More concerning is the potential on solar farms where a fire might spread from a malfunctioning panel to adjacent panels. Although this kind of event has not occurred so far, the widespread destruction of many panels could reach catastrophic proportions.

Carbon Capture

In carbon capture, carbon dioxide is separated from the gases, typically flue gases, in which it is contained and then compressed and piped as a gas into storage. This containment is typically conceived as underground storage such as abandoned oil fields. The key hazard is clearly a major carbon dioxide leakage from the pipeline and compression operations or the underground storage. Carbon dioxide is not in itself particularly toxic. After all, we breathe it out with each breath! However, it is an asphyxiant in large concentrations since it displaces the oxygen content in the air we are breathing. Moreover, carbon dioxide is heavier than air with the potential to gather in low-lying locations. It does not have an odour which might give an early warning. Since with carbon capture, carbon dioxide will be handled on a much larger scale than ever before, there is an increasing potential for catastrophe. Not to sound hysterical, but blowouts have occurred from oil fields extracting oil. Is there not the potential for a blowout from an oil field being used for carbon capture through some kind of kickback in the injection pipework? Or even leakage from the reservoir through ground faults. Those charged with carbon capture are, I am sure, taking such dangers quite seriously. Given the climate catastrophe which is to be prevented, this is a reasonable risk if it is well managed.

Wind Farms

I mention wind farms for completeness. From an occupational health perspective, there are many reasons to be concerned about wind farms. There have been injuries to workers carrying out erection and maintenance. Wind turbine fires. Collapsing of wind turbines, etc. However, from a catastrophic potential there seems little danger of any cascading effect. Sometimes hazards appear in unexpected places. See Derrybrien wind farm in the Incident Examples below.

Driverless Cars and Automated Roads/Motorways

Again, I mention this subject for completion. We have seen the outcome of software errors in the section on Air Transport in Chapter 9. I can imagine a mistake in programming or operating this system to have the potential to cause a massive motorway pile-up potentially involving trucks of hazardous materials such as gasoline and chemicals.

Commercial Space Exploration

Perhaps not of catastrophic potential. I am thinking more of the propensity for increased manufacture and storage of rocket fuel, not so much the remote possibility of commercial space rockets and components landing on the community. Some space satellites and vehicles carry radioactive materials for propulsion.

Incident Examples – Hydrogen

Fukushima Daiichi (2011)

In the Fukushima Daiichi Nuclear Accident (see Chapter 10) it was a hydrogen explosion which caused severe damage to three of the reactor buildings. The fuel rods became very hot due to the uncontrolled nuclear reaction and reacted with steam releasing hydrogen.

OneH2 Plant Explosion (2020)

This was a new fuel cell plant designed to produce hydrogen to power forklifts and small trucks. An explosion damaged 60 homes. Fortunately, there were no injuries (Vice.com: One of the country's only hydrogen fuel cell plants suffers huge explosion).

Incident Examples – Ammonia

Minot Train Derailment (2002) – One Fatality, Toxic Release

A train carrying five tank cars of anhydrous ammonia derailed and ruptured, releasing ammonia into the adjacent city. There was one fatality and 11 serious injuries. Derailment of freight trains carrying hazardous material is an issue of repeated concern. See Chapter 9 on Rail Transportation.

It was estimated that the accident caused over $2 million in damages and another $8 million in environmental clean-up costs (Wikipedia: Minot train derailment).

Potchefstroom, South Africa (1973) – 18 Fatalities, Toxic Release

An ammonia bullet tank failed, releasing 38 metric tonnes of anhydrous ammonia. The incident resulted in 18 fatalities, 6 of them outside of the installation's perimeter. Workers in the building just 80 m from the release survived, whereas people who had left their houses 180–200 m from it died. This particular catastrophe demonstrates the value of shelter-in-place versus evacuation for toxic gas leaks.

Incident Examples – Batteries

APS (2019)

A fire and subsequently an explosion took place at a Battery Energy Storage System. The incident caused the destruction of the battery facility and its storage container, and several firefighters were injured. The power supply was fitted with a fire suppressant system, but this proved inadequate to stop the cascading thermal runaway. The incident is thought to have been caused by a defect in one of the cells. The build of heat caused adjacent cells to malfunction in a cascade-like fashion. Some three hours after the thermal runaway was thought to have begun, firefighters opened up the door to the battery system, which agitated flammable gases that remained and brought the gases into contact with a spark or heat source – causing the explosion (Energy Storage News: Arizona battery fire's lessons can be learned by industry to prevent further incidents).

UPS Airlines Flight 6, Dubai (2010) – Two Fatalities

The cargo flight en route between Dubai and Cologne developed an inflight fire. The aircraft crashed, killing both crew members. Thick smoke appeared in the cockpit, and the pilot could not see the radio through the thick smoke. The pilot also discovered that he had no elevator control. This was subsequently found to be due to the fire burning through the protective covering and destroying the primary flight control system. The cargo hold contained a large quantity of lithium-type batteries. A number of improvements were made following the crash. These improvements included instructing pilots to climb to a higher altitude in the event of cargo hold fire to deprive the flames of oxygen and extinguish the fire. Improvement to ventilation and provision of smoke hoods in the cockpit were also recommended (Wikipedia: UPS Airlines Flight 6).

Incident Examples – Wind Farms

Derrybrien Wind Farm, Ireland (2003) – Widespread Environmental Damage

A landslide occurred during the construction of a wind farm. 450,000 cubic metres of peat were dislodged and travelled some 2.5 km initially and then after rains continued a further 20 km into a lough which was used as a supply of drinking water. More than 50% of the fish in the lake died. An environmental impact assessment should have been undertaken before the project was allowed to proceed (Wikipedia: Derrybrien Wind Farm).

CONCLUSION

New and Emerging Technologies – Potential for Future Catastrophes

LNG

The potential for a catastrophe is enormous, but facilities appear new and thoughtfully designed.

Not possible	Unlikely	Possible (50/50)	Likely, no more than one	Very likely/ maybe several
☐	☐	☒	☐	☐

Hydrogen

Whilst on the one hand it ignites so easily, on the other hand hydrogen disperses so quickly. A lot of research is going into making hydrogen safe when used as a fuel.

Not possible	Unlikely	Possible (50/50)	Likely, no more than one	Very likely/ maybe several
☐	☐	☒	☐	☐

Ammonia

This is a familiar hazard which has been well handled historically. The flight from fossil fuels may put pressure on speedy construction.

Not possible	Unlikely	Possible (50/50)	Likely, no more than one	Very likely/ maybe several
☐	☐	☒	☐	☐

Batteries

With increasing pressure to produce bigger and better battery storage there are going to be some incidents of infant mortality.

Not possible	Unlikely	Possible (50/50)	Likely, no more than one	Very likely/ maybe several
☐	☐	☐	☒	☐

Others

Solar farms, driverless cars, and even space exploration have limited potential.

Not possible	Unlikely	Possible (50/50)	Likely, no more than one	Very likely/ maybe several
☐	☐	☒	☐	☐

Section III

Catastrophe Prevention and Why It Is Failing

Chapter 18

Introduction

The number of catastrophes occurring worldwide suggests that methods of catastrophe prevention simply are not working. This conclusion is hard for this author to admit. As a process safety consultant and insurance survey risk engineer, my livelihood in the past 20+ years has been based on assessing process safety management systems, finding opportunities for improvement, and working with the client to implement those improvements.

I suggest it is the case that individual improvements do work, but unless all the links are securely 'joined up', the system fails. This subject, and its solution, is studied in the fifth and final section of this book.

Each chapter in this section addresses a particular group of safety management and catastrophe prevention aspects. Initial chapters cover issues at the 'front end' – operators, training, maintenance, contractors. We look at issues relating to equipment suitability. Emergency response is covered – the emergency response and fire brigade team are front-end personnel also. Then the section moves on to safety methodology, including incident investigation, risk assessments, Hazard and Operability Studies (HAZOPS), and other techniques. Finally, there are several chapters on the role of management, government, and regulators. At the end of each chapter, there is a brief summary of the ways in which this particular aspect of safety management can fail to prevent catastrophes.

So, let's examine some of the links. Whilst we all know the adage 'it starts at the top'. However, I am going to work in the opposite direction – from the operator up – simply because in many catastrophes operator error is erroneously seen as the root cause. Secondly, it's just too easy to pass the blame to the 'man at the top'. That does not provide a solution. Furthermore, the CEO or mainboard member may well know very little about operations management and safety. They may have brought to the main board excellent skills in other disciplines such as financial management, marketing and business development, etc.

In Section 2, a number of incidents, either catastrophes or incidents that could have become catastrophes, were described in some detail. For these specific aspects of safety management were identified, which had those aspects have been performed better or correctly, the incident may have been avoided. These contributing safety management factors for the following incidents (with their chapter numbers) are aggregated in Table 18.1:

- Piper Alpha (6)
- Deepwater Horizon (6)
- Richmond Refinery (6)
- Bhopal (7)
- La Porte (7)
- Brumadinho (8)
- Shoreham Air Show (9)
- Costa Concordia Cruise Ship (9)
- Transhuron ship (9)
- Ladbroke Grove Train Crash (9)
- Fukushima Daiichi (10)
- Chernobyl (10)
- Georgia Sugar (12)
- Beirut Warehouse (13)
- Freedom Industries (13)
- Buncefield (14)
- Port Talbot (14)
- DPC (15)
- Los Alamos (16)

Table 18.1 Frequency of Safety Management Issues in above Incidents

Safety Management Element	Total Mentions	Section 3 Chapter
Equipment Maintenance and Inspection	14	22
Emergency Response and Readiness	12	23
Operators and Front Line Personnel	11	19
Equipment Suitability and Reliability	9	21
Human Factors	8	20
Governance of Safety	8	31
Operating Procedures	7	19
Risk Assessment	7	25–27
Management of Change	7	31
Managing Legacy Issues	5	34
Laws, Regulations	4	32
Contractor Management	2	20
Control of Work	2	22
Assessing risk Outcomes	2	27
Consequence Analysis	2	28

All the other safety management techniques described in this section were either mentioned once or not at all in the example incidents analysis.

In the first chapter, we will talk about plant operators. In a different industrial sector, these could equally be the train driver, pilot, ship navigator, etc. without loss of strength of argument. After starting at 'the bottom end', this section then turns to safety techniques and finally to the structure of a holistic safety management system and its other elements.

For this reason, Section 3 does not follow the same structure as some other guidelines, for example, A.I.Ch.E. Risk-Based Process Safety. Non-technical readers can pass over Table 18.2, which maps the Risk-Based Process Safety Elements to the titles of the chapters in Section 3 of this book. It is included for readers who have a background in safety management systems.

Table 18.2 Chapter Mapping to Risk-Based Process Safety

Risk-Based Process Safety		'Catastrophe Prevention'	
Element Number	Element Title	Chapter Number	Chapter Title
Commit to Process Safety			
1	Process Safety Culture	33	Safety Culture and Other Social Structures
2	Compliance with Standards	19 32	Operators and Front Line Personnel Laws, Regulations, Standards
3	Process Safety Competency	24 25–27	Role of the Safety Department Risk Assessment #, HAZOP, and other techniques
4	Workforce Involvement	33	Safety Culture and Other Social Structures
5	Stakeholder Outreach	46	Common Factors with other 'Catastrophes' such as Climate Change
Understanding Hazards and Risk			
6	Process Knowledge Management	25–27	Risk Assessment, HAZOP, and other techniques
7	Hazard Identification and Risk Management	25	Risk Assessment, HAZOP, and other techniques
		27	Assessing Risk Outcomes and Risk Reduction
Manage Risk			
8	Operating Procedures	19	Operators and Front Line Personnel
9	Safe Work Practices	19 22	Operators and Front Line Personnel Equipment Maintenance and Inspection

(Continued)

Table 18.2 (Continued)

Risk-Based Process Safety		'Catastrophe Prevention'	
Element Number	Element Title	Chapter Number	Chapter Title
10	Asset Integrity and Reliability	22	Equipment Maintenance and Inspection
11	Contractor Management	20	Human Factors
12	Training and performance assurance	19	Operators and Front Line Personnel
13	Management of Change	31	Governance of Safety
14	Operational Readiness	22	Equipment Maintenance and Inspection
15	Conduct of Operations	19	Operators and Front Lone Operations
16	Emergency Management	23	Emergency Response and Readiness
Learn from Experience			
17	Incident Investigation	24	Incident Investigation
18	Measurement and Metrics	31	Governance of Safety
19	Auditing	31	Governance of Safety
20	Management Review and Improvement	31	Governance of Safety

Chapter 19

Operators and Front-Line Personnel

OPERATOR MISTAKES

Operators, pilots, train drivers make mistakes. You make mistakes. Think back to your last journeys in a car. Did you do anything wrong? Maybe drove through a traffic light at amber or worse, red. Failed to give way when you should have? Exceeded the speed limit? As is often said 'To err is human'.

Whilst an organisation needs to manage operators to reduce the potential for mistakes, we are mistaken if we think we can eliminate them. The field of human reliability deserves its own chapter. Therefore, we leave further discussion on this subject of mistakes until the next chapter on Human Factors, and concentrate first on how operators are instructed in what to do.

OPERATING PROCEDURES

In Section 2, eleven catastrophes were examined involving failures of the front-line person - operator, pilot, train driver. Seven of these involved problems with the operating procedures or the way those procedures were interpreted. These tragic events included the Bhopal chemical release, the Deepwater Horizon oil spill, the Ladbroke Grove train collision, and the Shoreham Air Show crash.

Simplistically, safety should be easy. Management just tells the operator what to do. So long as the operator does as instructed, all will be OK?

This simplistic assessment assumes two vital aspects:

1. Management knows what the operator should do.
2. The operator knows what management wants him to do and follows those instructions.

Why might either of these aspects not work?

In examining what is wrong with this approach, let's consider first who writes the operating procedures. Typically, the first operating procedures

DOI: 10.1201/9781003360759-22

are written when the facility has been designed and under construction. The person writing the procedures is, typically, not someone who has worked as a plant operator, and has certainly not been able to work as a plant operator on the plant being designed. This lack of practical experience is a serious void. Operating procedures can simply be wrong!

Illustrative Hypothetical Case – Whisky Distillery

Let's consider our Whisky Distillery again (see the Introduction and Appendix A). Distilleries typically have a kiln to dry the barley. This kiln halts germination and dries the barley in preparation for milling. The temperatures required are not high; the barley needs to be heated to around 60°C. Scottish malt whisky distilleries use peat to fire their kilns and claim that the peat smoke adds to the flavour of the resulting whisky. However, the use of peat is questionable from an environmental perspective, and some distilleries have been converted to natural gas-fired kilns. We will use this conversion to gas firing to demonstrate several aspects of safety management in this Section of the book. The diagrams and technology are kept at an elementary level to serve as examples for the non-technical reader.

The construction is, in principle, quite simple (Figure 19.1).

After the conversion, the peat floor is no longer in use. A simple natural gas burner has been installed in its place. An actual installation would be more complex than shown here

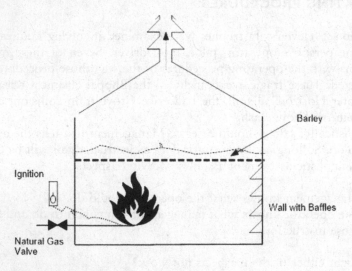

Figure 19.1 Whisky Distillery Gas Burner – Ignition.

The person assigned to write the procedure simply instructs the operator to open the natural gas valve and push the ignition button. Remember lighting a Bunsen burner at school – it can really be that simple. The writer remembers to include, in the procedure, an instruction for the operator to check that ignition has occurred. The operator is instructed to do this by looking through the flameproof window into the kiln.

For the operator, this works fine, at least for several ignitions. Then one day, the flame does not light. The operator tries again, and again until there is an explosion. The kiln is destroyed, the operator is injured and production ceases for several months. Why does this happen? Because by repeatedly unsuccessfully attempting to light the burner, natural gas accumulates within the kiln until it reaches a concentration at or above the lower explosive limit. The concentration of natural gas is such that an explosion occurs.

The person writing the procedure did not consider what to include if the burner fails to light. In practice, there are both procedural and engineered precautions required of a practical burner. We will return to this example later in Chapter 27.

Another example is a procedure where the operator is required to close valves A, B, and open C.

A and C are located physically close together, so it is more convenient for the operator to close A and C before closing B. An incident results because process fluids pass back through open valve B and out through opened valve C.

The procedure writer did not explain why the valves need to be operated in this order. Misunderstandings of the purpose of particular steps are a frequent cause of failure in operating procedures. These procedures should explain why as well as what.

Many other problems I have experienced with operating procedures:

- Out of date – equipment has been modified without updating the method.
- Conflicting – there are two procedures referring to the same operation. In the above example valves A and C might be parts of plant X1 and valve B is part of plant X2.
- Do not cover start-up or shutdown.
- Not practical to implement. Typically in such cases, the operator decides what to do.

Notes from Trevor's Files: Conflicting Procedures

Should a ship be earthed to the shore, or not? Earthing comprises connecting an electrically conducting cable between the vessel and the dockside. The Inter-Governmental Maritime Consultative Organization (IMCO) guidelines say that a large ship carrying, say fuel, should not be earthed since an earthing cable would not have the capacity to equalise any static potential generated between the ship and the shore. In contrast, standard procedure for unloading or loading a road tanker is to ensure that it is earthed. What is right or wrong in the case of the ship need not concern us here, other than to say there are some differences of opinion.

So, on several occasions when surveying dockside operations, I have seen procedures from the receiver of the raw materials, say the oil terminal, which specifically instruct the harbour personnel not to earth the ship. In contrast, the port authority, who owns the jetty and are responsible for harbour procedures, require the vessel to be earthed. I have found checklists where both earthed and unearthed check boxes have been ticked. What, to me, is essential is for the two parties to be consistent in their requirements, otherwise what respect do the harbour personnel have for procedures?

A further aspect of procedures concerns what the operator should do in abnormal situations. For example, what to do if the supply of electricity is lost, if suddenly there is no steam, etc. These procedures, sometimes referred to as emergency operating procedures (EOPs), are not standard because they are rarely, maybe never, carried out in the lifetime of a particular operator. However, a catastrophe could occur if the operator cannot quickly and fluently recall what to do. For example, with the loss of steam pressure, highly flammable process materials can back up into steam supply lines. A sudden cooling often leads to leaking flanges and the escape of process materials.

Whilst automation may cause some valves to close, etc. in the event of a sudden emergency, the operator should at least have a checklist of things to check! The nuclear industry has many such procedures – we have already discussed the implications of loss of cooling water to a nuclear reactor in Chapter 10! Emergency procedure training must be carried out and routinely refreshed – at least annually. A good practice is to do these as a tabletop 'what if' exercise. In addition, the training can be enhanced by operators physically going to the valves they would go to close or check and simulate that activity by hanging a tag on the valve. This type of drill gives a sense of real-time. Regrettably, retraining in emergency operating procedures is sometimes carried out by asking the operator to reread the procedure and sign that they understand. Such retraining is, I suggest, largely ineffective.

An excellent practice I have seen in the preparation of emergency practices such as loss of utilities is to have two standards for each utility. One of these procedures is for loss of utility 'with notice', and one procedure is for loss of utility 'without notice'. Many more precautions can be methodically carried out in the 'with notice' scenario compared with a shutdown 'without notice'.

Note that emergencies involving the emergency response team, e.g. fires, are covered separately, see Chapter 23.

Notes from Trevor's Files: Propane Unloading

During a short holiday, whilst taking a walk from my hotel room, I notice a tanker of propane arriving to replenish the tanks that supply gas for heating the hotel. Now I have watched many a tanker being filled at the terminal, but not discharged. I had a chance to view this activity 'undercover' with nothing to identify me as a process safety engineer! I just told the tanker driver I was being nosy, and he was fine with me observing what was going on. Now instructions for the operation were clear to read on the side of the road tanker. The process should start by connecting an earthing cable, and at least one set of wheels should be chocked. The chocks were available, but neither earthing cable nor chocks were used. Why not?

I think it is too easy to blame this on laziness. More likely, the tanker driver believed that the chocks were unnecessary, given that he was on totally level ground, and he felt that there was sufficient earthing provided by the electrically conductive hoses. This judgement is often a problem with operators and practices. The operators consider them 'over the top' and 'belt and braces'. However, when the risk assessment was done, it was done on the basis that there were multiple barriers; that both 'belt' and 'braces' were needed. What if the conductive braiding of the propane hose fails? What if the driver accidentally releases the brake, or the brakes fail, or the ground is not so flat ...? In Chapter 4 we discussed the benefits of multiple barriers. Operators typically don't grasp the importance of multiple barriers.

HOUSEKEEPING

My wife says my standard of housekeeping is poor. I look after my study. She looks after the rest of the house. My excuse that I have a 'busy office'. However, many would argue, rightly, that good house-keeping and safety are related. Let's not be concerned here about appearances. I am concerned about circumstances that can lead to catastrophes.

Notes from Trevor's Files: Dust and Rubble

On a recent insurance survey of a fertiliser plant, I found layers of dust inches deep in the operating area. This layer of dust covered all stairs and walkways and was both slippery and uneven. I was concerned on two grounds:

- The dust is flammable and could contribute to a dust explosion, or even secondary explosions if an explosion occurred elsewhere. For an example see Imperial Sugar incident described in Chapter 12.
- In the event of an incident, the slippery and uneven dust would delay response by personnel, or worse, cause a responder to the incident to fall and injure themselves. The fall and injury would delay the response to the incident. If the incident was a fire, a delay might cause the fire to spread and grow larger, potentially enveloping adjacent facilities directly due to the poor housekeeping.

While conducting some consultancy at deep coal mine above the Arctic Circle in Russia, I was concerned about the boulders and dust we had to climb over even when navigating the main mine tunnels, which are busy with many miners travelling in both directions. I suggested that the mine set a standard for these tunnels. For example, by setting a standard that at least one metres width of the walkway was free of rubble and other obstructions. This suggestion was widely rejected. 'We already have that standard', they said, 'it is in our regulations'. I had to leave frustrated. Here is a cultural issue – what is a regulation? We will return to this in Chapter 32.

TRAINING

Perhaps it is as simple as the operator should be trained appropriately.

Airline pilots have a rigorous training procedure which includes initial training, type training for specific aircraft, refresher training, emergency training, etc. One big assumption is that the training prescribed is fit for purpose. As described in Chapter 9 on Transportation, an example of inadequate training concerns the two Boeing 737 Max crashes.

Among other causes, it is clear that the pilots were not adequately trained in the new aircraft, particularly regarding the latest software which pitched the plane in response to certain conditions.

Many of the problems I have experienced with operator training include:

- Training being too generic; for example, training in how to operate a pump, as opposed to the specific type of pump required by the operator. A domestic example might be training in how to drive a car, and not how to drive a particular car with an automatic drive and other additional features.
- Training being too general; for example, training in handling hazardous materials saying to wear the "appropriate Personal Protective Equipment" – how does the operator find out what equipment?
- Conflicting training: for example, in the induction training, the operator is instructed to flush the material away using a water hose in the event of a spillage of material. In contrast, the site procedure where the operator works says to collect all spillage using an absorbent 'pig'.
- Excessive time gap between training and application; essential points are forgotten – but it's OK because the operator has been marked off as 'trained'!
- Training being all theoretical, with no practical content whereby the operator is required to demonstrate physically an understanding of the task in which the operator has been trained.
- Irrelevant training – training done 'just in case' in order to get the employee marked off as trained.
- Training not refreshed. Unless the operator practices the task regularly, information and sequences are readily forgotten. Examples are emergency procedures – which are rarely carried out unless there is an emergency. An example from my experience is First Aid Training. The training courses for First Responders are excellent, and I have done these courses at least twice. But I cannot remember much of the content. Hence it is recommended to repeat First Aid training each year.

Issues with training become more complicated where higher-level skills are required. An engineer may have qualified at university. Now, what local training does the engineer need to be able to carry out his/her duties at the plant?

Management and supervisory training can be controversial. Good managers may have become 'good' managers using totally different techniques.

Notes from Trevor's Files: Training Matrix Concept not accepted

Working with an international oil refining company in Central Europe, I asked the operating management about training. The response was that all questions about training be referred to Human Resources. We, therefore, requested to meet with HR, and we posed the question as to whether the organisation had a training matrix. Such a matrix would have a list of all the training required in any position, whether the training was needed before the person performed in the role, or whether it could be acquired whilst 'on the job', and whether the training had been completed. If not, a plan would be developed for the training to be completed.

A training matrix is an expectation of many of the ISO standards, including ISO 9000 (Quality). This particular organisation was certified in accordance with ISO 9000. Yet it seemed that there was no understanding or acceptance of the need for a training matrix.

Further issues can arise when the competence of an operator or group of operators is misunderstood.

This risk becomes more acute when work is contracted out. The contractor says that their people are qualified to a particular standard, and the operating organisation assumes contractor personnel knowledge accordingly, only to find serious voids.

SUMMARY: CAUSES OF FAILURE IN OPERATING PROCEDURES AND TRAINING

Reasons Why Procedures Sometimes Fail in Catastrophe Prevention Include:

☒ Person writing procedure lacks knowledge of the operation
☒ Person writing procedure does not consider all the ways the equipment might be misoperated and the consequence of this misoperation.
☒ The procedures are out of date.
☒ Two or more procedures are conflicting.
☒ Procedures do not cover abnormal situations
☒ Procedures involving abnormal situations are not rehearsed; re-training is done as a 'tick in the box exercise'.
☒ The procedures do not tell the operator what to do, leaving him/her to decide themselves.

Causes of Failure in Operator Training

☒ Training is too generic or general, not specific to the particular equipment being operated.
☒ Time between being trained and carrying out the task is too long.
☒ The training is either not relevant
☒ Training is too theoretical with little practical content
☒ Training not periodically refreshed and revalidated.

Chapter 20

Human Factors, Personnel Selection, and Contractors

HUMAN ERROR

When did you last make a mistake? The consequences may not have been severe – spilt the coffee, tripped on the stairs, pulled out of a road junction without clearly checking for oncoming vehicles. Those same mistakes might have caused burns to the hand, head injuries, or a traffic accident in slightly different circumstances.

There is a preoccupation in incident investigations to find the operator who made a mistake. And once that operator is found, some investigations are immediately curtailed, having found the guilty party. Finding the person who made a mistake is not the answer, at least in the context of incidents with catastrophic potential.

Attributing the accident solely to human error may prevent actions from being identified which would improve safety in the future. Human error is a frequently used explanation of incidents. Some research suggests that around 90% of industrial accident reports indicate a failure on the part of the injured person. Such reasons are convenient. Blaming individuals who directly cause accidents suggests that such accidents are unavoidable, which is hardly acceptable to those potentially affected by the accident. Blaming the individual is sometimes seen as absolving management of responsibility. This leads to incident recommendations that the individuals be disciplined, sacked, retrained, or told to be more careful (Davies and Adams. To Err is human: human error and workplace safety, 2015).

In Section 2, eight catastrophes were examined were identified as involving clear human factor failures. On closer examination probably all the incidents involved human factors to some degree. The most prominent events included Deepwater Horizon, Bhopal, the Shoreham air crash, and the Costa Concordia.

So, what are these human factors leading to error?

Human failures can be active or latent (HSE, Reducing Error and Influencing Behaviours 1999). There are three types of human error:

DOI: 10.1201/9781003360759-23

1. Slips and lapses – made inadvertently by experienced operators during routine tasks, perhaps because of distractions.
2. Mistakes – decisions subsequently found to be wrong, though the person making the mistake believed those actions to be correct at the time.
3. Violations – deliberate deviations from rules established for the safe operation of equipment.

Latent failures which contribute to human error include:

- Distractions, lack of time, inadequate procedures, poor lighting, or temperature extremes.
- Physical ability, competency, fatigue, stress, or drugs.
- Work pressure, long hours, or insufficient supervision; and
- Poor equipment design or workplace layout.

Failures can also be Active or Latent. With Active failures, there is an immediate consequence to the error. Active failures are typically made by front-line people such as control room staff or machine operators. Latent failures generally are failures in the health and safety management system and include errors in equipment design. Latent errors are made by people removed in time and space from operational activities, e.g. design engineers and managerial decision-makers.

Notes from Trevor's Files: 'Disused' railyard

If the perception of danger increases, people tend to behave more cautiously. On the other hand, people tend to act less cautiously when they feel safer. While conducting a survey at an alloy smelting works in Kazakhstan, we had passed carefully and safely through the noisy and dusty area in front of the furnaces, which were emitting roaring flames. Tripping hazards on the floor made us very careful. Once this section of the tour was completed, we had to traverse a large yard before reaching the environmental treatment works we wanted to visit. It was springtime. The large yard appeared to be disused, overgrown with vegetation. Birds were singing. We were relieved to have left the furnace area and were in deep conversation about what we had seen. I was at the head of the group along with my translator. Suddenly about one metre in front of us, a diesel locomotive passed by at high speed. Whereas the perception of danger in the furnace area was clear, in the large yard, we became relaxed. If we had been two seconds earlier me and my translator would have been dead, and it would have been my fault for not making myself aware of the hazards in the yard.

Human Error probability is disconcertingly large. For a reasonably simple task performed rapidly, reliability of only 90% is typically assumed in some kinds of risk assessment (see Chapter 27 on Layers of Protection Analysis). One in ten times, the operator gets it wrong! Now many factors might increase or decrease this reliability. For example, if the task is regularly practised, the reliability significantly improves. Reliability also improves if there is a second operator who might notice the error or raise the primary operator's awareness. Many factors can reduce reliability, including operator inexperience and ambiguity in expected performance standards.

In this author's experience, the necessity of accepting potential failure due to human factors is widely denied – see Notes from Trevor's Files: Organisations in the Former U.S.S.R. below. In my experience, oil field workers from Texas are generally very proud of the long hours they and their employees work, especially under the pressure of the drilling floor, even though such excessive hours are known to be error-inducing. Understandably management levels in an organisation are intolerant of mistakes from their staff.

The critical question is why human reliability does not seem to be widely understood; and why more is not done to reduce potential errors.

Of course, one often repeated approach to eliminating human errors is eliminating the human. Automating the system is seen as a solution to excluding errors. Indeed, automation can help, but as we shall see in future chapters, automated systems can produce many errors themselves.

First, some observations on human psychology and why errors might occur.

Notes from Trevor's Files: Coker Superintendent

I had the good fortune to work with several organisations in Russia, Ukraine, and Kazakhstan, where there was a similar approach to incident investigation. This approach seems to have its origins in the culture of the former Soviet Union. Incident investigations are typically focused on finding the operator or supervisor who had made a mistake. That person is then punished. This action was seen to be the end of the investigation.

In fairness, I also found that many organisations in the former U.S.S.R. had a high degree of trust in the professionalism and diligence of their employees.

I recall reviewing a particularly complex plant known as a Coker. In a Coker, tarry materials of little intrinsic value are fired into Coker cells at high temperature and pressure. This forms lighter, more valuable materials and some coke that separates out in the Coker cell. Periodically these vast Coker cells need to be opened to remove the 'petroleum coke', which is quite a valuable material. Opening a Coker cell is a complex operation requiring isolations of many pipes and the proper sequencing of cooling and depressurisation. Because of this complexity, many refineries have automated the process with automatic valves, interlocks, and software-controlled sequences. There have been incidents where this hazardous automated process has failed due, amongst other factors, to the failure of the large automatic valves to operate

Asked about the potential for such an incident on his Coker, my Russian-speaking colleague proudly replied that no such incident was possible on his plant. The operators always followed detailed procedures!

I was at least partially successful in convincing him that at least a checklist might be helpful as an aide-memoire for an otherwise very long procedure.

An interesting aspect of human behaviour often found evident in catastrophic incidents concerns 'confirmation bias'. Confirmation bias is the tendency to interpret and recall information in a way that confirms or supports one's own beliefs. People display this bias when they select information that supports their views, ignore contrary information, or interpret ambiguous evidence to support their existing attitudes. Examples of confirmation bias were identified in the discussion of the Deepwater Horizon incident in Chapter 6. Here the engineering team interpreted the results of a test as confirming that the well integrity test was satisfactory, and they were not open to other possible interpretations. At least part of the reason for the confirmation bias on this occasion is that no failures of well integrity tests had been ever experienced by the engineers involved. Therefore, a satisfactory well integrity test was considered a given and an almost redundant step in the activities (Hopkins (2012)).

A further aspect of human behaviours leading to human errors concerns the 'normalisation of warning signs'. There are often, although not always, warning signs before a catastrophe. However, if the people involved had frequently experienced such warning signs previously, and no significant incident resulted, the warning signs become normalised and not given much attention. A similar aspect is the 'normalisation of deviance', where people get familiar with deviating from a particular procedure with no significant consequences. The normalisation of deviance can be readily seen in driving habits where speed limits are routinely exceeded by at least 10%!

Notes from Trevor's Files: Normalisation of Deviance - Signage

Hazardous plants typically have many warning signs posted in appropriate locations. These might warn of the potential for toxic gases, flammable materials, etc. They may require specific Personal Protective Equipment (PPE) to be worn. They may restrict access only to authorised personnel.

On more than one occasion, when touring a plant, I have found myself beneath a warning sign with words to the effect "Toxic Gas – Protective Mask Must be Worn". But I have already walked into the area and had been for some time!

On enquiring of plant management, a typical explanation is that the sign I had observed was an old sign and no longer required. Personnel who work in such a plant must be confused about which signs they must obey and which not?

Construction organisations, at least in the U.K. are required to have signage indicating various PPE and other site safety requirements. I often observe people passing this sign without complying with the signage, for example, 'Hard Hat Required'. Construction staff, quite practically, explain that the requirements only apply to some parts of the site close to the actual construction work. So how do we know when to follow the sign and when not?

Another aspect of errors of judgement where teams of people are involved is termed 'risky shift'. Groups of people are often more inclined to make risky decisions than individual group members would make when acting alone. There is also a similar phenomenon known as 'groupthink'. When decisions are made by small groups, there can be an assumption that those decisions are unanimous. 'Groupthink' creates difficulties for those who might disagree with the dominant view of the group. We will return to risky shift and groupthink in Risk Assessment and Hazard Analysis.

So, having established that human factors are a frequent cause of incidents, how the risks associated with human factors be minimised. Human reliability is the opposite of human error.

There are established techniques for calculating this reliability. One respected method is the Human Error Assessment and Reduction Technique (H.E.A.R.T.) (Kirwan, 1994). It is a quantitative method whereby, at least in principle, the frequency of failure to complete a specific task can be estimated. The details need not involve us here, except to note that HEART and other such techniques require special training. However, it can be valuable to carry out a qualitative assessment. Such methods consider Performance Influencing Factors which can either exacerbate the potential for errors or reduce that potential.

Example performance influencing factors are shown in Table 20.1

Table 20.1 Performance Influencing Factors

Error Producing	Error Reducing
Poor lighting	Good lighting
A high number of alarms	Low frequency of alarms
Stressful environment	Unstressful environment.
Operator inexperience	Operator Experience
Absence of real-life training, e.g. by simulation	Real-life training, e.g. using a simulator
An incentive to use more hazardous procedures, e.g. another method would quickly complete that task.	Absence of alternative, more hazardous procedures which might be used as a shortcut.
Inconsistent meaning of displays	Consistent displays
etc.	

Notes from Trevor's Files: Red and Green lights

While touring an oil refinery in Kazakhstan, I noticed that field push button stations for pumps and other equipment had unlabelled red and green lights. On one portion of the plant, I asked the meaning of a red light and received the reply that Red meant 'Stopped'. In another part of the plant, the answer to the same question was that Red meant 'On' or 'Live'. Apparently, the discrepancy arises according to whether the design contractor for any portion of the refinery was European/US or Chinese!

The potential for operator error, especially when an operator is transferred from one plant to another, did not seem to be appreciated. Surely the situation could be improved by applying a label to each red or green light to indicate the meaning 'On'/'Off', etc.

Illustrative Hypothetical Case – Tank Filling

Let's consider a tank being filled with a hazardous material. In our hypothetical Fuels Terminal, it could be any of the gasoline or diesel tanks. In our Whisky distillery, it might be the Wastewater Tank which takes excess water from the Washback. This material is hazardous to the environment if discharged to the river directly (Figure 20.1).

The operator fills the tank through the manual valve until the tank is 'full', as indicated by the sight glass level indicator. The sight glass level indicator is simply a glass tube, typically protected inside a metal casing, through which the level can be viewed. Whilst this is an old-fashioned method of level indication, many are still to be seen in use today, in my experience, even in hazardous service.

The potential here for human error is apparent. Not noticing that the tank has reached its intended fill level, the tank overfills, and overflows. As the tank fills, the operator may become distracted.

The seriousness of this error, of course, depends on the consequences. If the material in the tank is rainwater, then there are no consequences except perhaps some local flooding if the filling rate is large enough! However, we are using examples where the filling fluids are moderately hazardous. In assessing risk (see Layers of Protection Analysis, Chapter 27), the probability of failing to perform this routine task requiring care correctly on demand would be between 1 in 10 and 1 in 100, depending on the operator's stress levels. This frequency of error is not a satisfactory situation given the potential environmental damage.

Error reduction methods here might include locating the sight glass level indicator adjacent to the manual valve, so the operator does not need to walk around the tank to see the level. Good access to the level indicator should be ensured and appropriate lighting. The task design should be considered. Whereas it might at first glance seem best if no other tasks are given to the operator during filling, errors due to boredom need to be considered. The time to overfill might be calculated and the operator is required to set a timer.

A checklist in which he enters the level at, say, every five minutes would also help.

There are other failure modes here connected with the level indicator, which we will come back to later when the benefits and drawbacks of automation are considered.

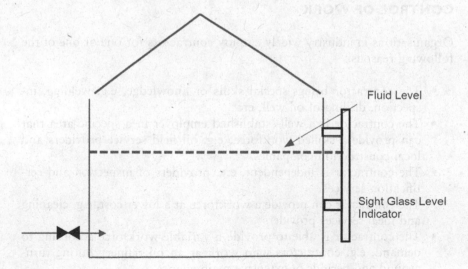

Fluid Level

Sight Glass Level Indicator

Figure 20.1 Tank Filling

PERSONNEL SELECTION

So, if an employee makes an error, it may be that the incident investigation questions the capability of that employee. Recruitment methods need to be examined to ensure that the employees have the required skills and competencies.

Competency comprises the:

- Skills
- Knowledge
- Behaviours that lead to successful performance in the job.

The most common method of personnel selection is a review of a CV followed by a job interview. More complex interview processes can include a selection panel of multiple interviewers and can consist of a variety of tests such as tests in:

Aptitude, intelligence, reasoning, memory, theoretical and practical knowledge, personality, etc.

Many selection processes are based on differentiating between candidates so that the best of the available candidates can be selected. There is reason to have confidence that the above techniques successfully achieve this differentiation. This need not concern us here. Our concern is that the personnel selected meet the **minimum** essential competence (skills, knowledge, behaviours) to do the job safely. Does a selection interview achieve this? I would suggest that it does not.

CONTRACTORS – SELECTION, PERFORMANCE, AND CONTROL OF WORK

Organisations in industry widely employ contractors for one or one of the following reasons:

- The contractor brings special skills or knowledge, e.g. welding, inspection, drilling of oil well, etc.
- The contractor is a well-established employer in a specific area that can provide a skilled workforce, e.g. oil field service providers and local construction companies.
- The contractor is independent, e.g. providers of inspection and certification services.
- The contractor can provide a workforce at a lower cost, e.g. cleaning and meal services providers.
- The contractor is able to provide a variable workforce according to demand, e.g. contractors who work at an oil refinery during turnaround and periods of extensive maintenance.

A widespread public misconception is that using the services of a contractor relieves the client organisation of responsibility for an incident in the contractor's field of responsibility. See Deepwater Horizon (Chapter 6), for a good example. The drilling rig on which the incident took place was wholly owned and operated by a drilling contractor. A separate specialist contractor was carrying out the well-cementing job that failed. Yet most of the 'blame' and subsequent fines were placed on the oil major directing the work, not the contracted organisations.

Notes from Trevor's Files: A dreadful contractor fatality

I had been promoted to Site Manager and was put in charge of a paint manufacturing facility that had recently suffered a fatality. Paint is manufactured in tanks where the constituents are mixed and tinted to the correct colour. When changing the colour to be produced, the tank needs to be cleaned. At this facility, cleaning was achieved by employing a contractor whose personnel entered the vessel and, using a solvent, cleaned the tank's walls.

On the day of the incident, a young man was inside a tank cleaning when there was a fire and explosion, which caused severe burns to the contractor's employee. Very sadly, he died after some weeks in hospital.

In the subsequent investigation, it became apparent that the incorrect solvent was being used. The solvent used had a particularly low flashpoint. The flashpoint of a material is the temperature at which a particular organic compound gives off sufficient vapour to ignite in air. This low flash point solvent cleaned much more quickly and efficiently than the higher flash point material, which should have been used. It was not established who decided to employ the incorrect solvent.

Quite rightly, it was the owner and operator of the paint manufacturing plant who was ultimately responsible, not the contractor.

There are good practices for contractor management, such as the following:

Contractor Selection
Contract Preparation
Contract Award
Orientation and Training
Managing the Contractors
Post-Contract Evaluation

(see dss+, formerly (Du Pont Sustainable Solutions) https://www.consultdss.com/contractor-management/)

The objective is to drive continuous improvement of the contractor regarding safety performance.

An excellent practice I have seen is the assignment of a contractor sponsor, specific to each contractor. The contractor sponsor regularly audits the contractor's activity and ensures users of the contractor carry out observational audits and feedback on performance. The contractor sponsor is key to the contract renewal process.

I have myself coached companies through the application of this excellent approach. However, it is not always applicable, maybe often not practical. To work successfully, the client has to have a genuine choice of contractors. Changing contractors can involve a lot of cost and effort. Sometimes, only one contractor has the necessary capability and available workforce, especially in more remote locations. Sometimes the client is obliged to employ a particular contractor – this is often the case where the client operates a facility on a site that is part of an industrial park and is obliged to use a contractor common to other companies.

Contractors and their employees are particularly vulnerable to decision-making, which puts financial considerations ahead of safety considerations.

A contractor can be 'deselected' by a client company for poor safety performance. This threat sounds like a positive feature at first sight. However, it can create a culture of fear within the contractor. A safety incident involving the contractor can lead to bankruptcy of the contractor and loss of employment for all the contractor's employees. Therefore, there is pressure on the contractor and its employees to do whatever the company employing the contractor is perceived to want.

Notes from Trevor's Files: Unjust Termination of Employment

When working on offshore oil platforms to help improve safety, I was asked to work with a particular general-purpose contractor. Note that the request came from the oil major, not the contractor itself. The oil major considered that this contractor had a problem with safety. I ran a workshop with the contractor's employees, which the general manager for the contractor in the region attended somewhat begrudgingly. His clear message to his employees was that any safety incident put in peril the contract and the employment of all those in the room. Further discussions with the contractor's employees showed that they were in fear of discussing any safety issue. They pointed to several examples, one of which involved a near miss when a tile fell from a ceiling close to a contractor's employee who reported the event. This had nothing to do with the employee. His employment was terminated!

Contractors' vulnerability towards safety also extends to the kind of safety advice offered. At the Brumadinho dam disaster reported in Chapter 8, the tailings dams were inspected by a well-respected global company providing inspection services. It is alleged that the inspection contractor had found that the dam failed to meet the necessary stability factor, which would prevent them from deeming it stable. Commissioned by the mining major, the employees looked instead for new calculation methods to achieve the desired results. In the end, the inspection contractor confirmed the dam's stability against its better judgement. As a result, neither the mine operator nor the authorities initiated dam stabilisation or evacuation measures.

The implication is that the local inspection contractor's employees feared giving the mining major 'bad news' because it might have affected their future employment stability. (European Center for Constitutional and Human Rights "The Safety Business: TUV SUDS role in the Brumadinho dam failure in Brazil").

Sometimes, the contractor's work is insufficiently distinguished from the work of the clients' direct personnel. Client and contractor procedures may be different and confused at the overlap. A good practice, especially in more complex situations, is to use a bridging document. This document should clarify who does what and whether under the contractors or clients safety procedures.

A particular example of this lack of clarity from my experience concerned a vessel entry requiring confined space entry precautions. Entering a tank is a highly hazardous activity. Residues within the tank may go on fire or cause toxic hazards, and the entry personnel may become asphyxiated if there is a lack of oxygen, etc. Whilst reviewing a vessel entry, which was being done by a contractor, I asked about the following:

- Means of emergency escape.
- Deployment of a standby man to wait at the manway of the vessel take action in the event of a problem.

Neither of these activities was apparent. It transpired that these requirements were part of the client's procedures but not that of the contractors.

SUMMARY: CAUSES OF FAILURE THROUGH HUMAN FACTORS AND IN PERSONNEL SELECTION AND USE OF CONTRACTORS

Reasons Why Human Factors Sometimes Cause or at Least Contribute to Catastrophes Include:

- ☒ Human reliability is overestimated, the risk of a mistake is underestimated.
- ☒ Ergonomic issues such as tiredness and stress.

⊠ Normalisation of deviance.
⊠ Risky shift.
⊠ Groupthink.

Reasons Why Good Personnel Selection Sometimes Fails to Prevent Catastrophes Include:

⊠ Selection methods differentiate between candidates but are not so good at testing for the minimum essential competence required for the job.

Reasons Why Good Contractor Management Sometimes Fails to Prevent Catastrophes Include:

⊠ Client of the contractor is tied to using that contractor, so contractor evaluation and reselection processes lack value.
⊠ Clients see the contractor as being solely responsible for safety.
⊠ Boundaries between the client's work and the contractor's work are poorly defined.

Chapter 21

Equipment Suitability

EQUIPMENT SELECTION AND DESIGN

In the incidents discussed in Section 2, nine were identified where equipment suitability was a significant contributory cause. These included the Deepwater Horizon oil spill, the Bhopal chemical release, the Brumadinho dam collapse, the Transhuron ship sinking, and the Chornobyl nuclear disaster.

Selecting a car is something most of us do several times in our lives. How do you go about this? Cost is one parameter. The number of seats – do we have to accommodate grandchildren, pet animals? Do we need the ability to transport heavy luggage during holiday times? Tow a caravan or boat?

As well as the cost of buying the car, there is also the cost of operating it. Such costs are affected by fuel efficiency, insurance, maintenance, etc. ...

Do we consider safety? We can refer to the vehicles' NCAP (European New Car Assessment Programme) rating in Europe. Are we prepared to pay for additional crash avoidance equipment and technology such as Automatic Braking Systems, Driver Alertness Detection Systems, Tyre pressure monitoring systems, etc.? These factors make purchasing a vehicle a complex task involving decisions as to what features your vehicle must have, or you would prefer to have, balanced against cost.

Is purchasing an aeroplane, ship, distillation column, pump, etc., different? Take the simple example of a pump. For sure, the pressure and flow rate to be achieved must be defined. The type of drive for the pump needs to be specified – electric, diesel, or steam turbine. What fluid will the pump contain? If materials are toxic, the seal arrangement should be considered. Might the seal fail, releasing toxic materials gradually or suddenly in a catastrophic manner? Should a special pump be purchased at additional cost, selecting one that is much less prone to leakages, such as a pump with a double mechanical seal or a canned pump with no seal?

Engineers specify all the features required of the equipment, and the list can be quite long. Then purchasing professionals carry out the task of finding prospective vendors. Typically, a tendering process will be carried out amongst several potential suppliers in order to determine the most suitable. Tendering

DOI. 10.1201/9781003360759-24

processes are required in most organisations and are usually compulsory in state-owned organisations in order to reduce the risk of bribery in any of its forms.

This process is prone to failure. Firstly, the engineer may forget some of the essential features required. Secondly, the purchasing professional may decide which bidder wins the tender based on cost without sufficient weight being given to safety and other features which the engineer has specified. The total lifetime cost of the equipment should be assessed, including the cost of maintenance, not only the initial purchase cost. However, the success of a purchasing professional is potentially influenced by the immediate savings the purchasing department can generate.

In the author's opinion, a further aspect is often overlooked. This aspect is consideration of Inherently Safer Design, or in the case of the oil, gas, and chemical industry, Inherently Safer Processes. The opportunity to consider this goes right back to the conceptual phase. Once the equipment is in the process of being specified, it is too late. Examples of inherently safer thinking include:

- Finding and using substitute materials, such as using water-based latex paint in place of one based on toxic or flammable solvents.
- Minimising the quantity of hazardous materials, such as reducing pipework sizes to the minimum size necessary, instead of using a standard, convenient size. This minimises the amount of hazardous material contained within the pipework.
- Simplification, such as using gravity flow to reduce the need for pumps.

Inherently safer design if not confined to the process industry. In the chapter on Mining (see Chapter 8), the hazards of tailings dams were considered. However, technology now exists and is being increasingly selected to produce dry stack tailings without creating hazards of the magnitude introduced by tailings dams. In the aeronautical industry, there are choices regarding inherently safer airport locations, in aeroplane size and propulsion, etc. ...

MATERIALS OF CONSTRUCTION AND CORROSION CONSIDERATIONS

A particular aspect of equipment selection concerns materials of construction.

It must be emphasised that all materials corrode at rates that depend on the conditions they are exposed to. If we take the example of a piece of pipework, the following considerations need to be considered:

- The fluid inside the pipework
- Temperature
- Pressure
- External atmospheric conditions.

Materials such as gold and platinum may have very slow corrosion rates. They are also costly and may not have the right physical qualities to make a robust piece of pipework of appropriate strength!

Kletz (2009) lists multiple occurrences where an inappropriate material has been provided and sometimes mislabelled. As well as selecting and ordering the correct material, it is sometimes the case that the wrong material is supplied. The resulting corrosion rates can be dramatic and catastrophic (see the DPC Incident example in Chapter 15).

Good practice here is Positive Material Identification (PMI). The equipment required, known as an XRF (X-ray Fluorescence), is available on 'e-bay' for less than Euro 10,000! Therefore, it surprises me that some organisations rely solely on delivery certificates. Of those organisations that use an analyser, many of them sample only 10% of metals supplied and then only special alloys. Welding rods and flanges should also be checked. Simply having a different material can cause the weld or flange to become a site for galvanic corrosion.

EQUIPMENT CONTROLS

Automation

A common solution to eliminating human errors is to automate the system. But automated systems fail also. The potential for failure of the automation system is often overlooked.

Illustrative Hypothetical Case – Tank Overfill Interlock

Let's consider again a tank being filled with the hazardous material. In our hypothetical Fuels Terminal, it could be any of the gasoline or diesel tanks; in our Whisky distillery, it might be the Waste Water Tank which takes excess water from the Washback. This material is hazardous to the environment if discharged into the river directly.

As noted previously, reliance on the operator filling the tank through the manual valve until the tank is 'full' as indicated by the sight glass level indicator is open to human error.

So, a decision is made to automate the filling with a level switch which automatically closes the filling valve at a predetermined level. However, in doing so we have replaced a vulnerability to human error with a vulnerability to an electronic or instrumentation error. Level switches typically of a float type, rather like the ballcock in the toilet cistern, are prone to failure due to corrosion or build-up of residues. Failures of tank level instruments were contributory causes to the major disasters at Capeco and

---

Buncefield Terminals (see Chapter 13 on Process industry infrastructure). Typically, a probability of failure on demand of such a level switch would be 1 in 10 to 1 in 100 demands, similar to the human failure example. This initially sounds OK. The human and the instrument are unlikely to fail simultaneously, so the probability is reduced by the instrument failure probability times the human failure probability. If we use 1 in 10 for the human failure and 1 in 10 for the instrument failure, then the combined probability is 1 failure in every 100 demands – a significant improvement.

But not so fast! Case studies show that the operator often considers this automation a time saver. Now that the task is automated, filling can commence. The operator can carry out other tasks or go to the restroom whilst filling continues and is automatically stopped on a high level. In this case, we have made no improvement since the human is no longer contributing to preventing overfill. Indeed, we may have made things worse. For the manual operation, the operator is likely to remember about the filling, perhaps a little too late or very late. But in the reliance in automation case, the operator is unlikely to check until much later – the potential for a large overflow is greater (Figure 21.1).

We will return to the risk of automation failure in general when we consider Layers of Protection Analysis and Safety Integrity Levels (see Chapter 27 on Assessing Risk Assessment Outcomes and Risk Reduction).

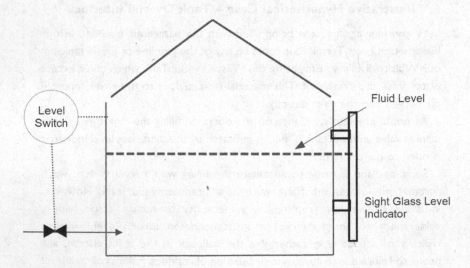

Figure 21.1 Tank with Level Switch.

Interlocks and Bypass Control

Most automated systems involve interlocks. Consider an elevator; some interlocks prevent operation if the elevator doors are not closed. A gas boiler is interlocked to close the fuel valve after a predetermined time if no flame is detected. The landing gear of a commercial aeroplane is usually interlocked so that it cannot be unfolded into a landing position if the aeroplane's airspeed is above a specific value to prevent damage to the undercarriage and aircraft instability.

As an example of a more advanced automation feature, vehicles can be supplied with a breathalyser interlock. The driver must breathe into the breathalyser. The car will not start unless the driver has passed the breathalyser test. Interestingly an internet search shows plenty of articles focused not on the benefits of such an interlock, but on how to bypass them!

A complex processing plant such as an oil refinery may have several hundred interlocks.

It is important to note that not all interlocks relate to safety or catastrophe prevention. An interlock may be present for quality or efficiency reasons. This fact gives rise to a source of potential error. Do the operating personnel know the purpose of the interlock? I have found that this is surprisingly often not the case and that the objective is often not documented for the operator to read nor included in their training.

Maybe this is not so important – if the interlock does its job, why does the operator need to know?

Unfortunately, the components of an interlock sometimes fail. The interlock will typically contain a detector, a signal wire, a controller that determines the action, a cable to cause the required event, and the device that carries out the desired event. In the case of the elevator, position switches detect whether the cabin doors are open or closed, the position of the cabin relative to the storeys of the building, wires to a control box which determines whether the cabin can or cannot be moved, and cabling to the motors and windings that move the elevator. Any one of these can fail.

The interlock needs to be bypassed in order to move the cabin to an appropriate storey to evacuate personnel and to carry out a repair. For the elevator, the purpose of the interlocks is straightforward and intuitive. But this is not necessarily the case in chemical processes, nor in more complex forms of transport, etc. ...

This act of bypassing presents a very significant risk because of the following:

- The purpose of the interlock may be misunderstood.
- Additional protection whilst the interlock is bypassed is not considered.
- The presence of the bypass is forgotten.
- The process has been designed and deemed safe on the basis that there is an interlock in place.

Notes from Trevor's Files: Interlocks Overlooked

While carrying out a process safety assessment on a large Middle Eastern petrochemical plant, one of my colleagues reviewed Process and Instrumentation Diagrams relating to a highly hazardous portion of the plant that handles ethylene. He noticed instrumentation intended to register the level of liquid ethylene in a separator vessel and trigger an interlock to shut down the plant if the level became too high. However, on further investigation, he found that this interlock had been inactivated long ago and that this fact was unknown to operating staff.

One imagines that, as the plant increased capacity, the separator level increased, and the interlocks started to restrict capacity. The engineers at the time considered them unnecessary and bypassed them. However, there is no record of the risk assessment they used to come to this conclusion. A high liquid level could result in liquid ethylene carrying over to portions of the plant designed to handle only gases. There were potentially severe consequences.

A good practice is to ensure that interlocks are classified according to their safety criticality (see SIL in Chapter 27 on Assessing Risk Outcomes and Risk Reduction for a potential classification method). Safety-critical interlocks should require higher-level authorisation above that of the shift supervisor. In the middle of the night, this may mean call out. An interlock bypass form should be easily visible, either electronically or in paper form. The form should include a requirement for 'mitigating measures' whilst the interlock bypass is in place. Mitigating measures may be to increase operator checks or require continuous monitoring using other instruments. All bypasses should include a risk assessment. Interlocks bypassed for extended periods should be more formally reviewed using a Management of Change Procedure (see Chapter 31 on the Governance of Safety).

Unfortunately, mitigation measures are often not considered. Shift supervision or even operators themselves are, in some organisations, permitted to bypass an interlock without further discussion. It should be recognised that operating a plant with a bypass is to operate it contrary to its design.

Software bypasses are permitted now and often used. The presence of the bypass is therefore not readily visible, and the bypass may go unobserved until the next test.

Human-Machine Interface and Alarms

A Human-Machine Interface (HMI) is a user interface or dashboard that connects a person to a machine, system, or device. The term can technically be applied to any screen that allows users to interact with a device. The dashboard on a car is an example of an HMI. The dashboards of many modern vehicles are at least partly software-controlled, as are aircraft controls. Process plants usually have a Distributed Control System, which has a software-controlled display. Some process plants have Supervisory Control and Data Acquisition systems.

My hydroponic greenhouse has an HMI, which comprises my laptop computer, which wirelessly relays commands to the greenhouse, brings back temperature and humidity, makes decisions, and sends out new instructions. This can be done using components readily available on the internet, plus some Python programming, which I learned as a hobby. Of course, it doesn't have the reliability needed for more serious applications!

Notes from Trevor's Files: Blank Display on Rental Car

Whilst travelling on business in Paris, I confessed to my Turkish travelling companion that I was nervous about renting a car and driving through Paris. He kindly volunteered to do the driving and rented a splendid car of a well-respected marque. He checked around the car and started off, only to find that the display screen was blank. No speedometer, fuel gauge, etc., were visible. We returned slowly and nervously to the car rental centre. They could not fix the issue. The car was changed to one of a slightly more traditional design. The absence of a speedometer is a serious issue with the potential consequences of misjudging speed, as is the lack of a fuel gauge with the possible consequences of running out of fuel and the car stopping at a dangerous location.

HMIs can fail. Often such systems have electrical backup systems so that the HMI continues in operation should the electricity supply fail. These can be based on diesel emergency electricity generators which can take a few minutes to start, and on batteries – an Uninterruptible Power Supply (UPS). As pointed out in Chapter 10, the Fukushima Daiichi reactors had diesel-powered electric generators but no UPS. Either system supplies enough power to maintain the control system for a duration. UPS systems are typically designed to maintain the control system and HMI for 30 mins to 1 hour. This assumes all other uses of UPS power have been switched off.

Despite this protection, there have been examples of control display screens going blank (black screen). This is a dire situation. In the event of a black screen, there are some actions that the operator should carry out. For example, to check if control in the field is being maintained. A checklist should be provided to the operator, and depending on the process, some units will need to be manually shut down. However, I have found that many organisations do not have a 'black screen' procedure arguing that their UPS system is sufficient.

UPS systems themselves sometimes fail. Instances in the author's experience of particular note have occurred when the power supply experiences a dip or disturbance but not a complete failure.

Causes of UPS failure include:

- The power outage lasted longer than the battery system of the UPS was able to provide power.
- Inadequate servicing of the UPS.
- UPS overloaded – UPS is asked to supply more power than it has capacity.
- Battery failure – typically die to old batteries or batteries which have not had enough time to be replenished from the last use.
- Overheating.
- Component failure.
- Etc. …

Further issues with HMIs arise when the screen representations of the plant being controlled are confusing or do not accurately represent how the process operates. Putting the entire Piping and Instrumentation diagram into a software HMI screen representation is not possible or appropriate. At the other extreme, simplification can mislead the control panel operator.

Until recently, Distributed Control System representations were highly coloured. More up-to-date technology uses screen representations that are in greyscale for all but the critical data such as alarms, valve positions, etc. … which are brightly coloured.

A particular issue with HMIs occurs in the context of alarms. During the design process, an alarm will be added once a particular parameter is identified as important. The reasons for the alarm may be safety, quality, production rate, and other economic factors. The result can often be a vast number of alarms which the operator much keep acknowledging. The danger, therefore, is that amongst these many alarms, there may be one alarm much more significant than the others. This all-important alarm may be overlooked. Alarm management is today an important issue. Efforts are made to ensure that the alarm load on each control panel operator is manageable under stress conditions.

Notes from Trevor's Files: Poorly Considered MOC – One of My Early Mistakes!

Here is something in my early years which I did poorly. The portion of the plant I was responsible for included some vertical pipes that had level alarms. These alarms were almost constantly going into and out of alarm. This was due to the frothy nature of the material passing down the vertical pipes. Operators quite rightly made the case that these alarms weren't doing any good. So, I had them disabled but only put in the paperwork Management of Change after the event. The Development Supervisor disagreed with my (proposed) action, and the MOC went unauthorised. However, I chose not to reinstate the alarms so as not to incur the wrath of the operators who had welcomed the change. No question, I should have had greater determination to prove my point and do things in the correct sequence.

FAILURES OF SAFETY DEVICES

Let's assume safety devices have been installed as recommended by adequate risk assessments.

Sometimes those safety devices themselves fail. Of course, if we have carried out our risk analysis properly, then the failure frequency of those devices will have been considered. But sometimes, safety devices fail with a frequency above that used in the assessment. Here are some examples:

Pressure Relief Valves

A pressure relief valve opens above a set pressure to prevent over-pressurising the equipment it is protecting. Such valves are routinely inspected at a frequency that is often defined by the laws of the country. On inspection, these valves should be tested before further work, such as cleaning, is done on the valve. This test is known as the 'Pre-Pop Test'. The objective is to determine if the valve was functional in protecting the equipment whilst it was in service. If it was not, then the equipment was not properly protected against overpressure. An investigation should then take place in order to understand why the valve failed. The failure might be due to corrosion, blockage with solids such as corrosion products, etc. Then decisions can be taken on upgrading the materials of construction, providing protection of the valve from solids, etc. ... However, some organisations do not do the pre-pop test. One reason sometimes given, apart from causing additional work, is that the relief valve must be handled in the contaminated state. Without the 'pre-pop test', the relief valve can only be cleaned and overhauled before a final test. This gives no information about the valve's performance in service. Without knowledge of in-service performance, the valve could conceivably fail shortly after overhaul and re-installation. It may have failed in service well before the test date, meaning that the vessel or pipework was not protected for a considerable period.

A pressure relief valve should be sized for a particular scenario with the potential to overpressure the equipment. It might be sized for fire engulfment or a runaway chemical reaction. There is an issue here because other scenarios might cause overpressure but at a lower relief rate of material. Pressure relief valves passing lower than design rates can fail. For example, they can 'chatter', a condition which causes wear. More seriously, the chattering may become so severe that the valve and adjacent pipework can fail. Examples are known where the bolts connecting the relief valve to the pipework and support steel have either loosened or sheared, resulting in the release of the hazardous materials into the atmosphere. Multiple relief valves having staggered pressure settings are a partial but expensive solution. This is particularly true if retrofitting them onto existing equipment.

Another cause of pressure relief valve failure is specific to relief valves which are paired so that each can be tested whilst the equipment is in operation. A three-way valve allows either valve to be connected to the process whilst the other is isolated. This is a common arrangement in oil refineries where the time between turnaround to allow relief valve inspection can be 5–6 years, and means to inspect and test the valve are required at more frequent intervals. Unless the three-way valve is specially designed, it can be inadvertently left in an intermediate position. This restricts the flow path for the process fluids in the event of an emergency release.

Incidental Case Study: Poorly Considered Management of Change

A petrochemical plant suffered a severe fire. This fire followed the release of material through a pressure relief valve due to a process deviation which need not concern us – the relief valve was doing its job. It was not one single relief valve, but several relief valves, all set to the same pressure. This arrangement was chosen to attain the maximum relief rate required for the particular scenario on which basis the process was designed. However, the rate of release was below the designed flow rate for all three valves, with the result that one of the valves 'chattered' – a process of repeatedly fractionally opening and then closing.

With 'chattering', the valve opens a little releasing some pressure and immediately reclosing, the pressure rises fractionally, and the valve opens again. This chattering of a heavy relief valve progressively weakened the bolts attaching the relief valve assembly to the pipework resulting in a significant release of flammable material and fire/explosion.

A recommendation was made that the multiple relief valve combination should have staggered set pressures so that only one valve opens first. The lowest set relief valve will take the load and operate properly without chatter if excess pressure is small.

This may sound simple. One needs to ask how even single relief valves are designed. They are usually sized according to the highest load expected such as the engulfment of the vessel by fire. But if the cause of overpressure is lower, is there also a risk that the relief valve will chatter? The author is familiar with other petrochemical operations with similar multiple relief valve arrangements. These other organisations were not convinced that it was necessary to stagger the relief valve settings. Staggering necessitated reducing the relief valve settings since the maximum pressure should not be raised. This meant that the lowest pressure valve setting was too close to the 'operating envelope'. If the proposed change were made, then inadvertent release of material would be more likely to occur due to operating excursions.

At the time of writing this book, this situation remains unresolved. It should be emphasised that the chattering relief valve case occurred on a relief valve set which discharged to a flare line. It is believed that relief valves which discharge to a flare line are particularly prone to chattering due to feedback of pressure surges on the downstream side of the relief valve.

Mechanical Interlocks

So far, we have considered interlocks mainly in the context of instrumented systems where the interlock has a mostly electronic content.

However, many interlocks are either totally mechanical or have a mostly mechanical content. An example is a simple lever on a gate on a barrier around moving equipment protecting personnel from harm by inadvertent entry and contacting the rapidly rotating parts. The mechanical latch may be connected directly to an electrical switch that powers the equipment. The electrical power is immediately disconnected if a person tries to enter with the equipment running.

An example closer to the catastrophic incidents of concern in this book is the control rods in a nuclear reactor. These are raised or lowered to control the nuclear reaction largely mechanically.

Mechanical devices suffer from failure modes related to stress failures of the materials of construction over time. Usually, they are designed to be fail-safe, i.e. to cause a shutdown should the mechanical element break. Fail-safe design is not always possible. Further, a mechanical device may fail due to breakage in either the open or closed position.

Electronic Interlocks

Electronic interlocks are typically less prone to failure since there are fewer moving parts. However, electrical/electronic interlocks still fail. Electromechanical relays are still used, as are overload devices such as those you may have in your home. These devices can stick. The interlocks may be improperly programmed (see Chapter 29 on Software Safety and Cyber Security).

The electronic interlock almost certainly relies on a mechanical component to achieve the desired interlock action such as the closure of an automatic valve. Reasons automatic valves can fail include:

- The gate or ball which stops the flow is corroded, and the valve sticks in the open position.
- If the valve is powered by instrument air, then the instrument air is critical to the activation.
- If the valve is powered by electricity, then the availability of electricity is essential to the activation.
- The valves have inadequate fire protection, so they have already failed due to heat exposure.
- There is a switch on the valve which can be used for manual operation, and the valve has been left in manual.

It is unfortunately quite common, in my experience, to find automatic valves switched to manual so as to avoid automatic shutdown, perhaps due to an erroneous signal.

The concept of 'fail safe' so that the valve will automatically go to the 'safe' position in the event of a failure is not a simple solution. If we assume that the safe position is 'closed', then the absence of instrument air or electricity, depending on the motive force, or the lack of an open signal will result in valve closure. But the sudden closure of a valve can create problems, and it is not always a simple case as to whether open or closed is the safe option.

It was the absence of motive power normally provided by DC electricity that significantly contributed to the Fukushima Daiichi Unit 1 reactor meltdown since the Isolation Condenser provided for just such a loss of cooling did not operate due to the associated valves being closed in the absence of electricity (see Chapter 10).

Fire and Gas Detectors

Fire and gas detectors are widely used to alert plant personnel to a release of material and associated fire. However, they are prone to failure through vibration, corrosion, or mishandling. Damage can occur through getting sprayed with water, steam, chemicals, etc. Gas detectors located close to equipment which habitually leak sound repeatedly and therefore become ignored, or worse bypassed. Often such systems are designed to auto-matically start fire protection equipment. These automatic fire protection devices are often switched to manual to prevent inadvertent operation.

SUMMARY: CAUSES OF FAILURE DUE TO PROBLEMS WITH EQUIPMENT SUITABILITY AND RELIABILITY

Reasons Why Issues in Equipment Selection and Design Sometimes Cause, or at least Contribute, to Catastrophes Include:

- ☒ Inadequate specification of all the critical features required of the equipment.
- ☒ Purchasing decisions made solely on price.
- ☒ Lack of consideration on inherently safer alternatives.

Reasons Why Materials of Construction and Corrosion Considerations Sometimes Cause, or at least Contribute, to Catastrophes Include:

- ☒ Lack of appreciation that almost all materials corrode over time.
- ☒ Reliance on the supplier supplying the correct material.
- ☒ Not using Positive Materials Identification.

Reasons Why Automation Sometimes Causes, or at least Contributes, to Catastrophes Include:

- ☒ Lack of recognition that automated systems fail.
- ☒ Operators can become reliant on automated systems and so do not do their own checks.

Reasons Why Interlocks and Bypass Controls Sometimes Cause or at least Contribute to Catastrophes Include:

- ☒ Lack of differentiation of interlocks which are critical safety from other interlocks.
- ☒ Operating staff do not know the purpose of the interlock, either because it is not covered in operating procedures, was not covered in operator training, or has been forgotten.
- ☒ Interlock bypass was put in place without consideration of mitigating measures.
- ☒ Interlock bypass requires minimal approval. Approval levels for more critical interlocks are not differentiated.

☒ Risk assessments not required for interlock bypasses.
☒ Long-term bypasses are not subjected to more in-depth risk assessment (see Management of Change).

Reasons Why the Human-Machine Interfaces and Alarms Sometimes Cause, or at least Contribute, to Catastrophes Include:

☒ Lack of appreciation that Human-Machine Interfaces can fail.
☒ HMI backup (uninterrupted power supply) maintains the HMI with electrical power for a limited duration.
☒ HMIs fail due to overload, poor maintenance, etc.
☒ Display not a sufficiently accurate representation of the operation.
☒ Display not ergonomically laid out for operator ease of recognition of critical features.
☒ Alarm overload.

Reasons Why Safety Devices Sometimes Cause, or at least Contribute, to Catastrophes Include:

☒ Pressure relief valves which fail in service or test are simply repaired without investigation.
☒ The design scenario on which the pressure relief valve is sized is incorrect.
☒ Multiple pressure relief valves are not staggered in set pressure resulting in 'chattering' and possible destruction of the valve when relieving.
☒ Mechanical interlocks break due to 'wear and tear'.
☒ 'Fail safe' cannot be achieved in some circumstances – failure in either direction causes a degree of risk to remain.
☒ Electrical/electronic interlocks can fail due to improper set-up in the control cabinet or the software.
☒ Bypasses on interlocks may have been forgotten and never returned to operation.
☒ The automatic device, typically a valve, may fail to operate due to corrosion.
☒ Electricity or instrument air providing the motive force for operating the device fails.
☒ The device is switched to manual, not automatic.
☒ The device activated by the interlock, typically a valve, is not fireproofed and so, in a fire, fails to operate in the desired direction.
☒ Fire and gas detectors fail due to corrosion, vibration, or mishandling.
☒ A gas detector frequently alarms, perhaps due to proximity to leaking equipment and becomes 'ignored' or is bypassed.

Chapter 22

Equipment Maintenance and Inspection

PREPARATION FOR MAINTENANCE

Before commencing maintenance activities, equipment needs to be adequately prepared. This includes isolating it from hazards which might arise during maintenance. In the context of the process industry, this involves isolating the equipment undergoing maintenance from the rest of the process. Pipework and the equipment need to be cleaned of hazardous materials as well as disconnecting sources of electricity and other utilities which might cause the equipment to start rotating or may put maintenance personnel in danger of electric shock. Similar considerations apply in other industries, such as the airline industry, for example, in the isolation of a jet engine for maintenance. Mistakes, in either not adequately completing the isolation, misunderstanding the isolation, or returning the equipment to service without fully restoring the equipment, are all too frequent sources of incidents.

In Section 2, 14 catastrophes were examined that involved failures in maintenance and inspection. These events included two of the most tragic – the Bhopal chemical release and the Piper Alpha oil platform disaster.

Typical faults in preparation for maintenance that have led to serious incidents include:

- Failure to isolate at all.
- Inadequate isolation. Whilst I have experienced closed valves being used, these are not considered adequate since a valve can leak materials. This can be due to the design of the valve, wear and tear, obstruction to the mechanism, etc. Isolation should not rely on gate valves in particular since the gates on these valves have a reputation for leaking material when closed (a phenomenon known as 'passing'). It is preferable to disconnect the pipe altogether, use slip plates (also known as blanks or spades), or a type of isolation known as double block and bleed arrangement. Two valves are closed in double block and bleed arrangements, and a drain valve between them is opened.

DOI. 10.1201/9781003360759-25

Thus any material which passes through the closed valves is released through the drain, not into the isolated equipment.

- Isolations removed too soon. Typically, this involves mis-communication through an inadequate understanding of the isolation or uncoordinated events. Good Lock-Out Tag-Out practices (see below) can help prevent this.
- Service lines, such as steam, water, and nitrogen are not isolated, leading to the process fluid to back up into service lines. Or of the service, such as steam being emitted during maintenance and en-dangering the maintenance personnel. Service lines are typically not corrosion-resistant to the process fluid. This situation can lead to flange leaks, etc. ...
- Isolations are not removed before the process is put back into service. This can lead to overpressure on the restart of the process, for example.
- Electrical isolation was carried out on the wrong equipment. This is, unfortunately, a serious concern in isolation. The electrical control room is typically remote from the equipment being isolated. Only correctly, which needs to be correctly interpreted, identifies that the correct equipment is isolated. Further, it is common that electrical control rooms have restricted access, usually only to specified electricians. The electrical isolation is sometimes considered as needing a separate work permit.
- Improper identification. In my opinion, hanging a tag saying 'Do Not Operate' on a valve or switch is far from satisfactory. The tag should be complete with additional essential information such as the date of isolation and the purpose of the isolation. This helps ensure that the purpose of the isolation is correctly identified.

Incidental Case Study: Ludwigshafen Pipeline

Three people died in Ludwigshafen, Germany, in 2016 when maintenance work was being carried out on pipelines connecting a chemical plant with unloading facilities in the harbour. It is believed that a worker using a cutting disc mistakenly started work on the wrong pipeline. Whereas he was intended to work on a decontaminated pipeline, he mistakenly started cutting on a pipeline containing highly flammable hydrocarbons.

Tracking the identity of pipelines over long distances can be difficult. Many companies have now introduced special permitting and marking systems to ensure positive verification by others before hazardous work starts around pipelines, to help ensure that the intended pipeline is selected by the mechanic for work.

Notes from Trevor's Files: Incomplete Decontamination – Another of My Early Mistakes!

I was responsible for organising the shutdown and dismantling of a process plant. The plant had existing shutdown procedures, which involved flushing the lines that had contained a flammable hydrocarbon. Some months after the shutdown, when dismantling was proceeding well, one of the contractors carrying out dismantling work involving a cutting torch reported a fire coming from one of the pipes. On further investigation, this particular pipework was on a bypass line, and clearly, the flushing of this line had been overlooked. There was a danger to the demolition operator from the unexpected flame. To compound my error, I assumed that now there was no remaining danger from this bypass. In practice, some hydrocarbon was retained, and a further fire occurred some days later. This is a case of inadequate procedures, inadequate incident investigation, and inadequate measures to prevent a recurrence on my part.

Returning equipment to service is another source of catastrophic incidents. One of the most fatal was the Piper Alpha disaster described earlier in this book (see Chapter 6). In brief, a pump had been shut down and a relief valve removed. However, the following shift did not know this and started the pump with the relief valve missing.

An excellent practice for equipment isolation is LOTO. No, this is not a gambling game, or at least it should not be! LOTO stands for Lock Out, Tag Out. This approach has been further extended with a verification step and is known as Lock Out, Tag Out, Try Out (LOTOTO). The verification step is confirmed that, for example, a locked-out motor does not start when the start button is pressed.

I grew up as an engineer in Du Pont practising what I consider to be one of the best LOTO procedures, and yet I have found many organisations saying that they practice LOTO and yet make some fundamental omissions. For example, when performing an isolation reliance is placed on locks. But a lock does not speak! If there is tagging of valves, pumps, compressor motors, etc. indicating 'Do Not Operate', a tag there is no other information as to the purpose of the tag. A tag should include the date the isolation commenced, the name of the person isolating, and briefly the purpose. A tag with its purpose identified is so much less likely to be confused. This is well worth the additional effort given the catastrophes that can follow from misunderstandings.

Not all authorities agree with me on this one, and some good texts show a standard tag where there is no provision of entering dates, etc. ...

Other deficiencies I have found include:

- Tags are in use, and there is a space for the date. But the date is not entered!
- Only one lock and tag are applied. The concept of ownership at any point in time is essential. So, the operator should be the first to apply the locks and tags, followed by the maintenance person or contractors, who may be multiple. The equipment is then considered turned over to maintenance/contractor. In restoring the equipment to operation, the maintenance/contractor removes their locks and tags first, and finally, the operator removes locks and tags before restarting the equipment.
- No other locks and tags are applied within the electrical control room. There is a reason for this – access to the electrical control room is often restricted to specific electricians only. I agree that this is a balance of risk, especially where special flame retardant overalls are required for wear in the electrical control. My preference is to make the effort to provide the right PPE and have the operator and maintenance personnel lockout in the electrical control room also.
- No verification method is used. A good practice is to verify the lockout of the normally energised part of the system by attempting to start the unit, e.g. pump or compressor motor.

A good practice is for all the lock keys to be placed in a master locked box, for which a supervisor holds the key.

Incidental Case Study: Phillips 66, Pasadena, October 1989

A chemical release occurred on a polyethylene plant, resulting in a massive vapour cloud explosion. There were 23 fatalities and upwards of 130 injuries, together with extensive damage to the plant facilities. It was determined in the subsequent investigation that satisfactory isolation employing good practices such as double-block systems or blind flanges was not carried out. A single isolating ball valve was accidentally opened at the time of the release due to an error in connecting air hoses. (Lees, 2014)

CONTROL OF WORK

The concept of a work permit is well established in industry in general. A work permit is required for all work being carried out on the facility other than the routine operator work for which there are established procedures. A work permit applies to all non-routine work such as repairs, inspection, testing, dismantling, and modification. There should be a work permit in both the above case of preparing for maintenance and many other types of work that may go on in a facility.

Essential features of a permit to work system include:

- Identification of the tasks to be carried out.
- A risk assessment (if not part of the permit) should be attached.
- Duration of the permitted tasks after which the permit is no longer valid.
- Any simultaneous activity.

Many errors are found during permit to work audits relating to unclear scope. Often generic permits are used without consideration of the conditions on the day.

Notes from Trevor's Files: Simultaneous Operations

Whilst carrying out observations in a petrochemical plant, we observed contractor personnel carrying out work reinsulating pipework. Work was also being carried out simultaneously, which involved opening a large filter that still contained the residues of highly toxic material. The operator was in full protective gear with an impermeable whole-body overall and self-contained breathing apparatus (SCBA) to ensure a supply of breathable air. The insulators were unaware of the potential exposure to toxic materials and were only equipped with regular overalls. Both the filter opening operation and the insulation work were covered by separate work permits, which on their own were quite adequate. The hazards of the simultaneous operations in close proximity had not been recognised. Just as the insulators were exposed to risks from the toxic material, the operator was exposed to dangers from the sharp metal edges and crowded workspace created by the insulators. The work permits and their risk assessments were generic and were not reviewed with regard to other work being carried out on the day.

Incidental Case Study: Hickson and Welch, Castleford, 1992

Maintenance work was being carried out, which involved softening the residue at the bottom of a vessel and raking out the residue. A spark occurred due to the use of a metal rake, and the subsequent fire killed five personnel. Whilst work permits had been raised, these permits were prepared for opening up the lid and for blanking off the vessel. The permits did not cover the procedure for softening and raking out the residue. Therefore, the risks of the raking step were not assessed. (Wise Global Training, 2015)

MAINTENANCE PLANNING

The connection between poor maintenance and catastrophe is clear. Our automobiles require regular maintenance to keep them safe, e.g. replacing worn brakes. The same is true of aeroplanes, ships, process plants, etc. ... In the absence of good maintenance, catastrophes can occur. Maintenance can be broken down into three main types:

- Breakdown or emergency maintenance. Something suddenly does not work. If it is vital to the operation, then an emergency repair needs to be expedited.
- Preventative maintenance is scheduled based on a specified period of time. The annual service on your automobile is an example.
- Predictive maintenance is scheduled on an as-needed basis, based on information about the condition of the equipment. If you plan some maintenance activity for your automobile based on hearing a persistent knocking sound, then you are carrying out predictive maintenance.

The percentage of emergency or breakdown maintenance versus preventative and predictive maintenance can be a revealing statistic. I tend to consider anything above 20% emergency or breakdown maintenance excessive, although it depends on the type of plant. Equipment less prone to sudden stoppage, which runs smoothly and continuously, is less likely to initiate an unexpected failure which could be the initiator of a serious incident.

Deferment or postponement of preventative maintenance should be avoided. If deferment is found unavoidable, it should be approved at an appropriate level that can consider the risk to safety and the business. A formal approval process for postponement of maintenance is, in my experience, a frequent recommendation of insurance surveys. Many organisations do not have such a procedure. Recognition of the risks associated with deferment should be included, together with any mitigating measures to be taken. For example, preventative maintenance may be planned on a storage tank. A postponement may be needed, for instance, because all tanks are full due to an unexpected decline in shipments. In the case of such a postponement, extra external inspection of the tank may be required, and the maximum filling level reduced.

MECHANICAL REPAIRS AND REPLACEMENT

'If it's not broken, don't fix it' is a dangerous idiom in considering catastrophe prevention. There are various reasons why something working adequately today might need repairs or replacement.

- Tests done as part of the predictive maintenance routine indicate that the equipment is likely to break down in the near future. It may be better to choose a time in advance, perhaps during a planned shutdown, to replace the equipment rather than have a forced unplanned shutdown when the equipment finally fails. Unexpected sudden shutdowns may also initiate a series of events resulting in a catastrophe. Many pieces of rotating equipment, such as pumps and compressors, are regularly checked in oil refineries, etc., by vibration measurements. An increase in the amount of vibration is a sign of wear or cracks in the equipment. Another good predictor is by taking samples of lubricating oil and sending them to a laboratory to assess the concentrations of metal particles.

- The equipment may be obsolete or spare parts not available. Strategically it may again be appropriate to choose the right moment to replace. Compressors, widely used in oil, gas, and chemical industries, are typical examples. These complex pieces of kit are regularly updated and modernised by the manufacturer. New models become available, just like the family car! In the process, obtaining spare parts for the older models becomes difficult.

- Newer models of the equipment may be more efficient, perhaps in terms of power consumption or environmental footprint. For example, furnaces that discharge exhaust gases with lower concentrations of nitrous oxide and sulphur dioxide emissions may be selected.

- Alternative models could be inherently safer. Pumps with a single seal could be replaced by pumps with a double mechanical seal. A further improvement is the use of canned pumps which have no seal and are therefore much less likely to leak.

Replacement of equipment by alternatives which are not exactly 'in-kind' has been a cause of many incidents. A replacement pump of a new model may have different seal materials, for example, which are not compatible with the fluids used in the particular process operated by the purchaser. Flanges may have been supplied in materials considered by the supplier to be an upgrade, but without the change being recognised and assessed, could result in the wrong welding methods being used. Therefore, it is essential to have a good Management of Change process (see Chapter 31 on the Governance of Safety). Staff must be alert to the dangers of such changes and have the willingness and energy to engage the Management of Change process when initiating such a replacement.

When equipment leaks, it is not always possible to shut down the equipment for repair. This is especially true of large-scale continuous processes such as oil refineries. Therefore, temporary repairs are sometimes used. These can be in the form of clamps or patches made of proprietary materials. A temporary repair is unlikely to be as robust as the repair which could be carried out with the equipment in a shutdown situation. Temporary repairs need to be periodically inspected and replaced with a permanent repair at the earliest opportunity. At insurance surveys, we typically ask about the number of temporary repairs in hazardous service.

ELECTRICAL REPAIRS AND REPLACEMENT

The degradation of electrical equipment is less easy to identify. Large-scale industry is typically fed with electricity through high voltage feeders, often over 100 kV. Electricity is less costly to transport if the voltages are high. Less energy is lost during transmission, and the cabling is more economical. For the electricity to be useful at the facility, its voltage must be reduced in substations, first to an intermediate voltage and then to the voltage which powers the motors and other equipment. This is usually 110V, 240V, or 440V. The reduction requires transformers. Transformers of different types. A common transformer construction comprises windings of wire, coated with insulation and immersed in oil. Progressive deterioration of the insulation and degradation of the oil may lead to short circuits within the transformer casing leading to overheating and ultimately to a transformer fire. A transformer fire is a quite spectacular event and may also cause substantial damage to equipment close to the transformer as well as destroying the transformer. In addition to property damage, combustion products from a transformer fire can be toxic, and the oil spills that result can be a significant pollution hazard. Insurance risk engineers look to see that there are physical fireproof barriers between transformers to help avoid fires spreading to adjacent transformers. They should be fitted with smoke detectors, preferably evening with water spray. Annual tests of the transformer oil are recommended to see if the oil is degrading or accumulating unwanted chemicals.

Electrical equipment in areas containing flammable or explosive gases should be of special quality and installation standards. In Europe, this standard is defined by the ATEX directives. Such equipment must be maintained appropriately; otherwise, the kit may produce sparks. Insulation checks and a visual check of such equipment need to be thoroughly carried out.

INSPECTION AND CORROSION

Inspection is unlikely to be the section of this book the reader is most interested to read. Indeed, it is a subject I knew least about for much of my career. My interest level significantly increased when The Lloyd's Markets Association published their study of the Major Causes of Losses in the Onshore Oil, Gas and Petrochemical Industries (2016). The Lloyd's Markets Association supports underwriters in understanding insurance risk and carries out research accordingly. In this study, financial losses above USD 50 million (property damage and business interruption only, liability losses are not included) were considered along with their causes. Mechanical integrity was found to be a cause of 43% of all losses. This includes 25% of losses due to piping internal corrosion alone. Of the 57% which were not mechanical integrity failures, issues to do with control of work (see above – work permit) contributed just 21% of the total, and operations and operating procedures just 25% of the total losses.

There are typically numerous pieces of rotating equipment such as pumps, compressors, and turbines at hazardous facilities. The rotors can degrade in time due to corrosion and often erosion due to particles within the process material being pumped or compressed. Frequent vibration measurements, sometimes with continuous vibration monitors, are necessary for the more critical pieces of rotating equipment to forego imminent failure and schedule maintenance.

> ### Incidental Case Study: Boiler Explosion, Connecticut 1854
>
> A boiler exploded, killing 21 people, and seriously injuring 50 personnel. Within a decade, the State of Connecticut passed a law requiring the annual inspection of boilers. (Wikipedia – Non-Destructive Testing)

In Equipment Specification (see Chapter 21), we considered materials of construction and corrosion.

Accepting that corrosion will occur, each piece of equipment will have an estimated life. In practice, the actual life may be shorter or longer, depending on the conditions the equipment experiences. Furthermore, many plants, especially oil refineries, vastly exceed their initial projected life. The distillation columns for the crude oil, for example, are typically replaced periodically because of anticipated corrosion. However, a more recent trend is to replace them using more corrosion-resistant materials.

Strange, therefore, that my training in chemical engineering paid little attention to inspection. Chemical engineering university curricula do not include dedicated courses to mechanical integrity. I guess that is left for the

Mechanical Engineers! Many texts on safety and loss prevention also rarely consider inspection in detail.

Inspection frequencies for pressure vessels are laid down by law in most countries. This goes back to the times of the industrial revolution when steam boilers were assembled to provide steam at high pressure to drive engines, including railway locomotives. Much was learned the hard way about making a suitable boiler and the errors that can be made, leading to a boiler explosion.

Boilers need to be periodically inspected to ensure that the necessary wall thickness of steel remains adequate, despite the inevitable corrosion, which will progressively reduce the life of the equipment.

Nowadays, such inspection requirement extends beyond the vessel to much of the associated pipework.

Many organisations quote the statutory frequency when justifying their inspection frequencies. The statutory frequency is a minimum. If circumstances merit, then a higher frequency can be elected. Furthermore, not all vessels and pipework fall into the categories where statutory inspection is mandated. Often some of this equipment which is not covered by statutory inspection could nevertheless cause a serious incident should it fail. Such equipment should be inspected as a prevention barrier against such failures.

Low-pressure pipework carrying hydrocarbons or toxic materials sometimes falls into this category. Consider the consequence of such pipework, which may have been in service for many years, suddenly starting to leak in the middle of the night! In such a case, a good practice is for the organisation to choose to inspect these items. Some organisations choose to do so; many do not.

Corrosion is rarely uniform. It is particularly prevalent wherever, for example, moisture might accumulate. Places where the materials differ even slightly lead to electrolytic corrosion such as at weld lines. There is a particular issue with 'dead legs', which are lengths of pipe perhaps only used infrequently, where there is little or no flow. A good practice is to inspect at least 4 points in the circumference of the pipe. This is not always done. Access is a further difficulty scaffolding may be needed – or the pipework may be too difficult to access.

The postponement of inspections should be avoided. If deferment is essential, it should be approved at an appropriate level that can consider the risk to safety and the business. Many organisations also do not have such a procedure. A formal approval process for postponement of inspection is, in my experience, a frequent recommendation of insurance surveys.

Case Study: Chevron Crude Column Pipe Failure

In Chapter 6, an incident at Chevron Richmond was discussed. In summary, a catastrophic pipe rupture occurred at a crude oil distillation column. A component of an 8" sidecut line ruptured, releasing a vapour cloud that engulfed 18 employees. The cloud subsequently ignited.

The investigation report notes that the failed component was made of carbon steel, which is known to be vulnerable to corrosion in this service, although commonly used since it is relatively inexpensive. This is OK so long as the carbon steel component is replaced before the end of its life. The type of corrosion which led to the rupture is known as sulphidation corrosion and is common to crude oil distillation columns. The corrosion rate varies according to its silicon content. Higher chromium steels are more resistant to this particular kind of corrosion. This circumstance provides an opportunity for Inherently Safer Design, by replacement with more corrosion resistant materials. However that had not been applied here.

In the ten years prior to the incident, recommendations had been made either to carry out a 100% inspection of the pipework or replace the line with an upgraded material.

However, in the four-yearly turnaround planning, the replacement of this pipe did not make the critical list. On the available thickness data, the pipe was calculated to have a life through to 2016. This is a good example where reliance on end-of-life calculations produced from a database without individual consideration by a knowledgeable inspection engineer can lead to erroneous decisions with catastrophic results.

(U.S. Chemical Safety and Hazard Investigation Board (2015). *Chevron Richmond Refinery Pipe Rupture and Fire*)

Special programmes are needed to detect corrosion under insulation (CUI) and corrosion under pipe supports (CUPS). These locations can trap moisture and lead to external corrosion at these locations. CUI programmes have recently become common. CUI is of particular concern in places close to the sea where wind-born salt can accelerate the deterioration of the pipe. CUPS is rarely practised. However, based on one organisation's experience, in particular, CUPS programmes should be accelerated.

Incidental Case Study: Good Practice for Detecting Corrosion under Pipe Supports

Following a leak of ethylene at a polymer plant, which was found to be due to corrosion under pipe supports, the organisation decided to initiate a programme to detect and repair corrosion under all pipe supports. Successful detection is a tough challenge. The pipework has to be lifted sufficiently to place the corrosion detector. Because of the hazards of lifting itself, pipework containing hazardous materials must be drained first. Where the inspection reveals sufficient corrosion, temporary repairs are carried out whilst the line is lifted. Despite the considerable downtime, cost and effort required, the corrosion under pipe supports programme is being replicated throughout this large multi-site organisation.

Risk-Based Inspection has been introduced in many organisations operating in countries that support this approach. The prescriptive approach where inspection frequencies are defined by statute has some shortcomings. In particular, it does not encourage the analysis of specific threats to equipment integrity, the consequences of failure at any specific locations, and the risks created by particular items of equipment. A further criticism is that statutory inspection intervals do not allow an organisation to benefit from good operating experience in specific areas. This would then allow the refocusing of inspection resources to other areas of concern. There has been a trend towards a risk-based approach supported by extensive plant operating experience, improved understanding of corrosion mechanisms, and the availability of fitness for service assessments (HSE 2001, Best practice for Risk-Based Inspection as a part of plant integrity management).

Thus, the application of RBI, where supported by statutory bodies, allows the choice of inspection intervals different from those in the statute. Shorter or longer intervals may be selected based on a risk assessment.

A fundamental cause of misuse of RBI systems, which I have personally witnessed, is that some organisations see RBI only for the cost-benefit of not having to inspect it as often. It is more constructive to see RBI as providing the tools to better focus inspection resources on higher priority risk areas.

Consider a typical oil refinery or other process plants. The number of vessels and pipework sections is huge. For inspection purposes, these are broken up into corrosion loops. In any corrosion loop, there will be multiple inspection points. At each inspection, this complexity generates a lot of data. Take any particular inspection location. Consider for the moment just pipework. At each inspection, the wall thickness is typically measured. It is then necessary to consider the starting wall thickness, the thickness at previous inspections, and the minimum safe wall thickness, which is usually specified in the code for the pipe. From this, the average corrosion rate can be calculated. This rate defines the time until the end-of-life of the pipe when the pipework thins to its minimum safe thickness. In many organisations, the inspectors say they calculate all of this manually or use spreadsheets on a pipe-by-pipe basis. In my opinion, this leaves considerable room for human error in that the inspector may overlook a critical corrosion issue amongst all the data and calculations.

A good practice is to use commercially available inspection database software, which automatically calculates end-of-life and issues reports alerting the inspector accordingly. This is not without dangers since the computerised database approach can be subject to data input errors. It is also 'impersonal' with no particular emphasis on items of known concern. It can lead to a lack of ownership of inspection results.

Notes from Trevor's Files: Personal Ownership for Inspection Data

When carrying out an insurance survey in one of the former countries of the USSR, I asked about inspection data for the specific section of the plant we were walking through.

We were introduced to the inspector who was carrying out work on the plant. He directed us to his locker, in the changing room, at the bottom of which was a handwritten book of inspection results. What was astounding was that he could talk about any particular piece of pipework and inspection result. He was the only inspector dedicated to a relatively small part of the plant. This enabled him to have such an informed approach and deep ownership of both the data and the plant for which he was assigned.

This approach does, of course, suffer the downside that the inspection process is mainly dependent on the health and well-being of just one person. Also, if anything happens to the handwritten records, much information would be lost.

It is generally recommended to hold hardcopy records in a fire safe and ensure that electronic records have a secure backup process.

Techniques for non-destructive testing continue to advance. Ultrasonic thickness testing is one of the most common among more than 70 other techniques, each of which has its benefits.

Inspectors require considerable training, which is often specific to particular techniques or groups of techniques. Often contracted services are used to ensure inspectors with the proper training and qualifications are available. In some organisations, therefore, the lead person for that organisation – the Inspection Head – does not have detailed knowledge of the inspection techniques and does not have appropriate qualifications. This potentially leaves the organisation exposed to misunderstandings regarding confidence or lack of confidence in inspection results. Techniques may be misapplied for particular applications.

AGEING AND END-OF-LIFE

Whether they be an oil processing operation, chemical plant, or nuclear plant, industrial facilities are constructed with a specified plant life. This is also true of aircraft and railway locomotives. This design life is part of the specification given to plant designers and constructors. Equipment and pipelines are selected with corrosion allowances accordingly. Assuming that the anticipated corrosion levels occur in actual operation as expected at design, then the pipework or equipment will retain sufficient thickness until the forecast end of life. Such design calculations are generally done conservatively, and much equipment retains sufficient thickness beyond its design life.

Whereas a plant might be designed with a 20–25 year design life, many plants extend their life to say 40–50 years. Similarly, the projected lifetime of aircraft can be extended. Of course, this means some parts of the plant or equipment will have to be replaced. Those parts needing to be replaced are identified by the inspectors based on their inspection results and calculations. As a result, they can forecast end-of-life dates for specific equipment and pipework.

This consideration of the design life of a plant and whether it can be extended is called 'ageing'.

FALSIFICATION OF INSPECTION RESULTS

The results of an inspector's work can severely impact the short-term output. In some circumstances, this can cause the inspector to either hide, ignore, falsify, or otherwise obfuscate the result. Or the result can disappear on the manager's desk. Perrow (2007, p. 143) points specifically to falsification in inspections at Japanese nuclear reactor:

'Japan's nuclear plants for some decades appeared to perform at higher levels of efficiency and safety than those in the United States. But in July 2000, four ominous unexpected shutdowns occurred, some releasing unacceptable radiation levels, in the plants run by Tokyo Electric Power Company (TEPCO), Japan's largest utility. In 2001, a whistle-blower triggered disclosures of falsified tests at some of the company's seventeen plants, and the government forced TEPCO to close some plants. In 2002, the company predicted that all of its seventeen plants might have to be shut down for inspection and repairs because falsified inspections and concealment of faults found in inspections that the government ordered; some of the faults were potentially catastrophic'.

···

Incident Case Study: Falsification of Inspection Data

A particular example of this concerns the case of a system engineer at a U.S. nuclear power plant. The engineer claims that in 2000 he found evidence of corrosion at the reactor head (boric acid) and attempted to clean them off. Before he could complete his task, he discovered that the scaffolding and equipment needed for his work had been removed and that management had signed a report saying that the reactor head had been thoroughly cleaned. Further examination had to wait until 2002, when he was able to complete a full cleaning of the head and found a 'pineapple sized hole' due to corrosion. It was subsequently determined that the boric acid had eaten away 70 pounds of carbon steel, leaving only half an inch of buckled stainless steel to contain the radioactive material inside of the vessel. After this discovery, the engineer was apparently transferred to another assignment where he found problems with reactor coolant pumps. He was then involved in another event where, after failing to convince the nuclear reactor operator that the problems of cracked shafts had to be fixed, he refused to sign a report saying the issue had been resolved. He was given the option of signing, resigning, or being terminated, although the operator of the plant says he was terminated due to poor job performance. In 2005 this same engineer was sued by the Nuclear Regulatory Commission (NRC) for failure to report the defects in the reactor.

It seems strange to me that the NRC sued the engineer directly, not the operating company.

The engineer was subsequently sentenced to three years on probation and fined $4,500 for helping to cover up the worst corrosion ever found at a U.S. reactor. He had faced up to five years in prison.

He was convicted of misleading regulators to delay a safety inspection at the nuclear power plant along Lake Erie. Defence attorneys have said the engineer was set up as a scapegoat because he spoke out about safety concerns.

World Nuclear Association – World Nuclear News 2008

In many organisations, the head of the Inspection department or even the inspector himself recommends what to do following inspection measurements and calculations. On the one hand, this makes sense. The inspector knows the corrosion mechanism, the past trends, and the specific location of corrosion. On the other hand, he/she is not responsible for assessing the risk.

In my insurance work, we pay close attention to the interaction between inspection and business management. Here is a recommendation I have made recently:

	Improve Authorisations of Completed Inspection Plans
Observations and Issues	The site has a good process for planning inspections, gathering the results, forecasting the end of useful life, determining any action to be taken, and setting a date for the next inspection. The inspection records are detailed and identify all the inspection points required. For example, measurements are needed for an atmospheric storage tank at all pipeline connections, various locations on the roof, wall sections, and floor. The database was observed to have a section marked 'Technical Review Complete'.
	When examining one example, we found that significant data to do with the lower section of the atmospheric tank was out-of-date. The database was automatically highlighting data calling for an overdue inspection. On further discussion, it became clear that this was due to the expansion of the inspection programme to cover this tank section to this degree of detail and that the data had not been required on previous inspections.
	It became apparent that some such inspection plans are reviewed by management, and some not with no obvious procedure to identify when an inspection plan should be authorised by business management.
	Some mechanical integrity failures have occurred where there have been misunderstandings between Inspection and Plant Management as to the risks resulting from an inspection and the subsequent forward inspection plan.
Recommendation (s)	It is recommended that the site improves the inspection procedure to define when an inspection plan requires business management approval in addition to the approval of the inspector. Criteria for management approval could be whenever any of the inspection data is incomplete or old or whenever the remaining forecast life is within a specified duration.

Thousands of inspections are made every year in a typical oil refinery or large petrochemical plant. Any particular pipework section may have ten or more inspection points. A fair amount of discernment is needed where any particular inspection point shows thinning. Is it real, representative? What are the criteria for the inspector to pass a decision on to business management? It is perhaps easier than it might at first seem to decide that any particular point could be overlooked – this time.

CORROSION MECHANISMS

Corrosion science is an evolving subject. New corrosion mechanisms are sometimes discovered in the process of investigating serious events.

In order to illustrate some of the challenges in identifying a quantifying corrosion mechanism we will select just one such mechanism – High-Temperature Hydrogen Attack (HTHA). HTHA is an irreversible deterioration of the mechanical properties of steel due to the reaction of hydrogen with carbon at temperatures above 200 °C. Atomic hydrogen may react with dissolved carbon within steel to form methane inside the material. This chemical reaction results in the deterioration of the mechanical properties of the steel, particularly its strength. The methane molecules are too large to diffuse out of the steel. Thus, they are trapped in and exert pressure on the grain boundaries. These voids grow through microscale creep deformation due to diffusion at the surface and grain boundaries or dislocations. Cracks result. The accumulation of methane may also cause blistering.

The issue led to the development of graphs for various carbon steel alloys, plotted against axes of temperature and hydrogen partial pressure. These 'Nelson' curves are detailed in the American Petroleum Institute (API) publication, Recommended Practice 941. The curves attempt to describe limit conditions for different grades of carbon and alloy steel, where exposure to adverse conditions above the curve will lead to HTHA effects on the microstructure.

The U.S. Chemical Safety and Hazard Investigation Board (CSB) determined that HTHA was the damage mechanism responsible for an incident in the USA in 2010, at the Tesoro Anacortes refinery, where the shell of a heat exchanger ruptured catastrophically, killing seven workers (CSB, Catastrophic Rupture of Heat Exchanger (Seven Fatalities)). The CSB were of the opinion that the heat exchanger was operating significantly below the relevant Nelson curve applicable at the time. Challenged by the CSB findings, the API added a curve for non-Post Weld Heat Treatment carbon steel in the 8th edition of API RP941 published in February 2016. The CSB concluded that this did not sufficiently address the issue, and in their Safety Alert of August 2016, as well as stating that operators should identify susceptible equipment and verify operating conditions, recommended that operators should:

'Replace carbon steel process equipment that operates above 400 °F and greater than 50 psia hydrogen partial pressure', and 'Use inherently safer materials, such as steels with higher chromium and molybdenum content'.

The temperature and pressure conditions stipulated by CSB differ considerably from those detailed in the API 'Nelson' curves.

SUMMARY: CAUSES OF FAILURE DUE TO IMPROPER MAINTENANCE AND INSPECTION

Reasons Why Preparation for Maintenance Sometimes Causes, or at least Contributes, to Catastrophes Include:

- ☒ Lack of recognition that a means of positive isolation is necessary.
- ☒ Inadequate isolation method.
- ☒ Purpose of isolation misunderstood leading to premature removal.
- ☒ Locks, with no informative tags, are employed.

☒ Tags in use but not completed.

☒ Locks and tags of only one party, typically operations, are applied – not with additional lock and tags from all the parties involved in doing the work.

☒ Locks and tags of only one party, typically operations, are applied – not with additional lock and tags from all the parties involved in doing the work.

☒ Only the electricians lock is applied in the control room.

☒ No 'Try' step to verify isolation is working.

Reasons Why Control of Work Sometimes Causes, or at least Contributes, to Catastrophes Include:

☒ Work permits are too general and not specific to the work situation of the day.

☒ Simultaneous work by two parties in close proximity is not recognised.

☒ Work permit scope does not recognise all the steps of the work.

Reasons Why Maintenance Activity Sometimes Causes, or at least Contributes, to Catastrophes Include:

☒ The advantages of Predictive and Preventative maintenance are not appreciated – excessive reliance on emergency or breakdown maintenance.

☒ Risk of deferring preventative maintenance not appreciated; no procedure applied for assessing the risks of deferment.

Reasons Why Repairs and Replacements Sometimes Cause, or at least Contribute, to Catastrophes Include:

☒ Lack of appreciation of the benefits or repairing or replacing equipment before it breaks down.

☒ Equipment replaced, but replacement not complete replica, resulting in unexpected failures.

☒ Temporary repairs are made, without regular inspection and replacement with permanent repairs within a reasonable time.

Reasons Why Inspection Sometimes Cause, or at least Contribute, to Catastrophes Include:

☒ Lack of appreciation of the extent to which corrosion causes catastrophic loss, especially that of pipework corrosion.

☒ Absence of training of other disciplines (operations engineers, process and chemical engineers, etc.) in Inspection techniques and programmes.

☒ Reliance on inspection intervals provided by statute.

☒ No inspection of pipework and equipment not required by statute.

☒ Use of Risk-Based Inspection to extend inspection intervals and reduce inspection cost, rather than refocusing the inspection effort.

☒ Risk of deferring inspections is not appreciated, no procedure applied for assessing the risks of deferment.

☒ Absence of a programme for detection of corrosion under insulation or corrosion under pipe supports.

☒ Excessive reliance on inspectors to spot trends in corrosion rates which indicate critical need for repair.

☒ Recognition of corrosion risk dependent on large impersonal software databases.

☒ Training of inspectors, especially inspection managers, in the many techniques available to carry out inspections.

☒ Falsification of inspection results, or not reporting them.

Chapter 23

Emergency Response and Readiness

EMERGENCY RESPONSE PLAN AND SCENARIOS

Suppose an emergency has occurred. In the first instance, this is rarely a catastrophe. The emergency may develop into a catastrophe if it is not managed well. For example, a small fire in overgrown vegetation around a pipe alley may overheat the pipe, resulting in a release, igniting and spreading fire to neighbouring equipment, etc. Even in the airline industry, a critical failure such as a failed undercarriage is not immediately a catastrophe. Pilots and ground staff are trained in emergency landings, and such emergency landings have been carried out without loss of life.

In Section 2, 12 of the incidents reviewed had issues with emergency response. These included Deepwater Horizon, Bhopal, Laporte, Brumadinho, and DPC.

In emergency response, we are dealing with controls which are placed to the right of the Bow Tie.

Most people associate emergency response as being the fire brigade. The scope is, however, broader. It includes:

- First response to the fire, both fire extinguishers, and personnel training in first response.
- Emergency rescue, with or without fire. Such as personnel overcome by fumes, including rescue from confined spaces inside vessels.
- Local civil defence support including fire brigade, ambulance and medical and police.
- Fixed means of fire extinguishing and suppression, such as deluge systems to cover process equipment and tanks with water or foam.
- Smoke and fire suppression inside control rooms and instrumentation buildings using automatic application with suppressant chemicals.

An emergency response plan is necessary for small events such as the fire in overgrown vegetation, through to fire engulfment of a hazardous process, aircraft crash, etc. ...

DOI: 10.1201/9781003360759-26

Note that a facility may choose to have minimal emergency response facilities in some limited cases. This philosophy accepts that a fire may destroy the affected units. This philosophy is only acceptable where there is minimal potential offsite impact. Examples encountered where the minimal approach is used include remote crude oil and gas extraction wells, oil well gathering and processing stations.

Incidental Case Study: Deepwater Horizon – Emergency Planning

In the Deepwater Horizon/Macondo catastrophe we reviewed in Chapter 6, it is remarkable that there was no viable emergency plan to deal with oil and gas flowing uncontrollably from the seafloor a mile below the sea surface, other than drilling a relief well. Drilling a relief well involves an extensive drilling period from a remotely located rig. The relief well must be directed precisely so as to intersect the blowout well. Once the relief well is drilled, it is then used to pump the leaking well full of cement. Drilling a relief well would leave the blowout well flowing uncontrollably for many weeks.

Regarding oil spill response, a plan did exist as required by the regulator. This document assessed that an accidental oil spill from the proposed activities could cause impacts to beaches. However, due to the distance from the shore (48 miles, 77 km) and the response capabilities that would be implemented, no significant impacts were expected! (Hopkins, 2012)

In Section 4, we will discuss catastrophic events caused by natural catastrophes, terrorism, and sabotage. The effects of such incidents are widespread and require well-prepared community action plans involving local governments and services and the operators of facilities containing hazardous materials or hazardous operations. Perrow questions how well the U.S.A. is equipped to deal with major disasters (Perrow, 2007, p. 99). There has been heavy criticism of the response to Hurricanes Katrina and Rita despite considerable investment and organisational restructuring to respond to such disasters in the wake of the 9/11 terrorist attacks. Statistics quoted by Perrow include the statement that only 11% of fire departments are adequately staffed. Only 11% have the equipment to deal with buildings of over 50 occupants. Only 13% would be able to deal with attacks by chemical and biological agents. This brings into question whether other countries are any better equipped. Indeed, in the U.K. the performance of the emergency services must be questioned after the Grenfell tower fire (see Chapter 15), at least when managing issues with high-rise buildings.

I have had the good fortune to witness personally co-ordinated drills between the emergency services and the site fire brigades at a major oil refinery and petrochemical plant in Germany. By contrast, I found the exercise exceptionally professionally conducted and coordinated, with emergency services arriving from many towns around this largely rural area.

Illustrative Hypothetical Cases – Emergency Planning

Let's consider our fuels terminal and whisky distillery. What kinds of emergency response plans should we develop?

In general, we would plan for:

- On-the-job injuries such as slips, trips, and falls.
- Medical issues such as heart attacks.
- Small fires of vegetation, boxes, etc. …
- Office fire.

Specific to the fuels terminal, we might plan for:

- Fire adjacent to a tank (within a retention basin if there is one).
- Fire inside a tank.
- Fire spreading from one tank to multiple tanks.
- Spillage of petrol and diesel requiring emergency retention and remediation.
- Fire at pumps.
- Fire in railroad tanks in transit to the terminal or whilst at the terminal.
- Fire in the road tankers either in transit from the terminal or whilst at the terminal.

Specific to the whisky distillery, we might plan for:

- Fire in the maltings house – dried barley.
- Fire/explosion on the peat floor or furnace replacing the peat floor.
- Fire/explosion in whisky warehouse (of ethanol vapour – otherwise known as the angel's share!).
- Spillage of any component of the whisky manufacturing process into the adjacent stream.

The steps in setting up an on-site emergency plan include:

- Identifying the significant hazards – source, type, scale, and consequences.
- Development of the scenarios which could arise and need an emergency response.
- Consideration of the plan of action to respond to each scenario.

- Establishing the procedures and frequencies to test and practise the emergency response.
- Determining the command structure.
- Deciding the competencies required (training and experience) to be able to conduct the plans of action for each scenario.
- Completing the training and staffing necessary to be able to meet the plan.
- Arrangements for any offsite emergency planning.

There are often difficulties in emergency response planning for incidents involving raw materials or products transported to and from the plant or warehouse. Is this the responsibility of the railway transport company or the contractor carrying materials in road tankers and trucks? These companies do not have the specialist knowledge of the particular materials being transported, whereas the company using them and producing them can be expected to have this expertise. There can be an undue reliance one on another. This is not so much of an issue for transporting well-known materials, including highly flammable materials such as gasoline. However, for less commonly used materials such as chlorine, the supplier and user should certainly have offsite emergency plans covering transportation.

It is usual for the site emergency response team to have the role of first response. That is controlling the situation until professional community emergency services can take over. Here lies a substantial source of error in emergency response handling. The site emergency response team will already be managing the situation with a knowledgeable on-scene commander. The on-scene commander will be either a shift supervisor or a full-time member of the site fire brigade. He or she then has to essentially hand over to the professional command from the local community brigade who does not have the same knowledge of the plant and the situation.

A particular issue, in my experience, concerning emergency response plans, is complexity. In former countries of the USSR, it is very typical to find long procedures with copious mathematical equations. As an insurance surveyor, I have repeatedly attempted to challenge this because the average firefighter does not have the competence or time to read such a document. A summary document would be more efficient for their use. Regrettably, it is still the case that the long-form documents are 'required by law'. At the other extreme short documents (otherwise known as emergency pre-plans) are often found not to contain some essential detail such as:

- Equipment description.
- Hazardous materials with quantities, and critical properties such as flash point and autoignition temperature.

- Toxic risk – including carcinogens such as Benzene and Asbestos.
- Explosion risk.
- Third-party locations at risk.
- Specials hazards such as the location of any instruments with radio-active elements, chemicals that react with water, etc. ...
- Key actions required, such as shutdowns, manual initiation of deluge, etc. ...
- Key tactical emergency response actions include activating firewater and foam fire extinguishing equipment.
- Cooling requirements for other adjacent equipment, as derived from radiation studies.
- Required mobile equipment response – vehicles required plus hydrants and monitors with firewater and foam capacity.
- Firewater runoff and any pollution interceptors. This is an aspect rarely included in practice.
- Maximum quantities of foam and firewater for major fire cases that are likely to test the capacity of the firefighting systems. These quantities are often not included, making it difficult to assess if the site has adequate firewater provision or not.

Excellent practice observed in industry is for each pre-plan for each scenario to be available on 2–3 A4 laminated sheets. The fire team can pick up the scenario on their way to the incident or available on each mobile response vehicle (Marsh Risk Engineering Position Paper No 2 2015 on Fire Pre-Plans).

There should be a major emergency response plan where there is potential for the emergency to have a significant community impact. There are specific regulations in Europe known as Seveso III, which have specific requirements regarding Major Emergency Plans. In the U.K., these are known as the COMAH (Control of Major Accident Hazards) regulations. These regulations give excellent advice on emergency response plans and whether or not an establishment meets the prescriptive criteria whereby Seveso III emergency plans are required. These criteria are based on the total amount of hazardous materials on-site. These are expressed as both generic categories of substances such as toxic, oxidising, explosive, etc. ... and for specific categories of hazardous materials. There are two categories of establishments, lower and upper tier. The requirements for emergency plans are mostly confined to upper-tier establishments.

Illustrative Hypothetical Case – Emergency Response Plans under Seveso III

Considering our hypothetical examples of the fuels terminal and whisky distillery we might be interested in, what quantities of hazardous material would require us to follow Seveso III regulations and therefore have major accident emergency plans?

The 2015 regulations are quite specific for the fuels terminal. The threshold is 25,000 tonnes for 'upper-tier' for both gasoline and diesel.

Ten thousand tonne tanks would be quite usual for a fuels terminal, so if our terminal has just such tanks, we need to develop major emergency response plans in accordance with the requirements of Seveso III. A 10,000 m^3 tank is something like 15 m tall and 30 m in diameter.

For whisky, the evaluation is only a little more complicated. The flammable component is ethyl alcohol, present at around 63.5% in a good Scotch whisky. For ethyl alcohol, the threshold for upper tier is 50,000 tonnes of ethanol or approximately 80,000 tonnes of whisky. A typical barrel holds around 180 kgs of whisky. Our establishment would qualify as 'upper tier' if it involved storage of some 440,000 barrels of whisky or more.

The swimming pool at my local gym is 25 m long × 12 m wide. This gives me a visualisation of such an area stacked three high with barrels of whisky – that's approximately 3,000 barrels of whisky.

Only a few whisky storage establishments are listed as upper tier in Scotland. Likely, our hypothetical venture is not upper tier, but if we send our product to a warehouse operated by more extensive distribution enterprise, then that warehouse may reach the upper tier.

What are the expectations of an operation meeting the requirements of the upper tier with regards to major emergency plans? There are additional requirements to communicate to local authorities. The following expectations are set for an offsite emergency plan:

- A plan prepared in conjunction with the local authority.
- Roles described to be carried out by the emergency services, local authorities, and other external organisations in the event of an accident.
- Provision for the restoration and clean-up of the environment following the incident.
- Consideration of 'domino effect' if the site is located close to other high-hazard sites.
- Arrangement for limiting risks to on-site personnel such as how warnings are to be given and the actions people are expected to take on receipt of the warning.
- Arrangements for providing early warning of the incident to local authorities.

The COMAH regulations contain what this author considers to be a quite extraordinary statement:

'The degree of planning should be proportional to the probability of an accident occurring'. Who assesses this probability and how? This seems to be inviting the kind of 'box ticking' exercise which we saw in Deepwater Horizon emergency plan preparation discussed above. Low probability can be satisfied by sparce planning details.

The potential for 'domino effect' has increased in recent years as large chemical majors have sold off part or all of their operations to different enterprises. The overall risk may not have changed, but there are now multiple plans instead of one offsite plan. This is due to the many operators on the industrial park. This has raised the complexity of emergency response handling. A good practice is for the industrial park to have one Emergency Response Team and one plan, with contributors from all the operations in the park. An excellent example is the Chemelot site in the southern Netherlands. This site was formerly a DSM operation and now contains over 150 companies and institutions.

EMERGENCY RESPONSE TEAM AND TRAINING

It is highly advisable that a hazardous facility has its own emergency response team.

Industrial parks often share a communal emergency response team. Smaller facilities of only moderate size may be dependent on the local community emergency response teams, so long as the response time is short.

A specialist contractor may also provide the on-site emergency response team. The contractor is responsible for maintaining a force of competent and trained personnel. This is advantageous since personnel can be moved between sites where the contractor provides a service. It also provides specialist and up-to-date techniques.

The minimum size of the team will be determined from the scenarios considered in the Emergency Response Plan. It is a common and acceptable practice for emergency response team members to be auxiliaries, i.e. usually conducting other tasks such as process operator, but can be immediately released from their normal function in an emergency.

At a minimum, at least part of the shift team would be trained in first response. First response might be to either a fire, a toxic gas release, a spill, or an injury.

Illustrative Hypothetical Case – Whisky Distillery and Fuels Terminal – Necessity for Emergency Response Team

In our hypothetical cases, of the whisky distillery and the fuels terminal, would you consider it necessary to have an emergency response team, or would you consider it acceptable to rely on the local community emergency response team? This rather depends on the response time of the local community response team and also on their equipment and training.

In the fuels terminal, we have significant volumes of highly flammable hydrocarbon, namely gasoline (petrol). Most fuels terminals would have a dedicated emergency response team. In the case of the whisky distillery, we are dealing with potential fires on which the local emergency response team are most likely to have both the equipment and training to handle, for example, in the maltings house or the drummed whisky storage area. It is unlikely that we need to have our own emergency response team here.

Incidental Case Study: Chevron Crude Column Failure – Incident Commander Briefing

In the case of the crude column pipe failure discussed earlier (see Chapter 6 on Oil and Gas and Chapter 22 on Equipment Maintenance and Inspection), the incident investigation also stressed the importance that the Incident Commander has the appropriate knowledge and is briefed by specialists as the incident unfolds.

The Incident Commander, on this occasion, had not been briefed on the potential for catastrophic failure of low-silicon carbon steel pipe. Consequently, he specified a 'hot zone' which was too small, and emergency responders approached more closely than was appropriate given the actual risk.

CSB (2015) Chevron Richmond Refinery Pipe Rupture and Fire

It is regrettable that, in many cases, personnel are chosen to be transferred to the emergency response team because they have proved less than competent elsewhere. I have found this, particularly in countries where it is extremely difficult to dismiss staff. This is typically the case in countries where the employment of local nationals is protected. Examples of this difficulty are found in the Middle East, Kazakhstan, and other regions.

Specialised training is needed for the more senior firefighters. Notable examples are the Moreton-in-the-Marsh Fire Service College and Rotterdam International Safety Center (RISC), now known as RelyOn Nutec Fire Academy. It is common practice for the emergency response chief and his or her deputy to do at least one week's training at such facilities every year.

EMERGENCY PRACTICES AND DRILLS

Emergencies are, hopefully, rare events. Without regular practices and drills, any training in handling emergency scenarios will likely be progressively forgotten or not recalled with the necessary fluency. Therefore, routine exercises are essential. Typically, a facility will have one drill each week. In my opinion, this is barely sufficient. For a five-shift operation, that's less than one drill per month for each shift team. Suppose each scenario is going to be practised once per year. The number of scenarios practised is limited to just 12. Processes more complex than our simple fuels terminal example would have many more, which means some plans are practised less than once per annum.

Drills should be followed by an 'after-action review' to see what went well and what needs improvement. Actions should be assigned and tracked to ensure those actions are completed.

Sometimes, unfortunately, shortcuts are taken in carrying out drills. At one extreme, they can be cut down to merely 'tabletop' exercises where the drill actions are discussed. Whilst such drills are not without value, they give little exposure to the time, or lack of time, actual required to complete a task, and limited recollection of the physical placement of equipment, hoses, hydrants, etc.

Another limitation to drills, in many cases unavoidable, is the need to carry out a 'dry drill'.

In a dry drill, no water is applied to the equipment, as opposed to a wet exercise. Dry drills can be necessary on, for example, equipment such as distillation columns where the application of large amounts of water externally would cause cooling upsets to the manufacturing process.

Illustrative Hypothetical Case – Fuels Terminal – Emergency Response Drills – Scenarios to Practice

What emergency scenarios would you consider for regular practice for the fuels terminal?

Suggestions are:

- Tank fire with cooling of adjacent tanks.
- Rail tank car fire.
- Rescue of a collapsed person on top of a gasoline/diesel tank.
- Recur of a collapsed person on top of an LPG tank.
- Road tanker fire on site.
- Road tanker fire in transit.
- Containment of a gasoline/diesel leakage.

- Fire at product pumps.
- Fire at LPG tank.

Emergency drills should also be practised with the local community emergency services. For Seveso III sites, this must be done once every three years, but some organisations recognise the benefits and have such a major practice annually.

FIRE AND GAS DETECTION

It is necessary to have fire and gas detectors with alarms. You almost certainly have them in your house, and if you do not, then please get equipped right away! Smoke detectors alert inhabitants to the first stages of fire, such as might be caused by an overheated frying pan, and carbon monoxide alarms are fitted to alert people to a malfunctioning gas boiler or solid fuel burner.

Illustrative Hypothetical Case – Distillery – Fire and Gas Alarms

What provision of fire and gas alarms would you consider for the fuels terminal?

Suggestions:

- Office areas – fire alarm and carbon monoxide alarm if there is a gas boiler.
- Maltings house above the barley floor – fire detection.
- Maltings house above barley floor and outside of the peat floor – carbon monoxide alarms.
- Warehouses – fire alarms.
- Product warehouse – gas alarms to detect explosive accumulations of ethyl alcohol.

Fire detectors can be misplaced so that they are obstructed from seeing the fire, especially if they are of the optical type. Smoke alarms take time to respond, especially if they are placed distant from the potential ignition unless they are of the Very Early Smoke Detection Apparatus (VESDA) type.

Gas detectors can be misplaced. The density of the potential gas release will determine if the gas rises or falls if it is released. Methane and hydrogen rise, most other gases fall. The placement of the detector with respect to the prevailing wind direction is also a consideration.

Detectors and their alarms require frequent testing, typically once every three months. Sometimes these tests are omitted. Detectors sometimes fail or start to flip into and out of alarm due to plant issues, giving rise to alarm overload (see Chapter 6). Hence, detectors are sometimes temporarily bypassed, with the potential to overlook the bypass and lose the alarm's protection for an extended period.

FIREFIGHTING WATER AND FOAM

Facilities handling flammable materials normally have a source of firefighting water. This applies even if reliance is on local community emergency response – the amount of water that can be transported in a mobile firefighting vehicle is limited.

Firefighting water storage may be in atmospheric tanks or in storage ponds. The water may be abstracted from natural watercourses and wells. Firefighting water normally requires little pre-treatment, although filtration is recommended to prevent organisms from making home in the pipework and tanks. Saline water is not recommended since the salt content makes it corrosive to steel components. In an emergency, the water can be directly abstracted from lakes, wells, and even from the sea.

Normally water for firefighting is held in dedicated storage. However, some organisations have storage tanks that are shared with process water needs. I always feel somewhat uncomfortable with this arrangement. Such shared storages are operated with a rule that a certain proportion of the storage is reserved for firewater. In the event of the storage tank level approaching the level reserved for firefighting water the processes taking process water must be shut down. I imagine that taking this decision to shut down, when there is no fire, is a difficult one!

How much firefighting water to store?

Illustrative Hypothetical Cases – How Much Firefighting Water Storage?

Let's consider our Whisky Distillery and Fuels Terminal again.

How much firefighting water storage is sufficient?

First, we note from Appendix A that both sites are adjacent to streams. Although we don't know much about these streams, let's assume that they are of sufficient volume and flow rate to allow the community fire brigade to abstract water from them if there is a fire. The on-site storage should be sufficient for the site fire brigade to carry out their first response.

For the fuels terminal, if the tanks are a typical 10,000 m³ volume, a typical fire water demand is around 1500 m³/hr. Firewater storage for approximately 30–45 mins supply is commonly provided. However, this must be questioned, especially where natural watercourses are remote, or tank separations mean increased potential for a fire in one tank to spread to adjacent tanks.

For the whisky distillery, we have the potential for fire in the warehouses, product barrel storage and in the maltings house. We should carry out a fire risk assessment, a required step in the U.K. It is likely that a fire in any of these areas would be limited, and reconstruction practical within say 2–3 months. One exception is the potential for a vapour cloud explosion considered in Chapters 25–27 on Process Hazards Reviews. A vapour cloud explosion would not be contained by fire water, although some of the follow-on damage could be reduced.

The amount of firefighting water that needs to be stored is something of a contentious issue. It is often assumed that firefighting water can be reclaimed. Firefighting water sprayed onto equipment creates a runoff. Most sites have graded concrete with drains directed to a rainwater/firewater retention pond or tank. The claim is that runoff can be pumped out of this retention pond/tank whilst fighting the fire and reused by the firefighting appliances. However, I have never seen this tested. The extent of the runoff water contamination is difficult to predict in advance.

Emergency Response Equipment – Fixed and Mobile

Fast first response can be critical in preventing a catastrophe, especially from domino effect fires travelling from one piece of equipment to other adjacent facilities.

Many installations have a fire water main which would feed hydrants, where firefighting trucks can hook up their equipment to, and monitors, which are fixed installations able to spray water or foam onto equipment. Now there is a big problem with firewater mains – they corrode! So many refineries and terminals are engaged in an expensive process of replacing the original carbon steel pipework with plastic alternatives. Meanwhile the fire water main cannot be fully pressurised and there is a risk of rupture in the event that it needs to be used at full pressure!

Firewater pumps are needed to pump the water from storage into the firewater main at sufficient pressure and flow rate to deliver water at the needed location.

It is also good practice to have a mixture of pumps with both electrical and diesel motors, so that fire water can be supplied in the event of an electrical failure. A failure in electrical supply can indeed be a potential cause of severe loss of process control and fire.

Notes from Trevor's Files: All Fire Water Pumps with Electrical Drives

At an oil refinery in Kazakhstan, I was reassured about the continuity of the electrical supplies, with many power stations being visible around the refinery. The refinery is equipped with firewater pumps with electric drives. It is a good feature of this refinery in that it has two separate independent fire water pump houses.

However, we learnt that in 2011 a general power outage did occur and that it took some two days to fully restore the power to the refinery. The power to the firewater pumps was restored after 6 hours. Whilst diesel generators have been subsequently installed; they would not be able to supply the firewater pumps, which run off a different electrical supply. In the event of a general power outage, the refinery would be unable to fight the fire using water from the firewater main.

It is usual for firewater pumps to be regularly tested, and a member of the emergency response team typically goes out to the pumps once a week to run each of them. This is a simple test to see that the pumps do indeed run. The test sounds like it should be a straightforward practice, but if not done correctly, it can result in progressive deterioration of the diesel pumps due to 'wet stacking'. When the pump motor starts from cold, unburned fuel is exhausted from the combustion chamber. Carbon build-up can occur on motor components such as the exhaust valves. A good practice is to run each diesel motor for at least 30 minutes.

Annual tests on such fire water pumps should be carried out. Impellor wear can occur with time, so a capacity check is needed, and comparisons made with the design data for the pump. Often such tests are not done because of the absence of suitable flow measurement instruments. Without these tests, there can be no certainty that the required amount of firefighting water will be delivered in the event of an emergency. A further test is to check the flow delivered from one of the hydrants furthest from the firewater pumps. This tests the ability of the pumps to move the required amount of water. It also checks that the firewater main is not blocked and has sufficient capacity. Such a test is 'messy' in that a lot of water is discharged at the hydrant. Again, means of flow measurement are required. Such tests are sometimes omitted.

Means of automatically generating and applying firefighting foam are sometimes provided, especially at critical equipment. A typical issue with firefighting foam is storage. Foams have maximum and minimum storage temperatures and should be tested at least annually to ensure that they remain of good quality. In addition, some equipment can be protected by special techniques such as steam curtains which disperse flammable gasses and help isolate releases. Water curtains can be used to knock down toxic releases. Therefore, a key feature is that such equipment can be initiated remotely and quickly. If an operator must go to the facility to start the operation of the steam or water curtain, he or she may be drawn into the area at risk of fire or explosion. Regrettably, many such installations are found with manual valves shut, meaning that if they are required, an operator would indeed have to enter the area at risk. Worse, if it is not appreciated that a manual valve is closed at the time of the emergency, the protective curtain is not available.

Mobile firefighting equipment is generally well catered for, and we will not attempt to replicate additional data in this book. Other than to point out the importance of the availability of a turntable ladder sufficient to get to the higher platforms of the equipment on the site, especially distillation columns. This ladder may be for personnel rescue as well as fire.

Other means of fire suppression may be employed, especially where water is not permitted due to the damage it would cause. These typically involve the release of a fire suppressant gas within an enclosure where the equipment is located. Large compressors often have an enclosure around the motor with such a suppressant system. Electrical control rooms, instrumentation rooms, and even control rooms can be protected by such means. Sometimes the suppressant gas is an inert gas such as Nitrogen or Carbon Dioxide. In this case, the enclosed space or room needs to be uninhabited; otherwise, asphyxiation would occur. If the room needs to be entered, the fire suppression must be securely disabled first. The danger here is that re-enabling of the system is forgotten. Other fire suppression chemicals work in small concentrations to extinguish the fire and are much less likely to be hazardous to people.

Most fire suppressant systems are intended to operate in automatic, i.e. on detecting smoke or fire, the suppressant is released automatically. Keys are always provided, which allow the suppressant system to be switched from automatic to manual. I have found that the keys in such systems are almost always switched to manual. The reason is simple enough – sometimes spurious alarms occur, and rather than permit expensive suppressant to be released, possibly unnecessarily, the operating staff prefer to make a manual check first!

IMPAIRMENT OF EMERGENCY EQUIPMENT

Sometimes emergency equipment needs to be taken out of service for inspection, repair, or modification. If the site is still operating or still containing the materials which might cause a need for the emergency equipment, then additional risks are involved. The normal access to the equipment is not available. There should be an 'impairment procedure' in which the equipment to be taken out of service is described, the risks involved in terms of access to equipment, and any mitigations available.

Often such procedures are not available. It can be, of course, that more general procedures are used by a different name. This is fine. Consider the example of impairment not working in the example below.

Notes from Trevor's Files: Fire Water System Impaired

When being shown around a large terminal storing gasoline, diesel, etc., we approached a facility to load road tankers. Some distance away, I noticed a tag on the fire water supply line to the facility that said that the firewater system was not operational. A valve in the firewater line had been tagged closed. However, when we reached the tanker loading facility, operations were ongoing. The personnel involved did not know that there was an issue with the firewater system. It was later explained that there was a small corrosion hole in one of the firewater pipes, and that if they had needed the system, it would have worked fine, albeit with a leak at the hole. The reader will understand that I was not happy with this explanation. In the event of a fire, who would have known to go and open this remote valve? Would that person also have known opening this tagged valve was OK?

SUMMARY: CAUSES OF FAILURE DUE TO ISSUES WITH EMERGENCY RESPONSE AND READINESS

Reasons why the errors in the development of an Emergency Response Plan and incident scenarios sometimes cause, or at least contribute, to catastrophes include:

- ☒ Plan is considered a 'tick in the box' exercise with unsubstantiated optimism regarding the effectiveness of specific actions.
- ☒ The scope of the plan is too narrow, e.g. may consider fire but not toxic release or medical incidents.
- ☒ Neglecting offsite scenarios resulting from incidents to pipeline, incoming material shipments and product shipments.

☒ Blurred responsibilities between the organisations own emergency response team and that of the local community/civil defence once the local community service arrives and takes over.
☒ Plan too complex, including scientific calculations, etc.
☒ Emergency plan is too simple, and does not include essential data such as required firefighting water and coolant water flows.
☒ Major emergency response plan does not consider domino effect from or on neighbouring installations.

Reasons Why Problems in the Creation of an Emergency Response Team and Their Training Sometimes Cause, or at least Contribute, to Catastrophes Include:

☒ Incident Commander lacks sufficient training or understanding of the technical process or does not receive adequate support or is provided with essential information early in handling the incident.
☒ Emergency response team used as a place to reassign personnel who have proved less than competent elsewhere.

Reasons Why Deficiencies in Emergency Practices and Drills Sometimes Cause, or at least Contribute, to Catastrophes Include:

☒ Drills are carried out only as desktop exercises without any real-time simulation.
☒ Insufficient practice during the year considering the number of scenarios requiring rehearsal.

Reasons Why Errors in the Provision of Fire and Gas Detectors and Alarms Sometimes Cause, or at least Contribute, to Catastrophes Include:

☒ Fire detectors placed incorrectly so that they do not see the fire.
☒ Smoke detectors are too remote from the equipment being protected.
☒ Gas detectors are not placed in the correct position after taking into account whether the gas will rise or sink on release.
☒ Inadequate test frequency.
☒ Troublesome fire and gas alarms become bypassed and forgotten.

Reasons Why Errors in the Deployment of Firefighting Water and Foam Systems Sometimes Cause, or at least Contribute, to Catastrophes Include:

☒ Insufficient firefighting water storage.
☒ Corroded firefighting water main.

☒ Reliance on unproven recycling and reuse of firefighting water in the course of fighting the fire.

Reasons Why Errors in the Set-up of Emergency Response Equipment Sometimes Cause, or at least Contribute, to Catastrophes Include:

☒ Reliance on firewater pumps with only electrical drives, not alternatives such as diesel.
☒ Regular firewater pump run tests are too short leading to pump damage.
☒ Firefighting foam is improperly stored and not retested for quality.
☒ Special equipment such as foam applicators, steam and water curtains isolated by a manual valve – unavailable for remote activation.
☒ Fire suppressant release system not in automatic, either bypassed or in manual.

Chapter 24

The Role of the Safety Department and Incident Investigation

THE SAFETY DEPARTMENT

Safety is the responsibility of the safety department, isn't it? Well … no, not really. Only line management can be responsible for the safety within their department. So then, what is the role of the safety department? Almost every organisation has a safety department. Even in small organisations, this may be just one safety officer. Indeed, the role can be combined with other functions. I suggest that the role of the safety department includes the following:

- Monitoring the health and safety risks and hazards in the workplace, or at least oversee the process for doing so.
- Advising managers and employees on minimising or ultimately avoiding risks and hazards in the workplace.
- Ensuring the business is legally compliant with all health and safety legislation, or at least identifying gaps needing correction.
- Working with and training employees to manage, monitor, and improve the health and safety standards in the workplace.
- Making sure that all safety inspections in the workplace (for example, monitoring noise levels) are carried out correctly. Note that this does not necessarily mean being responsible for carrying them out. There may be a separate Occupational Health department covering, for example, noise measurement, and an Inspection department including highly skilled people who carry out lifting equipment inspections, or indeed such tasks may be outsourced.
- Assisting with the creation and management of health and safety monitoring systems and policies in the workplace and overseeing the health of these systems taking corrective action when necessary.
- Offering general health and safety advice to all employees.

Depending on the organisational structure emergency response, occupational health, and medical first aid may or may not be integrated into the

DOI: 10.1201/9781003360759-27

safety department role. (Ref. Occupational Safety and Health Consultants Register: What does a health and safety manager do?)

Some other references talk about the safety department's role in enforcement, but I don't see that the safety department can enforce things. It has a role in persuading line management when things are wrong and has some strong tools to do so.

Notes from Trevor's Files: Corrupting the Safety Department?

During an international assignment, I was invited to join in the morning calls where each department on an oil refining and petrochemical complex reported on the previous day's activities and production. The call, rightly or wrongly, was always led by the CEO, the overall site manager. There had been an ongoing problem following a leak, and staff were wrestling with ways to get a restart in the shortest possible time. There was just a hint in the conversation that maybe the risk assessment requirements might be getting in the way of the creative temporary arrangements being proposed to fast track the restart of the complex.

The CEO then turned his attention to the HSE Department. 'What is Health and Safety doing to help us out of this crisis?' he demanded. The implication was clear to all on the call. Risk assessments related to the restart were approved with only minor upgrades, and I suspect not the scrutiny that might have taken place without the implicit threat.

I am not a fan of morning calls where all can participate. I am from the old school where turnover was first done from the shift supervisor to the superintendent and then passed up the line to the unit manager and overall site manager. Although this was all done rapidly in a morning, it did allow some degree of thought as to how to pass on the message and pass down feedback to help ensure the correct interpretation of instructions.

The safety department is often perceived as being interested only in the prevention of slips, trips, and falls. Perhaps also it indicates some interest in occupational health issues and fire prevention. Such perceptions arise from audit activity and participation in the safety observation programme. Slips, trips, and falls issues are easily visible. So is the wearing of hearing protection in noisy areas. So too is the accumulation of rubbish which might create a fire hazard. However, most issues that might form part of the causal tree on the way to a catastrophe are not directly in view of the safety officer 'on walkaround'. Occupational safety issues are likely to be more evident and crowd his or her mind.

Unfortunately, the safety department is not widely regarded as a good posting during an aspiring manager's career. In some organisations, it appears to be used as a place for people who have been found wanting when in production or other 'sharp end' roles. Fortunately, this is, in my opinion, now changing for the better. Safety is being better regarded as a career in its own right. More people are entering industry with qualifications in safety and seeing their careers perhaps not necessarily as moving up within an organisation but moving up by moving between organisations achieving higher levels within the safety profession. Disadvantage – safety professionals are seen as not being people who have experienced the real world.

INCIDENT INVESTIGATION

If incident investigation were working well, then one would expect that, from each incident, causes would be found and means to prevent recurrence identified. If means to prevent recurrence were acted upon within the organisation, then at least similar incidents would be prevented. Further, if the incident merited sharing with other organisations with potentially similar issues, most incidents would be eliminated. This is clearly not the case, at least for incidents on a catastrophic scale as discussed in Section 2. However, there is good evidence that at a slips, trips and falls level, incident investigation and other techniques are having a beneficial effect (HSE Statistics – non-fatal injuries at work in Great Britain).

Incident Recognition and Reporting

It may seem strange to need to discuss incident recognition. For occupational safety events, this is relatively straightforward; something happens which did cause harm to a person or might have done.

However, in the context of industrial safety, events which might lead to a catastrophe (but fortunately did not) can be less visible and more challenging to define.

....

Incidental Case Study: BP Texas City

In Chapter 6, the incident at BP Texas City in 2005, where there were 15 fatalities, was discussed. Underlying factors included:

- The distillation tower level indicator showed that the tower level was declining when it was overfilling.
- The pressure relief devices on the column lifted.

- Lack of supervisory oversight and technically trained personnel during the start-up.
- Operator fatigue from working 12 hr shifts for 29 or more consecutive days.
- Occupied trailers were sited too close to process units handling hazardous materials.

Which, if any of these factors, if identified in isolation, would have resulted in an investigation? My experience is that the malfunction of an instrument would not necessarily lead to an investigation, even if it is recognised as safety-critical. We will return to this disappointing state of affairs in the section on Safety Integrity Levels (SIL) in Chapter 27 on Risk Reduction.

Furthermore, a pressure relief device's activation would not always result in an investigation, although this is increasingly becoming recognised as good practice.

The ergonomic and welfare issues and the temporary plant layout issues raised would not be the subject of incident investigation in isolation.

The point being that no one of the above factors, on its own, would have caused an investigation.

'Near miss' reporting has considerable value and is a common feature in the safety management systems of many organisations. It is mainly considered in the context of occupational health injuries rather than in process safety, where catastrophes may be a knock-on effect from the near-miss.

Near misses are typically defined as unplanned events which could have resulted in injury or property damage but didn't. This is sadly all too simplistic. Taken literally, an organisation could be overwhelmed with the investigation of near-miss incidents. My opinion is that near-miss incidents investigation should be limited to investigating those incidents where, if just one critical item were different, injury or property damage would have occurred. There is no consensus on this. Clearly, investigation of all near-miss incidents is an excellent practice, just so long as the organisation has the resources to do this properly.

Notes from Trevor's Files: One of My Gardening Mistakes!

Whilst this would not have resulted in a catastrophe, let me try to illustrate the distinction between near-miss events meriting an investigation. When dismantling an old wooden structure in our garden, I was conscious that some of the roof supports might be rotten. I had successfully removed one roof panel when I heard a loud cracking sound. I immediately moved away without stopping to see the cause. Had I not done so, the heavy roof panels would have fallen on my head, quite likely causing concussion or worse. I would categorise this in a work context as a near-miss meriting investigation. Only my quick reactions at the time saved me from injury.

There are somewhat complex rules regarding the statutory requirements for reporting of incidents. For occupational health injuries, requirements depend on the severity of the injury. Still, since we are considering catastrophes in this book, these need not concern us here other than for multiple fatality situations.

Requirements for reporting releases of toxic and flammable materials are somewhat hard to unravel since there are two interesting statutory bodies – one associated with occupational health and the other with the environment. In the U.K., there is a requirement to report to the HSE (Health and Safety Executive) dangerous occurrences under a regulation known as RIDDOR, which includes:

- The collapse, overturning, or failure of load-bearing parts of lifts and lifting equipment.
- The failure of any closed vessel or of any associated pipework forming part of a pressure system.
- Plant or equipment coming into contact with overhead power lines.
- The accidental release of any substance which could cause injury to any person.
- Electrical incidents causing explosion and fire.
- Incidents with explosives, Biological Agents, Radiation.
- Collapse of scaffolding.
- Train collisions.
- Incidents relating to wells, including blow-out (excluding water wells).
- Pipeline incidents.

(HSE, 2013, Leaflet INDG453 Reporting accidents and incidents at work)

The Environmental Agency and DEFRA (Department for the Environment, Food and Rural Affairs) require a report of any incident with the potential to cause environmental damage.

Note that the requirements in the U.K. are not prescriptive in the sense of defining the amount of released material; a judgement has to be made as to the potential environmental damage.

The regulations for offshore oil spills are more precisely defined – for example, greater than 500 kg of flammable liquid or gas. Offshore spills are dealt with by a different regulation, which is often interpreted as requiring reporting as little as one drop of oil.

Incident Investigation Process

Incident investigation involves discrete steps, such as:

1. Make the initial response and complete an initial report.
2. Form an investigation team.
3. Determine the facts.
4. Determine all contributing factors.
5. Determine systems to be strengthened.
6. Recommend corrective and preventive actions.
7. Document and communicate findings.
8. Follow up.

(Du Pont Sustainable Solutions – Incident Investigations: Getting Started)

An honest, thorough, and systematic approach is needed for effective investigations (Lees, 2014). Well, here is our first problem. No matter how much we try to avoid it, someone or some department end's up feeling that they were to blame. Incident investigation training may help us, rightly, to see incident investigation for its goal of improvement and finding means to prevent a recurrence. But local cultures prevail. Management may want to apportion blame even if they have been trained otherwise. This can happen inadvertently. An investigation that has ten recommendations, all to be actioned by the maintenance department, is a sure indicator as to who will feel the blame!

No one wants to feel blame. In the real world, it happens over all sorts of daily activities. It is a deeply distressing emotion, which can and does result sometimes in psychological illness.

Since we want to avoid the bad feelings that blame, real or imagined, brings, it is natural to try and find ways to avoid situations that might arise. This tendency may be on an individual or a departmental level – 'maintenance messed up on this one again!'. Of course, not all of the blame game is controllable within the incident investigations room. Outside, people are talking, as they do, often based on the flimsiest of evidence. In larger organisations, an incident can start an almost political debate where opinions become immediately polarised in a way that becomes difficult to unravel during the investigation process. Suppose the maintenance department representatives in the investigation all come with a feeling of commitment to protect the maintenance department from blame 'at all costs'. In that case, it won't be easy to establish that there are indeed opportunities to prevent recurrence within the maintenance department.

Notes from Trevor's Files: Incident Investigations and Internal Politics!

My colleagues in safety consultancy will be very disappointed to see me writing about blame. Therefore, I reflect here on an incident investigation concerning an event on a large site, as it happens in Saudi Arabia. Perhaps more significantly, the incident being investigated took place at a joint venture between companies having on the one side an Arabic culture and on the other side a Far Eastern culture. I was leading a team of consultants to improve safety at the location. Incident investigation training had been carried out. On the team, we had one of our best incident investigation coaches.

The incident involved smoke being emitted from a particular piece of equipment. The site had a good policy that all fires be investigated, even when confined to a small amount of smoke. Maintenance work involving some hot work and insulation had recently been completed. People coming to the investigation were polarised in their views. Production people were convinced that there was no process leak. Maintenance personnel were equally convinced that their hot work and insulation work was completed correctly.

There is a certain amount of tolerable time which can be allocated to an incident investigation. After one hour, the investigation team leader announced that the investigation was closed and left the room, although no consensus had been achieved. He published his report a few weeks later, with his own conclusions. It probably did not help that joint venture responsibilities were divided. Production was led by a director with a Far Eastern culture and Maintenance by a director of Arabic culture. Since the investigation leader was Arabic, I think the reader can guess how the investigation was reported!

I don't mean here to be critical of any particular national culture. If roles are reversed, and the site was on Far Eastern soil, I can imagine a reverse outcome.

The excellent incident investigation coach was so frustrated he left the project, and I dared to voice my frustrations during the monthly safety meeting, which included site directors. This distinctly English behaviour did not go down well with the local management, and I was close to being barred from the project. Fortunately, some careful political work by my colleagues allowed me to continue the work.

Creating a blame-free investigation atmosphere is critical to success. However, that does not mean that individuals should not be held accountable.

One of the values of investigating near-miss and minor incidents is that this creates the opportunity for incident investigation skills to be practised. In the author's opinion, a common mistake in extending incident investigation to near-miss or minor incidents is that organisations use the same protocol and investigation team membership. More senior people who should be involved in a more serious incident do not have the time or degree of interest to conduct lesser incident investigations properly. As a young engineer, I felt deeply proud to be given the responsibility of leading investigations into minor incidents.

For very serious incidents, state authorities will be involved. In many countries, they conduct the investigation. I have always encouraged the organisation to carry out its own investigation in parallel where they can. This allows at least interim measures to prevent recurrence to be determined in a more helpful timeline. However, sometimes even the organisation itself does not have access to the same evidence that the authorities have.

In this regard, I must also mention the primary responsibilities immediately after an incident – to render the accident scene safe. After this, for a major incident, the site needs to be secured to assemble key evidence. Rather startlingly, this is sometimes referred to as securing the 'scene of the crime'.

There are many excellent texts and training courses on incident investigation conduct and methods, such as Why-Tree, which we will not attempt to reproduce here, other than emphasising the importance of discovering the root cause. We are trained to ask why multiple times in order to develop a causal tree until we come to a root cause. Root causes determined in this way are often system causes.

For example, a leak of toxic material occurred from a piece of pipework. During the investigation, it was determined that this was due to a leaking flange. Why was the flange leaking – because the mechanic had not properly torqued the bolts. Why was that – because he was moved to another emergency job before this one was fully completed. Why was that – because work scheduling procedures were inadequate.

System causes are often related to procedures, training, and work scheduling issues. We will come back to the challenges of root cause analysis and dealing with root causes later.

Case Study: Space Shuttle Challenger Disaster

All seven crew members of the Space Shuttle Challenger died when the assembly broke apart just 73 seconds into its flight. The Rogers Commission investigated the disaster with many interesting conclusions regarding both the technology and, most importantly, the way of working of the NASA organisation. The latter was found to have significantly contributed to the event. Most notably, the decision to launch was flawed, but communication structures impeded the discussion of critical technical concerns. Plenty is written elsewhere on this. But I would like to highlight a particularly striking comment in the context of safety professionals. 'The Rogers Commission was surprised that NASA's safety staff were never mentioned during many hours of testimony. They also discovered that there were no safety representatives on the mission management team that made key launch decisions. The commission also concluded that the safety team was not sufficiently independent to be able to create the necessary checks and balances in the decision-making system'.

Bahr, N. (2015, p. 74)

There are a number of software aids for incident investigation. On the one hand, such software helps maintain records and discipline. It is also great at tracking recommendations and actions. On the other hand, it can be seen as taking thinking power away. The duty of the investigation committee becomes that of populating the software model and seeing what 'it' says.

There is a particularly new and insightful approach to incident investigation. It is based on the reasonable assumption that, in most cases, the parties involved in the incident – operators, engineers, managers, CEOs, etc. were doing what they thought was right. The question then is to find out how people's assessment and actions made sense at the time, given the circumstances that surrounded the event (for more on this approach see Kernick (2022): Catastrophe and Systemic Change – Learnings for the Grenfell Tower Fire and Other Disasters). I don't believe any organisation plans to have disasters or knowingly takes great risks. However, the approach proposed requires an organisation prepared for considerable introspection. It is likely that such an incident investigation would take considerably longer than the traditional approaches, and that the systemic changes that might be proposed as recommendations to prevent recurrence might be difficult to implement. Having endorsed this approach, I have not seen any actual incident investigations carried out in this manner. In Chapter 42 on Improvements to Risk Management methodology I will attempt to do so for an incident of particular personal interest.

SUMMARY: CAUSES OF FAILURE IN THE ROLE AND ACTIONS OF THE SAFETY DEPARTMENT, AND OF DEFICIENCIES IN INCIDENT INVESTIGATIONS

Reasons Why the Activities of the Safety Department Can Fail to Prevent Catastrophes Include:

☒ Real or perceived pressure from more powerful parts of the organisational hierarchy can influence the actions of the safety department.

☒ The safety department is sometimes used as a dumping ground for those who cannot make it in other parts of the organisation.

☒ The safety department is sometimes inappropriately perceived as a department with personnel who were unable to make it elsewhere.

Reasons Why Deficiencies in Incident Investigation Can Fail to Prevent Catastrophes Prevention Include:

☒ The incident investigation is preoccupied with finding someone to blame or placing blame on an individual department.

☒ There is little value for investigating 'near miss' investigations where no injury occurred.

☒ Conversely, too many 'near miss' events are put forward for investigation, to the extent that the organisation is overloaded with investigations.

☒ The investigation comes to the wrong conclusion under pressure from more powerful members of the hierarchy.

☒ The skills of the safety department are not valued in facilitation or guiding the incident investigation process.

Chapter 25

Risk Assessment Basics and Preparation

HAZARD IDENTIFICATION

In the context of catastrophes in the process industries, we usually first think of chemical hazards – the chemical's physical, chemical, and toxic properties. It is essential also to consider any biological and physical risks such as electricity, radiation, and high pressures. Although not likely to result in a catastrophe, other hazards such as ergonomic and basic safety such as slipping/tripping hazards and machinery guarding are also considered in this step. Hazard identification is the first step in evaluating the risks of a current or future facility. For some reason, in some organisations, this seems to create undue difficulty.

The methods of Hazard Identification and Risk Assessment are of course not restricted to the process industries. They are equally applicable to transportation, etc. ... Indeed, they are applicable to any task where there are hazards. The complexity may vary according to the task. The task of cleaning windows should be subject to a risk assessment, although clearly not as complex a risk assessment on flying a jet airliner! The application of risk assessment in this book will focus on its application to processes. Very similar processes apply to other industries such as aeronautical, rail and space! Bahr's (2015) System Safety Engineering and Risk Assessment gives excellent descriptions of these methods applied in a variety of industries.

In Section 2, a number of incidents were reviewed where the risk assessment was clearly flawed or had not been carried out at all. Bhopal, the Shoreham air crash, and Chornobyl are most notable. However it is likely that there are deficiencies in risk assessment involved in every incident. Therefore, this book dedicates several chapters to the subject:

- Chapter 25: Risk Assessment Basics and Preparation
- Chapter 26: Risk Assessment Techniques, HAZOP, Fault Tree, and other methods
- Chapter 27: Risk Assessment Outcomes, Layers of Protection Analysis, and Safety Integrity Levels

DOI: 10.1201/9781003360759-28

- Chapter 28: Consequence Analysis and Consequence Mitigation
- Chapter 29: Software Safety and Cyber Security
- Chapter 30: Risk Assessment Conclusions and Risk Reduction

'The hazard identification study should be creative and dynamic' (Wise, 2015). While I have found in the U.K. and most of Europe that this is largely true, there is either a lack of knowledge or confidence to do this in Middle Eastern and former countries of the U.S.S.R. I doubt that this has to do with education. Instead, it comes from the need to get their instruction from more senior people. And, of course, more senior people are less accessible. In the former U.S.S.R, these countries insist that such information comes from state-approved design houses. This takes time and money, and it is not always clear if the design houses, some of which have a disproportionate number of academics, really have the competence to carry out this task properly.

In most cases, the hazards can be promptly listed using a simple hazard checklist. However, it is imperative to look for additional risks that might not be readily apparent to the people creating the list. For chemical risks it is not necessary to get an MSDS from the actual supplier you intend to use! Many are MSDSs are readily available on the internet. A chemical is a chemical, although there may be some minor differences depending on impurities.

An MSDS will show amongst other data:

- Product identification.
- Relevant uses.
- Hazard Classification and precautionary statements.
- First-aid measures.
- Acute Toxicity – the ability to cause harm to the body through absorption, ingestion, or inhalation through a single exposure.
- Chronic toxicity – the ability to cause harm to the body through absorption, ingestion, or inhalation through repeated exposures.
- Firefighting measures.
- Accidental release measures.
- Environmental precautions.
- Safe handling.
- Occupational exposure limits.
- Exposure controls.
- Physical and Chemical Properties, including flammability – flash point and autoignition temperature.
- Stability and Reactivity.
- Ecological information.
- Disposal considerations.
- Transport information.
- Regulatory information.

This is a lot of data. As an example, the sheet for ethyl alcohol runs to 11 pages. There are MSDSs for water running to six pages. Interestingly an MSDS for *water* includes the following statements:

- Precautions for safe handling – Wash hands and other exposed areas with mild soap and *water* before eating, drinking, or smoking and when leaving work!
- Fire hazard – not flammable
- Firefighting instructions – use a *water* spray or fog for cooling exposed containers!

Strangely, carcinogenic and toxic materials such as benzene and nicotine are also of six pages duration.

Notes from Trevor's Files: Unravelling Material Safety Data Sheets

During my early days as a process engineer, we had a practice at a particular site of producing chemical information sheets, which were handy summaries of the hazards of a specific material. In most organisations, there are extensive books of Material Safety Data Sheets.

I am a proponent of assembling the shorter chemical information sheets with data relevant only to the specific hazards relating to the usage of the material in the process.

The MSDS is, of course, the basis for such chemical information sheets, but with all the unnecessary information stripped out. Line items such as 'no data available' and 'not classified' and irrelevant physical data also need not be replicated. Further, the chemical information sheet should include explanations suitable for all employees to understand.

In the U.K., employers are required to carry out a COSHH assessment.

COSHH stands for Control of Substances Hazardous to Health. This requires risk assessments of the operations where the hazardous materials are used. These risk assessments are usually more informative and easier to read than the MSDSs. Further, an MSDS can cause unwarranted concern, for example, if the word carcinogenicity is mentioned. The COSHH assessment puts the hazard into the context of the work to be performed and the resulting risk.

Hazards may not be immediately apparent to those responsible. For example, at the Beirut warehouse where the massive explosion occurred (see Chapter 13). Here the ammonium nitrate was considered a fertilizer. Its explosion potential, especially the risks from contaminated material, was not appreciated.

A particular example of hazards, which may not be immediately apparent, concerns the chemical compatibility of materials used in the process. Du Pont Sustainable Solutions (now dss+) recommends creating a matrix in which all the chemicals used in the plant are listed in columns both across and down the matrix. The consequences of each material being mixed are entered in each box of the matrix. It appears that often this is sometimes conceived as a problematic and unnecessary piece of work for two reasons:

1. The reactivity of the mixture is not known – This seems to me highly unlikely. Most chemists will know this even if you can't find it on Google – call them! If this does not work, get access to Bretherick's Handbook of Reactive Chemical Hazards. If it is still not known – show this on the matrix. Tests are available such as accelerating rate calorimetry. The importance of not to the risk assessment will become more apparent later.
2. There is no credible chance of the two materials mixing – Always challenge this assertion. Some remarkable unplanned mixings with catastrophic results have occurred.

Chemical hazards may only result if two materials are mixed, and this mixing may not be part of the normal process.

Incidental Case Study: Unplanned Mixing

An incident in 2016 due to the unplanned mixing of sulphuric acid and sodium hypochlorite has been discussed in Chapter 12. The incident occurred when the material was inadvertently unloaded into the wrong tank. This case study by the CSB includes an incompatibility chart showing that of the five chemicals unloaded at this facility, all but 2 of the 20 possible combinations were incompatible and could result in severe consequences if mixed.

Further research showed that other incidents of chemicals being unloaded into the wrong tank had occurred on multiple occasions. The consequences, of course, depend on the substances, but the potential for this error needs to be appreciated.

(U.S. Chemical Safety and Hazard Investigation Board (2018) Key Lessons for Preventing Inadvertent Mixing During Chemical Unloading Operations – MGPI Processing)

In another incident in 1996, a chemical believed to be epichlorohydrin began off-loading at the Albright and Wilson site in Avonmouth, U.K. Shortly afterwards, a series of explosions destroyed both the storage tank and the road tanker. Later, it was discovered that the material actually discharged contained sodium chlorite, which reacts explosively with epichlorohydrin. This was again due to a communications error.

In carrying out hazard identification, problems can arise due to lack of knowledge but also lack of will.

Bearers of bad news are rarely well appreciated at the time. Unearthing some hazards that had not been previously fully appreciated can sometimes be a brave step for a young engineer.

Rarely are established management open to hearing about a previously unrecognised risk.

Notes from Trevor's Files: Unforeseen Risk – Unwelcome News

During my early days as a safety consultant working with oil exploration and production companies in the North Sea, I established good relations with a company with several installations offshore. In addition to my regular coaching in behavioural safety, I was asked if I could review the safety of a project approaching managerial approval. Of course, I agreed to this exciting assignment. The project involved an established offshore oil platform that was currently extracting oil from a sweet (i.e. low hydrogen sulphide) reservoir. Production from the sweet reservoir was in decline. In order to extend the life of the platform, it was proposed to start extracting from an adjacent field and doing a tie-back from this adjacent field to the existing platform using subsea wellheads and manifolds installed on the seabed.

However, there was one big obstacle. The adjacent field was sour (i.e. the extracted oil contained a significant amount of hydrogen sulphide). The existing platform had been engineered for sweet oil and gas, not sour. The proposed solution was to inject formaldehyde at the sub-sea wellheads in sufficient quantities to scavenge out the hydrogen sulphide before arriving at the platform. Substantial experimental work, including a full-size prototype built onshore, had been carried out. The objective of the experimental work was to verify that the formaldehyde could knock out the hydrogen sulphide to the levels expected.

The organisation was, therefore, already mentally committed to this project. As well as being economically attractive, the project would also extend platform life and thus provide more extended job security for the organisation's employees.

When reviewing the detailed project documentation, which included process hazards reviews, unfortunately, an aspect that had not been considered. A simple mass balance consideration should lead even the non-technical reader to wonder what happens to the formaldehyde and hydrogen sulphide. They are no longer in the oil! The answer is that they form reaction products that then appear in the water phase. This water phase is separated out on the platform and discharged to the sea. I needed to understand what chemicals were released due to the formaldehyde/hydrogen sulphide reaction. Had any studies been done of their toxicity to the marine environment? As I came to understand, the reaction products can be quite complex. My understanding is that the project was later abandoned.

I was not invited to continue my work with the organisation. The offshore world has a wonderful abbreviation for unwelcome workers – NRB – Not Required Back!

INTRODUCTION TO RISK ASSESSMENT

Risk assessment, risk assessment, risk assessment … Wherever and whatever nowadays it seems, needs a risk assessment. During the pandemic, even going to church required a risk assessment. Taking a group of scouts on a mountain trip requires a risk assessment. My surgeon needed to do a risk assessment before carrying out a surgical procedure.

Whatever type of risk assessment, the risk assessment is typically described as needing:

1. Identification of the hazards.
2. Determining who might be harmed and how.
3. Evaluation of the risks and deciding on the precautions.
4. Recording the findings of the risk assessment and implementing its conclusions.
5. Reviewing and updating the assessment regularly.
   (Wise, 2015)

In my opinion, this five-step process misses one vital step, which is perhaps implied. After the risk assessment is completed and any recommendations implemented, who decides that the risk is now acceptable? This decision-making is often absent as a discrete and recorded step.

In the U.K. there is a legal requirement that a risk assessment should be 'suitable and sufficient', a requirement which is echoed in the legislation of many other countries with similar intent.

In the context of this book on catastrophes, we are interested only in risk assessments that might prevent disasters. In the hazard identification work

Table 25.1 Time Taken to Carry out an Assessment

	What-if/Checklist	HAZOP
Simple, small systems	12–25 hrs	6–12 days
Complex/large systems	2–4 weeks	3–12 weeks

Source: Bahr (2015, p. 263).

above, we will have identified some kind of catastrophic potential. So how to carry out a risk assessment?

One source says that over 100 different safety analysis methodologies have been documented (Bahr, 2015, p. 253). In my opinion, the shear number of techniques presents a significant problem. How do I choose the appropriate method? How many techniques do I need to train my people to use each of them competently? There is a broad spectrum of types of risk assessment to be carried out, and the methodology selected should reflect this.

There are many excellent texts on selecting and conducting risk assessments using different assessment methodologies e.g. (Bahr, 2017, p. 261). Marvin Rausand, 2011, in Risk Assessment: Theory, Methods, and Applications, dedicates an entire book of over 600 pages to the subject. Thankfully this text includes against each technique a list of the main advantages and the main limitations. Table 25.1 gives an estimate of the time that it would take to carry out an assessment using each technique:

Whatever technique, risk assessment requires a significant amount of time and resources.

PREPARING FOR THE RISK ASSESSMENT

Before carrying out a Risk Assessment, several preparatory steps are necessary, and these steps are, for the most part, required irrespective of the Risk Assessment Method selected.

1. Assembling of information that will be required in the assessment. In the context of the process industry, this would include the items shown in Table 25.2. Difficulties typically found in collecting and using this equipment are also listed in this table.

   In addition to the above 10 example items, I have seen lists of preparatory information totalling 28 types to be collected before the risk analysis can start.
2. Putting together the risk assessment team and resource planning.
3. Chartering the team.

Table 25.2 Information to Be Prepared before the Risk Assessment

Information Required	Purpose	Difficulties Typically Encountered
Any previous risk assessments	To gain efficiency with the benefit of previous work.	The previous risk assessment might have been at the design stage and not include some of the additions or changes made in construction and commissioning. Previous studies may not have had the same scope or purpose.
Plot Plan of location of process to be studied	To assess the distance between process equipment of particular value in determining the consequences of an event especially fire in affecting adjacent equipment.	Plot plans are typically not updated with modifications unless the modification is significant. So a new pump may not show, although a new distillation column would almost certainly show.
Piping and Instrumentation Diagrams	To define each node of study on which to carry out the failure mode questions – guidewords if using HAZOP.	Are the Piping and Instrumentation Diagrams (P&IDs) up to date? Have the drawings been updated with all modifications made, such as additional instruments, drain valves, etc., often added during final build and commissioning?
Operating Instructions	To evaluate what could go wrong when the procedure is followed and what might go wrong if the procedure is not followed.	Although this is probably the most straightforward information to collect, there can be issues in getting the most up-to-date instruction. Sometimes conflicting instructions are uncovered, a valuable finding in itself.
Emergency Shutdown Instructions	To evaluate failure modes during unplanned shutdowns.	Sometimes these don't exist. Should be separated for various causes of shutdown or partial shutdown, including loss of steam, loss of electricity, etc.
Design Data of Safety Valves	Primarily to know the scenario they were designed to protect against e.g. fire engulfment of the vessel.	Sometimes this data cannot be found, especially for an older plant. Or calculations show volumes, but the scenario on which they are based is not evident.
Design data of the equipment. Including, for example, pump curves	For example, pump curves – to determine how high might pump material or at what flow rate.	Design data is a very broad description. A lot of work can result in assembling data which is ultimately not used by the risk assessment team.

and hold-up of hazardous materials	Hold-up of hazardous materials to estimate the maximum potential extent of a spill of material released from the equipment in an uncontrolled manner.	For an older facility, it is also often the case that the original data is no longer available.
Logic descriptions and logic diagrams or safety instrumented systems, together with Safety Integrity Level evaluations	To enable the study of the way automated systems react in certain scenarios.	Logic diagrams are sometimes not documented with written descriptions of their intents. Interpretation of the software takes considerable time and skill. Where software is documented, it is sometimes the case that the documented intent does not match the software as written.
Information about incidents in similar processes. This would include incidents and near misses on the unit itself plus other relevant incidents within the organisation or globally.	To facilitate learning from other events. To contribute to the discussion about the probability and consequence of a particular scenario.	'Similar processes' can be interpreted in a broad or narrow way. Many relevant case studies occur in distinctly different facilities. Often insufficient detail is available in the public domain to be valuable.
Chemical Reaction matrix (see Risk Identification above)	To enable the assessment of consequences of inadvertent mixing of two or more incompatible materials.	Requires some work to assemble, work that is sometimes rejected by those tasked with completing on the basis that, in their mind, mixing of these chemicals is not feasible (but it is not their place to judge).

Typically, there are issues getting both the people with the right mixture of skills and knowledge available, trained and motivated to carry out the risk assessment. This is usually additional work to someone's 'day job'.

Issues arise when information is not available. In older plants, the need to preserve some of this information may not have been as evident when the plant was designed as it is today. Original equipment manuals on, say, rotating equipment can be hard to locate. If some information cannot be found, does that mean the risk assessment cannot be carried out? I argue that the team should press ahead with the risk assessment, making conservative assumptions concerning issues related to the missing information. Others may say that the assessment needs to be deferred until all the information is collated.

BOW TIE

Bow Tie is an excellent method of communicating process safety information to staff, contractors, regulators, senior management, the public, and other stakeholders (CCPS, 2018, 'Bow Ties in Risk Management – A Concept Book for Process Safety').

It is not a risk assessment method itself, but in this author's opinion, an excellent way of preparing for a risk assessment, whether it be by HAZOP or many other available techniques. We introduced Bow Tie methodology in Chapter 1 and have used Bow-Tie diagrams throughout Section 2 to demonstrate some of the barrier failures, which enabled a number of the catastrophes described to occur.

Some of the issues with HAZOP and other risk assessment techniques occur since they get immediately involved in detail. Participants can be 'unable to see the wood for the trees'. Bow Tie, I recommend, as an introduction to a risk assessment such as HAZOP to help avoid this blindness.

However, there are dangers with Bow Tie. Central to the Bow Tie is the existence of a Top Event. Any particular hazardous facility may have several top events. Whereas HAZOP, if done well, will 'spy out' all top events, Bow Tie assumes, not unreasonably, that the Top Events will be apparent 'by inspection'. However, one single Bow Tie is unlikely to characterise any one facility – it takes several (Figure 25.1).

*For process safety, read catastrophe prevention or major accident prevention methods for non-process industries such as transportation.

I suggest the best way to develop an understanding of Bow Tie Diagrams is through the creation of some diagrams relating to our hypothetical cases regarding the whisky distillery and the fuels terminal.

CCPS (2018) 'Bow Ties in Risk Management – A Concept Book for Process Safety' contains excellent explanations as to how to construct Bow Ties and gives examples of Well-Worded and Poorly-Worded Hazards, Top

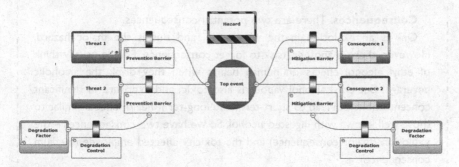

Figure 25.1 Generic Bow Tie Diagram.

Created using BowTieXP Software © CGE Risk Management Solutions B.V. 2021. BowTieXP is a registered trademark of CGE Risk Management Solutions B.V.

Events, Threats, etc. which I will not attempt to replicate here other than to commend this book to the interested reader.

At first, I started drawing Bow Ties using regular drawing software. But this is tiresome and can lead to inconsistencies. Throughout this book, I have used BowTie XP from CGE Risk Management Solutions to construct Bow Ties. Whilst this or other Bow Tie drawing software is not essential, those who draw Bow Tie diagrams regularly will benefit from its use. The software also has many other features, such as the ability to link to risk assessments and is mainly self-documenting.

Illustrative Hypothetical Case – Whisky Distillery – Application of Bow Tie

Two severe consequences of incidents in the distillery should be apparent from a review of the hazards see Chapter 2.

- Fire in the Kiln area of the distillery.
- Fire/Explosion in the Warehouse.

Let's prepare a Bow Tie for the Fire/Explosion in the warehouse.

Hazard. First what is the Hazard? The ethanol (ethyl alcohol) is the hazard. It itself is not a problem. The problem is that the ethyl alcohol permeates the oak cask. This is a deliberate part of the whisky maturation process and is not a mistake! So, this is set as the hazard.

Top Event. The top event here is not the explosion. The top event should not be the consequence. It is an event in which control of the hazard is lost. So, the top event of interest is the accumulation of ethanol vapours in the warehouse.

Consequences. There are two potential consequences.

One is an explosion in the warehouse, and this is the major hazard. However, there is another, easy to forget consequence. People usually think of ethyl alcohol effects on human beings when they drink the alcoholic beverage. But ethyl alcohol vapour is also toxic, and exposure in significant concentrations can lead to short-term and long-term health effects similar to those well known with ingested alcohol. So we have two consequences – the explosion (major consequence) and the toxicity affected employee (medium consequence).

Threats. Identifying the threats is more challenging, and it is essential to distinguish between threats and barriers. Bow tie experts advise selecting threats that have a direct causation and should be specific. In this case, I have chosen the threat to be stagnant air in the warehouse since this will directly lead to the top event, given that the alcohol being lost from the barrel is constant and predictable. Another threat is that the warehouse becomes overloaded beyond design capacity, and therefore the ethyl alcohol emitted is beyond that which normal air extraction can manage.

So far, the Bow Tie looks as in Figure 25.2.

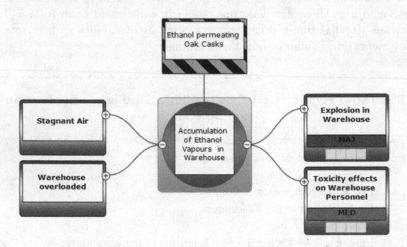

Figure 25.2 Bow Time of Whisky Warehouse Showing Threats, Top Event, and Consequences.

Created using BowTieXP Software © CGE Risk Management Solutions B.V. 2021. BowTieXP is a registered trademark of CGE Risk Management Solutions B.V.

There are several barriers that prevent the threats from becoming a top event:

PREVENTION BARRIER

A barrier to loss of air extraction might be ventilation systems, either forced or natural draught. A barrier to Warehouse overload would be warehouse stocking procedures.

Barriers to an explosion could include an explosimeter alarm or periodic gas analysis.

Prevention of toxic effects could include monitoring air in the warehouse or other types of occupational health monitoring.

The Bow Tie now looks as follows with the barriers added (Figure 25.3).

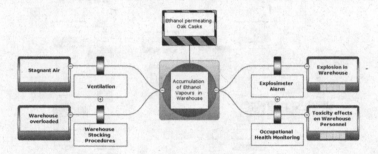

Figure 25.3 Bow Tie of Whisky Warehouse with Barriers Shown.

Created using BowTieXP Software © CGE Risk Management Solutions B.V. 2021. BowTieXP is a registered trademark of CGE Risk Management Solutions B.V.

There are some conditions that could affect the effectiveness of the barriers identified.

For example, the ventilation could be inadequate (the degradation factor), and this can be controlled by features such as ventilator alarms, ventilator maintenance, and the conducting of airflow checks (degradation controls). To the right-hand side of the bow tie there may be inadequate alarm maintenance with the control of alarm checks (Figure 25.4).

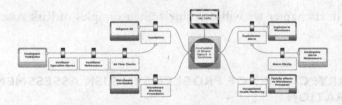

Figure 25.4 Bow Tie of Whisky Warehouse with Barriers and Degradation Factors and Degradation Controls.

Created using BowTieP software © CGE Risk Management Solutions B.V. 2021. BowTieXP is a registered trademark of CGE Risk Management Solutions B.V.

It is very useful to annotate the final diagram, which can be achieved neatly with Bow Tie software such as BowTieXP. I have chosen to annotate this diagram with the following aspects of the barriers and degradation controls:

- Effectiveness (Very Good to Very Poor)
- Barrier type (Active hardware, Behavioural, etc.)
- Brf code (Maintenance Management, Operating Procedures, etc.)
- Accountable (Operations, Maintenance, HSE, etc.)

So the final Bow Tie for this top event looks like in Figure 25.5.

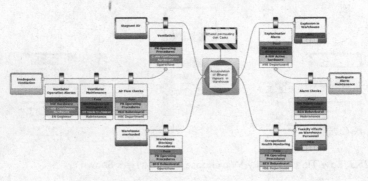

Figure 25.5 Bow Tie of Whisky Warehouse with Barriers and Degradation Controls Annotated.

Created using BowTieXP software © CGE Risk Management Solutions B.V. 2021. BowTieXP is a registered trademark of CGE Risk Management Solutions B.V.

Using Bow Tie the risk assessment team now has an excellent understanding in preparation for the risk assessment.

In the next chapter we will examine some examples of Risk Assessment methods.

SUMMARY: CAUSES OF PROBLEMS IN RISK ASSESSMENT PREPARATION

Reasons Why the Risk Identification Step Sometimes Fails to Prevent Catastrophes Include:

☒ People completing the Identification lack confidence in identifying the risks, making this step unduly long.

☒ Data overload results, hard to see what is important and what is not important. Material Safety Data Sheets are a good example.

☒ Chemical Reactivity Matrix not completed, either because those carrying out the Risk Identification jump ahead, and see no opportunity for the materials to mix, or lack of confidence that they can come to a conclusion without consulting a Chemist – use Bretherick!

☒ The outcome of the Risk Identification may be already giving 'unwelcome news'; pressure is conceived from the organisation's hierarchy.

Reasons Why the Risk Assessment Preparation Step Sometimes Fails to Prevent Catastrophes Include:

☒ Overload of information to be studied.

☒ Choice or risk assessment method inappropriate to the type of facility to be assessed.

☒ Choosing method unnecessarily complex for the operation to be studied – Consider What-If/Checklist first!

☒ Availability of team members with the right skills and knowledge.

☒ Training of team members in the technique selected.

☒ Motivation of team members for whom the risk assessment is conceived as additional work.

Chapter 26

Risk Assessment Techniques – HAZOP, Fault Tree, Event Tree, and Other Methods

SELECTING A RISK ASSESSMENT METHOD

A risk assessment method needs to be chosen. Often organisations have specific guidance on this. A simple five-step process is typically used for occupational safety type risk assessments, but this is not the type of risk assessment intended to prevent the catastrophes that are the subject of this book.

In the context of the processing industry, including oil and gas, chemicals, and the nuclear industry, HAZOP is the technique of choice. Insurance industry risk engineers always look to see the HAZOP. Before we get into HAZOP and its limitations, let me dwell for a moment on life before HAZOP. Although first developed in 1963 by ICI, HAZOP has only become the method of choice throughout the process industry in the last two decades.

There are many other methods of Risk Assessment. HAZOP is particularly suited for the process industries but not so well suited for other industries such as air, rail, and ship transportation. Failure Mode and Effect or other techniques may be more appropriate in these industries. However, these different techniques suffer from many of the same pitfalls as HAZOP. So, in this chapter we will focus on HAZOP as an example of the types of techniques which might be used.

Before HAZOP, I recall using Checklist methods and What-If. Bahr (2017, p. 263) lists amongst the strengths of What-If/Checklist:

- Very inexpensive
- Does not need a lot of data
- User friendly
- Can be used quickly and easily

Why use anything else, other than a What-If/Checklist? Well, the same reference lists the weaknesses of What-If/Checklist as:

- Not as complete as other analyses
- Not good for complex systems
- Not good for identifying interdependency

Once I have described below the mess which some organisations get into over HAZOP and other techniques, it is worth reflecting on whether What-If/Checklist would have been a better approach. My own view is that a What-If/Checklist should precede a HAZOP or other technique since it can help see the big picture before becoming drowned in the detail. I would also argue that for some simpler processes, a good What-If is all that is appropriate, recognising resource constraints that all organisations suffer.

HAZOP does not, in general, deal well with mechanical integrity issues – what if a pipe corrodes and leaks hazardous material. Mechanical integrity issues are not brought out well by typical HAZOP guidewords.

HAZOP

Most, if not all, of the techniques are recommended to be done by a team of people. The team should be multi-disciplinary. A team who, among its members, have expert knowledge of every area of the process plant and its operations. For HAZOP, a group of between five and eight people is necessary, including the following (Wise, 2015, p. 27):

- Design Engineer
- Process Engineer
- Electrical Engineer
- Instrument Engineer
- Operations Manager
- Maintenance and/or Inspection Engineer

The team should also include a facilitator who has experience in Hazard and Operability Studies. Many organisations also specify that the facilitator is not directly involved in the design or operation of the chemical plant. A scribe who is very familiar with HAZOP work is also a valuable addition. The scribe records all the activities of the HAZOP.

A large plant is normally broken down into particular process blocks to improve the focus of the work. The team requires between 3 and 12 weeks of work for each process block.

Essential though the teamwork is to the successful outcome of the HAZOP, it raises some considerable difficulties:

- All of the above highly skilled personnel who are an essential part of the plant's daily operation must be released for the HAZOP.
- Success depends on the effectiveness of the chairperson.
- Success also depends on the knowledge and experience of each team member.

There is considerable 'dead time' when some specialist team members. They may not be involved in the discussion for significant periods. For example, the electrical engineer may not be involved whilst process issues are discussed.

The team output is vulnerable to influence from human behaviour issues such as risky shift and groupthink (see Chapter 20 on Human Factors).

Let's delve into the workings of HAZOP in just a bit more detail. HAZOP is a systematic hazard identification process to explore how the system, or a plant may deviate from the design intent and create hazards and operability problems. The analysis is done in a series of meetings as a guided brainstorming based on a set of guidewords.

Examples of guidewords are:

- No/None
- More of
- Less of
- Reverse
- Before
- After
- Etc. …

So, the team asks itself questions such as:

1. Could there be 'no flow'?
2. If so, how could this happen?
3. What are the consequences of 'no flow'?
4. Are the consequences hazardous, or do they prevent efficient operation?
5. Can 'no flow' be prevented by changing the design or operational procedures?
6. Can the consequences of 'no flow' be prevented by changing the design or operational procedure?
7. Does the severity of the hazard or problem justify the extra expense?

Now imagine the team constructed as above, and let's examine which of the above questions the team can be reasonably expected to answer.

Taking questions 1 and 2:

1. Could there be 'no flow'?
2. If so, how could this happen?

These are reasonable questions to ask the team. They might respond with pluggage, closed valve, pump failure, etc. …

Taking questions 3 and 4:

3. What are the consequences of 'no flow'?
4. Are the consequences hazardous, or do they prevent efficient operation?

Does the team have the skills to reply to this question? In qualitative terms, they might be able to respond, for example, that overheating might occur. They might even be able to speculate that an explosion might result, but not be able to estimate the size of the explosion, the financial impact, or the number of potential fatalities or estimate the off-site impact.

Additionally, the team is expected to identify if this condition prevents efficient operation.

Taking questions 5 and 6:

5. Can 'no flow' be prevented by changing the design or operational procedures?
6. Can the consequences be prevented by changing the design or operational procedures?

I suggest that the most a HAZOP team can do is to make a few helpful suggestions. In the 'no flow' case, instrumentation might be added to detect 'no flow' and to shut down the process if this is detected automatically. But there might be several potential solutions.

Taking question 7:

7. Does the severity of the hazard or problem justify the extra expense?

I suggest that the team is not equipped to determine if the severity of the hazard or problem justifies the additional expense. We will review later when considering how to resolve this in a systematic way (see Chapter 27).

Illustrative Hypothetical Case – Whisky Distillery – HAZOP Extract

Let's consider again our Whisky Distillery (see Chapter 1 – Introduction and Appendix A), where the peat used to heat the barley has been replaced by a natural gas-fired burner for environmental reasons.

The construction is quite simple and is shown in Figure 26.1.

The peat floor is no longer in use. A simple natural gas burner has been installed in its place. An actual installation would be much more complex than shown here, with a commercially available burner management system. However, for this operating procedure example, let's keep it really simple.

In a HAZOP, we would progress through all the guidewords – which would cause a HAZOP team to consider that there was a significant risk.

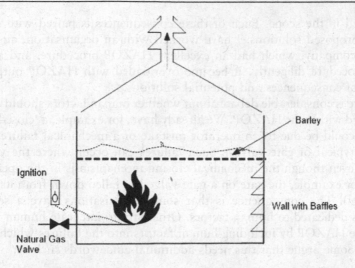

Figure 26.1 Whisky Kiln Ignition.

The guideword 'No flow' in the context of the gas supply is a fail-safe result. However, using the guidewords 'more of' and 'less of' should give rise to concern about explosive mixtures of gas forming within the kiln before the ignition button is pressed. The guideword 'As well as' might also raise concerns about repeat ignition attempts leading to a backlog of gas in the kiln, potentially forming an explosive mixture.

So, the HAZOP team might suggest procedures to ensure that the ignition is carried out at the same time as the opening of the natural gas valve. Or they might suggest that pressing the ignition button opens the natural gas valve. Or even better, they might suggest that pressing the ignition button causes the natural gas valve to open, and after a predetermined time, ignition occurs. Additionally, they might suggest that a temperature detector, and/or a flame scanner be installed within the kiln to detect if ignition has occurred so that if there is a failure to ignite, gas does not continue to flow into the kiln, etc. Just from one relatively simple part of the process, we have generated many improvements. Which ones are justified, and which are not?

From this short immersion into the detail of one kind of risk assessment technique, we can see that it doesn't quite work in the way foreseen initially. After a promising start, we get to questions that the team is ill-equipped to answer, requiring specialist knowledge and input. Further, the team is then asked to make judgements about affordability. Is this an appropriate question for a HAZOP team? I suggest not.

In practice, the team comes up with literally hundreds of potential significant consequences, especially considering that operational efficiency is included in the scope. Each of these consequences is paired with one or more proposed solutions. I have worked with an organisation, an oil refining company, which had an excellent HAZOP procedure, and applied this procedure diligently. It became overloaded with HAZOP outputs in terms of consequences and potential solutions.

There is considerable debate about whether human factors should also be included within the HAZOP. We already have, for example, a 'closed valve' which could be due to an operator mistake or a mechanical failure of the valve (typical of gate valves). Such failures can occur where the valve is closed even though the mechanical closing mechanism is in the open position (for example, the gate on a gate valve has fallen down from its raised position). The consequence is that some organisations have a separate process dedicated to human factors. Others try to integrate human factors into the HAZOP by including human factors into the failures which might occur. Some argue that this needs additional guidewords such as:

- Omission – not done
- Too little or too short
- Too early or too late
- Wrong order
- Etc.

Others argue that these extra guidewords are unnecessary and implicit in the regular HAZOP guidewords above.

So, we have a picture of complexity and potential confusion, with some arguing this, and some arguing that, about the HAZOP process. There are career safety engineers who have been dedicated to HAZOP and who have strong and often contradicting views on how HAZOP should be conducted.

For example, extracting from one textbook, the following overall steps are recommended for performing a HAZOP (Rausand, 2011, p. 252):

1. Plan and prepare.
2. Identify possible deviations.
3. Determine causes of deviations.
4. Determine consequences of deviations.
5. Identify existing barriers/safeguards.
6. Assess risk.
7. Propose improvements.
8. Report the analysis.

The same textbook then comments – regarding Step 6 on assessing risk, that this step is not part of all HAZOP studies. No alternative approach is offered. Other experts on state emphatically that 'When assessing the consequences, it is important not to take any credit for protective systems or instruments already included in the design'. They do not explain why (see, for example, Rockwell Automation, 2013, Process Safebook 1: Functional Safety in the Process Industry – Principles, standards and implementation). I find the resulting assessment of risk somewhat confused. If there is a column on the HAZOP worksheet marked 'Risk' or RPN number (see below), is this before or after consideration of existing safeguards?

I prefer an approach that shows:

- Probability, consequence, and risk with no safeguards
- Probability, consequence, and risk with existing safeguards
- Probability, consequence, and risk after proposed additional safeguards

This approach would considerably help clarity, but it is not practised.

There is a column marked risk in almost all examples of HAZOP worksheets that I reviewed in the preparation of this book. This column is subdivided into Frequency (or probability), Severity (or consequence), and a Risk Priority Number or similar. The risk priority number may also be expressed as a critical risk, very high risk, high risk, elevated risk, low risk, and very low risk. Categorising the risk in this way greatly assists in prioritising output from HAZOPs, but it does not reduce that output.

With organisations who are struggling with the sheer volume of output, I consider this itself to be an unsafe situation and have proposed:

1. Considering only HAZOP output relating to safety only, not operability. This is considered to run contrary to HAZOP principles – I argue that it can be essential if top priorities are to be identified.
2. Considering only risks that are of 'high risk' or greater. Elevated risks can be 'stored' until the higher risks are dealt with. Some also consider this to not be in line with best practices. I argue that a degree of practicality is needed in an organisation with many risks classified high or greater.

There is a further element of complexity and workload to be considered. A HAZOP should consider not just normal operation but also the processes of start-up, shutdown, and other abnormal modes such as loss of steam, electricity, etc. ... This is huge work that is, in my opinion, rarely done to completion. My experience is that the team's energy level declines as the work progresses. These additional operating modes may have been covered but not with the same intensity and scrutiny.

Some actual examples of problems with HAZOP are described by Trevor Kletz (Kletz, 2009) as summarised below:

1. Suppose a HAZOP is being carried out on a storage tank with a floating roof. The floating roof has a drain to manually discharge rainwater from the roof into the bunded area surrounding the tank.

The bund, which is intended to catch any release from the tank, has a pipe and valve penetrating the wall of the bund so that the operator can drain rainwater accumulating within the bund. This pipe with valve discharges into a waterway. In the HAZOP, the team considers the possibility of material from the tank getting into the waterway. The material could conceivably pass to the floating roof through seal failure. However, no other consequences are foreseeable. Even if the hose from the floating roof fails, the spill would be located within the bund. Unless someone in the HAZOP team has much experience and credibility, the team will miss the fact that the bund drain valve may well be left open. These valves often are left open in my experience. After heavy rainfall, the operator will go around the tanks to empty the bunds of water. But this takes time for each tank. Perhaps there is a change of shifts. The operator forgets to go back and close the valves.

2. A HAZOP team were reviewing the means of transfer of an explosive powder using a scoop. The team were concerned about the possibility of an electrostatic charge accumulating on the powder and scoop. They decided that a metal scoop would be better than a plastic one. No one realised that a conducting scoop would increase the risk of ignition if the operator were not grounded. A conducting scoop would pass a spark from the scoop to the ground. A spark from a plastic scoop would be less likely to occur and would be less energetic.

3. During a purification step a small amount of an oxidising agent is added to a much larger volume of hydrocarbon. The reaction between the two substances is known to be highly exothermic. However, none of the team knew this or consulted an expert. Neither did they consult Bretherick's Handbook of Reactive Chemical Hazards, which is a standard text on reactive mixtures. An explosion occurred after a few months of operation.

To assess the validity of HAZOP, we need to turn first to the assessment of probability and consequence. These factors are fundamental in determining the acceptability or unacceptability of risk (see Chapter 3).

ASSESSING PROBABILITY

HAZOP instructions, in my experience, give very little guidance on how a HAZOP team should assess probability. There is a reasonably robust way to be found in Layers of Protection Analysis which we will discuss later (see Chapter 27 on Assessing Risk Outcomes and Risk Reduction), but LOPA is not part of HAZOP. Having said that, I have experience of one large organisation that prefers to carry out a 'mini-LOPA' within the HAZOP to complete the probability section. They choose to do this in order to reduce the subjectivity involved in assessing probability.

Most established organisations, I have found, have developed their own tables of probability. These are based on some data sources that can be purchased, combined with the organisation's own experience. My favourite source is PGS 3/CPR18 E 'Guidelines for quantitative risk assessment'. This document is freely available on the internet.

Some example failure rates typically used are shown in Table 26.1:

Table 26.1 Typical Failure Rates

Failure	Frequency
Pump failure resulting in loss of flow	Once every 10 years
Failure of a double mechanical seal pump with loss of containment	Once every 100 years
Heat exchanger tube leak (smaller exchangers with <100 tubes)	Once every 10 years
Lightning strike	Once every 100 years

However, these sources are not readily understandable by the average operating technician or by senior management trying to understand the source data for the costly recommendations coming their way!

Probabilities 'taken from books' should be adjusted with actual plant experience otherwise the assessment is inadequate. Take the incidents of Buncefield and CAPECO described in Chapter 13, where, in both cases, the level instruments were known to be unreliable. Would the maintenance personnel have been present in the team and been willing to point out their actual experience in a risk assessment? More likely the 'book value' for probability would have been used. The maintenance personnel might have been told to do a better job.

Illustrative Hypothetical Case – Whisky Distillery – HAZOP Extract – Assessing Probability

Let's consider again our Whisky Distillery (Chapter 1 – Introduction and Appendix A), where the peat used to heat the barley has been replaced by a natural gas-fired burner for environmental reasons.

The peat floor is no longer in use. A simple natural gas burner has been installed in its place. An actual installation would be more complex than shown here, with a commercially available burner management system. See Figure 26.1 above.

Using the guidewords 'more of' and 'less of' gave the HAZOP team concerns about explosive mixtures of gas forming within the kiln before the ignition button is pressed. 'As well as' also gave rise to concerns about repeat ignition attempts leading to a backlog of gas in the kiln, which might form an explosive mixture. 'None of' gave concern about flame out, a phenomenon observed by some operators in the team, might also lead to an explosive mixture of gases. The HAZOP team proposed a number of potential improvements. To help determine the severity of the risk, we now need to assess the probability.

'More of' and 'less of' are human factors issues relating in the system, which is not yet automated, to the actions of the operator lighting the kiln. In Chapter 20 on Human Factors, we have commented that human reliability is often overestimated. Although there are no precise instructions, an error of one in every ten attempts seems reasonable.

The HAZOP team have 'guesstimated' based on some experience there is a problem lighting the kiln once in every ten ignitions. This is where the kiln does not light the first time and needs to be relit. The kiln is lit once every day at start of shift. The kiln operates 48 weeks per year during weekdays only.

Therefore, the event frequency of failing to ignite is around 24 events per year. However, every failure to ignite does not necessarily lead to a dangerous backlog of gas. The operator should have a procedure and will also use his or her common sense to wait a while for any unignited gas to disperse before attempting to reignite. The HAZOP team assumes that the operator fails to wait the required time on one in every ten occasions, giving an event frequency of 2.4 times per year. Not every backlog will lead to an explosive mixture, but we will assume so to be 'on the safe side'.

ASSESSING CONSEQUENCE

HAZOP instructions, in my experience, also give very little guidance on how a HAZOP team should assess consequences. Typical guidance is to assess the worst credible consequence of the deviation, determined by the degree of injury, environmental damage, and material and financial damage that may result. There are techniques for calculating consequences from, for example, releases of flammable and/or toxic vapours. Such techniques, based on computer modelling, are useful but require a lot of input data about the precise configuration of the site and the equipment within it. Different computer models can give different results.

I am used to applying such a model to estimate financial loss from a vapour cloud explosion. This kind of calculation is frequently required for property damage insurance. The results from the particular model I am asked to use can be as much as twice that of other models. Modelling a different release may be less than 0.75 that of an alternative model. Most importantly, such models require training and time – they are not available to a HAZOP team looking to assess consequences. The HAZOP team is left to consider qualitatively what the consequence might be (see Chapter 28 on Consequence Analysis and Dispersion modelling).

Illustrative Hypothetical Case – Whisky Distillery – HAZOP Extract – Assessing Consequence

Let's consider our Whisky Distillery again (see Chapter 1 – Introduction and Appendix A). For environmental reasons, the peat used to heat the barley has been replaced by a natural gas-fired burner.

HAZOP team is concerned about explosive mixtures of gas forming within the kiln before the ignition button is pressed. The guideword 'As well as' also raised concerns about repeat ignition attempts leading to a backlog of gas in the kiln, which might form an explosive mixture. The HAZOP team proposed several potential improvements.

The HAZOP team have determined a potential event frequency of twice per year. They now need to assess the potential consequences.

Maltings houses, by inspection of photographs, are not robust buildings. It is highly likely that an explosion would destroy most of the maltings house. Since there are no blast-proof walls, at least not so far in our design, a fatality or severe injury to the operator igniting the kiln seems likely.

We need to contact a project engineer to assess the cost of rebuilding the maltings house.

Since we don't have a project engineer to hand, let's see if we can come to a 'ball park' estimate. There are internet references to the cost of starting a craft distillery as EUR 3–5 million.

Considering that a substantial amount of the cost will be in the distillation equipment itself, being built of expensive metals like copper and stainless steel, we estimate approximately EUR 500,000 to reconstruct the malt house. We will use this figure later in our analysis.

INTRODUCTION TO QUANTITATIVE METHODS

HAZOP is a qualitative method. It meticulously analyses a process and comes up with a list of potential failures and recommendations to reduce risk. HAZOP often requires an additional and separate stage, called Layers of Protection Analysis (LOPA), to help quantify these consequences and prioritise the many HAZOP outputs. LOPA is discussed in Chapter 27 on Risk Outcomes and Risk Reduction.

In the following paragraphs, two risk assessment techniques are discussed, which lead directly to quantified outcomes. Quantified outcomes are much more meaningful to senior management. So why not use quantified methods all the time? They are very costly in terms of time and require a high level of expertise. A colleague of mine used to say that for a quantitative risk assessment, the clock starts ticking at USD 100,000, even for the simplest systems. Furthermore, because of their mathematical nature, these methods do not lend themselves well to the participation of the operations or maintenance personnel. However, the exclusion of such front-line personnel runs the considerable risk of not identifying practicalities in plant operation.

What kinds of quantitative information are we referring to? Fatalities or more specifically, the probability of fatalities from a given event, is certainly a valuable piece of information that is well understood (see Chapter 28 – Consequence Analysis – on FN Curves). The monetary value of the damage resulting is another piece of useful information for consideration in the boardroom. The probability of damage following a particular event is sometimes reduced to the risk expectation value, which is defined as the probability of the damaged state occurring times the monetary value of the damaged state (Bahr, 2015, p. 336).

It should be noted that quantitative methods are typically focused around just one event. In practice, a facility may have several scenarios, each with a specific event that requires analysis. Society is not concerned about the risks from one piece of plant in isolation; they are concerned with the total risk from a facility.

The absence of 'summing' the various risks from a facility is a major void and can lead to deliberate or inadvertent misinformation. Having said that, it is sometimes the case that one major risk overshadows many other less significant risks.

Illustrative Hypothetical Cases: Emergency Scenarios

In our hypothetical illustrative examples, the whisky distillery would need risk assessments regarding catastrophic potential around both the fire or explosions in the kiln area and a fire or explosion in the whisky maturation warehouse. In the fuels terminal, there are multiple scenarios:

- Overflow of gasoline/diesel/fuel oil tanks
- Leak from gasoline/diesel/fuel oil tanks
- Overpressure of LPG sphere
- Leak from LPG sphere
- Fire at rail unloading operations
- Fire at road loading operations
- Fire/explosion at the LPG bottling plant

Each of these would need its own quantitative analysis.

These quantitative techniques are very much reliant on the estimates made for probability. As we have seen, the likelihood of a failure of a piece of equipment, instrumentation or human being is only a very rough estimate. Hence the output number of the quantitative method can easily be given an inappropriate accuracy, given the uncertainty over the input values. A way to maintain sight of the uncertainties throughout the quantitative process is the use of 'Fuzzy Logic'. This requires the rather amusingly entitled methods of fuzzification, inference, and defuzzification. The 'fuzzy logic approach' is all very mathematical and therefore not easily understood.

FAULT TREE

Fault Tree Analysis is a method that can be used in the process industries but is also widely used in a broad range of activities, including aeronautical and space engineering and railway engineering. It is used to model particular failure modes leading to a Top Event. In the context of the Bow Tie diagram (see Chapter 4), Fault Tree analysis is used to analyse the sequences of events that could lead to the Top Event. It considers Threats that must come together to lead to the Top Event. This is gathered together as a logic tree with various 'AND' and 'OR' gates. The logic tree can grow very quickly and become very cumbersome to manipulate. Bow Tie has the strength that it is very illustrative but has to be a simplification. Fault Tree Analysis is very thorough and can include every detail. 'Cut Sets', which lead to the top event, can be identified from the logic tree. A fault tree can be quantified by placing probabilities on the initiating events. The output is a probability for the top event.

EVENT TREE

Event trees start with an initiating event and then determine the potential consequences and probabilities. It is a quantitative technique that enables risk to be assessed in terms of probability numbers like fault trees. It is often thought of as the right-hand side of the Bow Tie, from Initiating Event to Consequences.

Event trees have been used to examine many process and non-process scenarios. Abdussamie (2018) examines the risks involved with ships carrying Liquified Natural Gas, and floating LNG vessels (used as terminals – see Chapter 9 on Transportation Industry). The authors apply the risk matrix and use hazard identification to calculate multiple consequences in a mathematically exacting manner. A table of probabilities is used. For example, 'failure of thrusters' is assigned a probability index of 4 – Likely to occur once per year in a fleet of 100 ships a figure referenced from another text (Vanem, E. et al. SAFEDOR Project, 2005). Drilling down, I found that this data was based on an expert panel analysing past incidents to LNG ships and assigning probabilities. This is a reasonable approach, but it may not represent the actual probability. There have been so few incidents involving LNG tankers that one questions if sufficient statistical data is available to justify the analysis. In a relatively new and growing industry, we will not know the actual probabilities until significantly more data has been accumulated. Sadly, the conclusions do not mention a bottom-line consequence from all this detailed work. Work of the pilot boat is identified as the highest hazard, then emphasises the importance of adopting a safe procedure, an argument that sounds a little circuitous.

Surely the availability of a safe procedure is assumed from the start. Otherwise, LNG operations would not be allowed at all!

OTHER TECHNIQUES

There are many other risk assessment techniques applied to catastrophe prevention, such as:

- Structured What-If Technique (SWIFT)
- Failure Mode and Effect (FMEA)
- Management Oversight and Risk Tree (MORT)
- Petri-Net
- Bayesian Network Modelling

And many more. For the purposes of this book, I will not try to explain them nor comment on their benefits and disadvantages. This whole subject gets extraordinarily complex and academic. Imagine making a presentation to the company's main board on the outcome from a risk assessment entitled 'Bayesian Stochastic Petri Nets using Fuzzy Logic'. What would be your reaction if you lived in the vicinity of a hazardous plant and received some risk data from such a process. Would you trust it? Or would you consider that you were just being browbeaten by science?

SUMMARY: CAUSES OF FAILURE OF RISK ASSESSMENT TECHNIQUES INCLUDING HAZOP

Reasons Why HAZOP and Other Risk Assessment Techniques Sometimes Fail to Prevent Catastrophes Include

- ☒ Dependant on the skills of the Chairperson.
- ☒ Dependant on the professional experience of the Scribe, otherwise the whole process gets held up, or important aspects of the discussion are missed.

☒ The team is vulnerable to human factors issues making the risk assessment unreliable. The factors include 'risky shift' and 'group think'.

☒ HAZOP requires consideration of Operability issues too. This can be a distraction from the core task of containing hazards.

☒ Some assessment of the consequences of a particular failure is required by the team. The extent of the consequences is not always clear for a HAZOP team.

☒ The team is asked to make recommendations. Sometimes issues of expense are perceived at this step, and therefore some costly recommendations excluded.

☒ HAZOP is an extremely detailed process. Key practical information may be missing if the right people are not in the room or know what actually goes on but are unwilling to say it.

☒ There can be confusion as to whether human factors – human errors – should or should not be considered in the HAZOP. Of course, they should, otherwise, it is incomplete. Some organisations have separate human factors studies.

☒ The residual risk can be unclear – residual risk without controls, residual risk with current controls, residual risk when recommendation is implemented.

☒ For some participants, especially maintenance and instrumentation specialists there is considerable 'dead time' when their knowledge and experience are not being called upon.

Reasons Why the Assessment of Probability and Consequence Assessment Steps Sometimes Fail to Prevent Catastrophes Include

☒ Confusion as to whether this is within the scope of the HAZOP team or outside of it.

☒ Availability of Probability of Failure data.

☒ Unwillingness to adjust probability of failure data based on actual plant experience.

☒ Underestimating or overestimating the consequence, on which basis recommendations may be made.

Chapter 27

Assessing Risk Outcomes, Layers of Protection Analysis, and Safety Integrity Levels

ASSESSING RISK

Once probability and consequence have been determined then risk can be assessed. We need to turn those probability assessments expressed in terms of times per year, and consequences in terms of injury and cost, into a 'risk'. This risk is normally expressed as:

> Critical risk, very high risk, high risk, elevated risk, low risk, and very low risk or similar.

It is good practice for an organisation to rank its risks according to a matrix such as the matrix shown in Figure 27.1.

Probabilities might be described quantitatively (Table 27.1):

Note a difficulty here. Are incident frequencies to be considered per site, group of sites, total organisation, in any particular industry, continent, or globally? I have found that the industry and scope are often not made clear.

Consequences might also be described quantitatively, either in terms of occupational safety, environmental, or financial impact. See Table 27.2.

Note that the financial impact consequence is also very dependent on the size of the site.

Other columns are sometimes usefully added such as:

- Reputation
- Legal/Regulator
- Operational Impact (in terms of interruption to production)

In order to create such a matrix, the tolerability of risks needs to be quantified. Taking the example of personal safety criteria, quantification of the tolerability of risk depends on how those risks are perceived. This depends on several factors, including:

DOI: 10.1201/9701003360759-30

Consequence

Probability	Very Minor	Minor	Serious	Very Serious	Catastrophic
Frequent	Elevated	Elevated	High	Very High	Critical
Occasional	Low	Elevated	Elevated	High	Very High
Seldom	Low	Low	Elevated	High	High
Unlikely	Very Low	Low	Low	Elevated	High
Very Unlikely	Very Low	Very Low	Low	Low	Elevated

Figure 27.1 Risk Matrix.

Table 27.1 Probability Classifications in Risk Matrices

Probability Classification	Frequency
Frequently	Once per annum
Occasional	Once every ten years
Seldom	Once every 100 years
Unlikely	Once every 1000 years
Very Unlikely	Once every 10000 years

Table 27.2 Consequences

Consequence Classification	Personal Safety	Environmental	Financial Impact
Catastrophic	Multiple deaths	Serious ecological with long-term pollution or health risks to the local community	Euro 100 million
Very Serious	One fatality or permanently disabling injuries	Serious excursion to environmental permit. Corrective measures outside of the site are necessary	Eur 20–100million
Serious	Serious injury, work disability, lost workday case	Serious excursion to environmental permit. Corrective measures outside limited to actions within the site boundary	Eur 5–20 million
Minor	Temporary limited work including injuries due to irritation	Slight permit exceedance. No environmental damage outside of site boundaries	Eur 100,000 to 500,000
Very Minor	Medical Treatment, first aid, short illness	Limited damage or emission within site boundaries	Eur 10,000 to 100,000

- Personal experience of adverse effects
- Social or cultural background and beliefs
- The degree of control one has over a particular risk
- The extent to which information is gained from different sources, e.g. the media.

The HSE has made a valuable assessment of the acceptability of risk, affectionately known as R2P2 – HSE – Reducing Risks, Protecting People (2001). This study was mentioned earlier in Chapter 3 on What Makes Risk Tolerable or Intolerable. The organisation creating a matrix such as that above would do well to benchmark their matrix versus R2P2. We will return to this in discussing the concept of 'As Low as Reasonably Practicable (ALARP)', in Chapter 28. R2P2 tolerability criteria deal mostly with the risk of death. According to R2P2, a risk is intolerable if a worker is exposed to a danger of death more frequent than 1 in 1000 years. In the above matrix, the columns 'Very serious' and 'Catastrophic' include fatalities. The row 'Unlikely' corresponds to intolerable frequency and so is set to the Category 'Elevated Risk' for Very Serious. Other categories are developed by judgement, but at least this matrix has an anchor point within R2P2.

Illustrative Hypothetical Case – Whisky Distillery – HAZOP extract – Assessing Risk Exposure Frequency

In Chapter 26 the HAZOP team, developing a HAZOP for kiln, were concerned about explosive mixtures.

They came up with an event frequency of 2.4 times per year.

For the consequences, they assessed that a fatality or severe injury was a credible outcome, either to the kiln operator either on the barley floor or at the kiln. The economic cost was considered to be around EUR 500,000, based on reconstruction of the kiln.

The probability is 'Frequent' and the Consequence 'Catastrophic'.

If the organisation was using the above sample matrix, we can see that the personal safety impact is 'Very Serious', the financial impact 'Minor', and the probability 'Frequent'. This gives an overall Risk described as 'Very High'.

It seems like something needs to be done!

QUANTITATIVE RISK ASSESSMENT

In quantitative risk assessment (QRA), we combine:

- quantitative methods (or probabilistic) methods of assessing the likelihood of an incident (the top event)

- quantitative methods of assessing the effectiveness of mitigating the consequences

The result is a probability figure for a particular consequence, such as 'one or more fatalities' or 'fatalities to members of the public'.

QRA is often included with the Safety Cases required under Seveso legislation, although I understand it is not mandated that it be included.

Further, as noted by Lord Cullen, in his report on the Pipe Alpha disaster, when safety cases were being proposed for introduction offshore: '*QRA is an element that cannot* be ignored in decision-making about risk since it is the only discipline capable, however imperfectly, of enabling a number to be applied and comparisons of a sort made, other than in purely qualitative kind. That said, the numerical element must be viewed with great caution and treated as only one parameter in an essentially judgmental exercise' (Haddon-Cave 2009, p. 167).

There are three main reasons for being dubious about QRA:

1. It is costly, time-consuming and requires specialist engineers to accomplish.
2. It depends on the probability of failure data that go into the calculations. Whereas data going into LOPA (see below) is only 'an order of magnitude', the failure data probabilities used in QRA are expected to be more accurate. The outcome can be significantly swayed in either direction depending on the source of data. Since the data is usually expressed as a range, the user can choose whether to use the lower failure frequency or the highest, with considerable effect on the final figure.
3. QRA is based on a particular scenario (consequence). Is this the worst consequence? In practice, various scenarios may need to be analysed to find out the worst consequence.

An improvement would be to carry out a sensitivity analysis, using optimistic and pessimistic probability data, and determine the outcome on the incident frequency deduced. However, I have never seen this carried out in practice.

Prior to the Fukushima Daiichi incident (see Chapter 10), probabilistic assessments were not carried out in Japan for incidents affecting personnel outside of the site boundary (external events) (The Atomic Energy Society of Japan (2014) The Fukushima Daiichi Nuclear Accident – Final Report). For probabilistic risk assessments, read QRA. The reason given in the report is that there was considerable doubt about the meaning and role of probabilistic risk assessment in Japan. The Atomic Energy Authority of Japan continues to note that 'One of the excuses is that methods of assessing external events were still immature, or that there was no reliable data'. Following Fukushima Daiichi, it seems that probabilistic risk assessments will be now pursued in Japan. Nevertheless, I understand their initial doubts about the value of probabilistic methods. I would also agree, however, that the practice has considerably matured over recent years.

RISK REDUCTION HIERARCHY

The following hierarchy should be used in considering methods of risk reduction:

Step 1. **Design or Reorganise to eliminate hazards.** For example, using a catalyst that reduces the temperature and pressure of a process.

Step 2. **Substitute the hazard with something safer.** For example, the replacement of flammable solvents with ones that are not flammable. Sometimes powder that creates toxic or dust explosion risks can be replaced by a pelletised form of the material.

Step 3. **Isolate the hazard from people.** For example, in mining, deploy robotic cutting machines at the rock face so that miners are remote from the area of risk of rockfall. This also applies to relocating control rooms and office blocks to more remote distances from the process and ensuring residential housing does not encroach on the potentially hazardous zones created by the plant.

Step 4. **Use Engineering controls.** This can include devices such as pressure safety relief valves, mechanical interlocks, and electrical interlocks. This category also includes the designing and implementation of Safety Instrumented Systems. Ventilation can be used to reduce the risk of build-up of toxic or flammable gases indoors. Spill containment can be used to limit the extent of a release.

Step 5. **Use Administrative controls.** These are primarily in the form of procedures that the operator must carry out, combined with the training to help ensure that they are understood. Control of work using work permits is an administrative control, as is interlock bypass (see Chapter 22 on Equipment Maintenance and Inspection). Housekeeping, labelling, signage are other means of administrative control.

Step 6. **Use Personal Protective Equipment.** Although the last in the hierarchy, this is often the item of first resort. Examples are the provision of HAZMAT suits, breathing air sets, etc.

Whilst this is the hierarchy, financial pressures often cause them to be considered in reverse order since steps 5 and 6 are likely to be the least expensive and/or require the least effort to implement. It is sometimes not possible to carry out some of these measures retrospectively on existing facilities.

DESIGNING A SAFETY INSTRUMENTED SYSTEM

In the risk reduction hierarchy above Step 4: using Engineering Controls is the most common type of risk reduction action.

Many recommendations which result from a Process Hazards Analysis concern the provision of additional instrumentation. Further, if there is already instrumentation in the process, the extent to which this can help prevent the catastrophic event needs to be evaluated. As we have seen, the probability that it will reduce the consequence needs to be assessed, both for existing instrumentation and planned instrumentation. Some of this instrumentation will simply be indicators, like the speed indicator in your car. Some will be interlocked. For example, if your car lights are switched to automatic, they come on once the light has dimmed.

Back in the 80s it was common practice to list all the instruments which are safety-critical. With the advent of electronic process control systems, it was a recommended best practice during that era that safety-critical interlocks be hard-wired (i.e. not electronic) so as to not be vulnerable to software corruption. As electronics control systems became more reliable, and the desire to add more and more instrumentation, the need to use electronic systems to provide safety-critical actions became beneficial. This was no less than a revolution; a revolution that is still not fully appreciated. The resulting dependence on electronics brings more complexity, the need for more specialist knowledge, increased vulnerability to software programming errors and corruption, and exposure to cyber-attacks.

We have to start this discussion by referencing an international standard. I'm not too fond of name-dropping international standards, and so far, I have mostly succeeded in not mentioning standards and regulations in this book. Once somebody mentions ASME ..., API ..., NFPA ... such, they cut off everyone in the discussion except those familiar with that code/standard/regulation, and the non-technical reader loses interest. The implication, of course, is that people involved in the discussion should know the standard and know it intimately. Knowing these codes standards, and regulations require intensive reading and understanding of how they are implemented. Suddenly catastrophe prevention becomes the realm of the specialist to the exclusion of everyone else.

At this point in our discussion of risk assessment I have no choice but to introduce the standard IEC61508, together with the use of the technical methods Layers of Protection Analysis (LOPA) and the methodology of Safety Instrumented Systems (SIS).

LOPA AND SIS

Whilst IEC61508 is recognised as a best practice, it is not, at least in the U.K., a legal requirement. However, in the event of an incident, an organisation would be vulnerable to legal action for not applying best practices unless it was implementing a similar or better approach. It is also implied that IEC61508 is retrospective. Many processing plants were built before IEC61508 was introduced. It is expected that a plant would measure itself using the approaches in IEC61508 and recognise its risks accordingly. Where the site did not have instrumentation meeting the expectations of the standard, it would be expected to have a plan to achieve this.

The principle is that we have a safety goal, which arises from our risk analysis above. We might, for example, have a 'Very High' risk which we want to instrument so that it becomes a 'Low Risk'. To achieve this, we need to reduce the probability of the event occurring and if possible, reduce the consequences also.

For the following discussion, we need a few definitions:

An Independent Layer of Protection (IPL) is:

A layer of protection that will prevent an unsafe scenario from progressing regardless of the initiating event or the performance of another layer of protection.

A Safety Instrumented System (SIS) is:

A combination of sensors, logic servers and final elements that detects an abnormal or out of limit condition and brings it to a safe condition.

The Safety integrity level (SIL) is defined as a relative level of risk reduction provided by a safety function or to specify a target level of risk reduction. In simple terms, SIL is a measurement of performance required for a safety instrumented function (SIF). SIL 1 requires a lower standard of instrumentation than SIL 2 and higher levels which require progressively higher levels of reliability.

A conditional modifier is defined as enabling events or conditions which have to occur or to be present before the initiating event can result in the consequence.

As an example of a conditional modifier, consider the overflow of a tank containing hydrocarbons, e.g. gasoline. There is a risk that the vapours from the hydrocarbon will ignite. The potential ignition sources may be in a particular direction that is different from the prevailing wind direction. If a potential ignition source is to the north west, and the wind only blows in this direction for 20% of the year, the probability may be reduced by this conditional modifier. If the operator is present only for a certain number of minutes each hour, this may also be a conditional modifier, however in the latter case we need to be careful – are we studying the injury potential or the financial damage. The former is affected by the operators presence, the latter is not. Conditional modifiers may thus reduce the SIL rating required.

Notes from Trevor's Files: We Are a SIL3 Plant!

During a visit to a large petrochemical complex, as it happens in Uzbekistan, the Technical Director announced that we need not ask about SIL levels. The plant, which was of recent construction, was all built to SIL3 standard he proudly declared. The plant had indeed been built to a very high standard of technology, and I have little doubt that the process control systems and logic solvers were specified and labelled as SIL3. Quite possibly other parts of the instrumentation loop, including the detectors and actuated devices, were also purchased to the same standard. But the statement that we are 'SIL3' shows a lack of understanding at a senior technical level of the intent and workings of IEC61511. Some control loops may indeed be over-engineered. But the actual SIL level of the loop needs to be assessed in the context of the risks of the process involved. This is needed, amongst other reasons, to assess the frequency at which the instrumentation loop needs to be tested.

Let's assume, by way of example, that we have a 'Very Serious' risk which needs to be reduced from 'High' to 'Low'. Any instrumentation we introduce to improve the risk has its own probability of failure, which depends on the design and quality of construction. The nomenclature used is shown in Table 27.3:

Table 27.3 Probability of Failure Targets

Demand Mode SIL Targets Probability of Failure to Perform Its Desired Function on Demand	Continuous Mode SIL Targets Probability of a Dangerous Failure Per Year	Safety Integrity Level
Better than 1 failure every 10,000 demands	Better than 1 failure every 10,000 years	4
Better than 1 failure every 1000 demands	Better than 1 failure every 1000 years	3
Better than 1 failure every 100 demands	Better than 1 failure every 100 years	2
Better than 1 failure every 10 demands	Better than 1 failure every 10 years	1

Illustrative Hypothetical Case: Whisky Distillery – LOPA SIL

Let's consider our Whisky Distillery again. We established under Illustrative Hypothetical Case – Whisky Distillery – HAZOP extract – Risk above, that there is a potential event frequency of 2.4 backlogs of flammable gas per year following an ignition failure. For consequences, we assessed that a fatality or severe injury was a credible outcome, together with a financial cost of around EUR 500,000. We concluded that something needs to be done to improve the risk.

A glance at your domestic gas boiler will indeed show that there is instrumentation to protect against the development of an explosive mixture due to failure to light or 'flame out'. But for the purpose of this simple demonstration of SIL, let's try to design something very basic.

Let's propose the installation of a flame scanner – a device which detects radiation from a flame. If the flame scanner does not detect a flame, then the proposed instrumentation automatically closes the gas feed valve. We need a logic solver here because we need this flame scanner interlock to be disabled for the first, say, 10 seconds after ignition to allow the flame to become established (Figure 27.2).

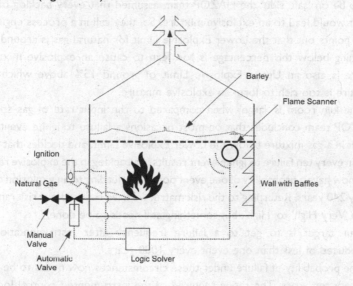

Figure 27.2 Kiln with Logic Solver.

We want to improve the probability of the incident from 'Frequent' (> 1 per year) to 'Unlikely' (<1 per 1000 years). Should we use the demand mode or continuous mode figures from Table 27.3?

The hazardous condition occurs during the ignitions sequence so we should use the demand mode on this occasion.

Therefore we are looking for the additional instrumentation to have a low probability of failure on demand $(1/1000)*(1/2.4)$ or 1 in every 2400 years. From the above table, we can see that this requires a SIL3 rating.

SIL3 is a very demanding rating that would require expensive equipment and probably require two or more flame scanners 'voting'. The interlock closing the gas valve would activate if any one of the flame scanners detected 'flame out'.

It is time to re-evaluate our assumptions.

We assumed that the operator would make a mistake and not allow gas to disperse on one in every ten occasions. One in every ten is a common failure frequency used in Layers of Protection Analysis for human failure. Nevertheless, I imagine that, in going to the plant manager saying that operators had a failure frequency on one in every ten, would result in the proposal to get better operators! So let's reinforce the procedure and use one in every 100 – there is some merit in this – the operator is not in a stressed situation at the start of the shift. Instead of a fatality frequency of one event 2.4 per year, we now have one event every 24 years.

To be on 'safe side' the HAZOP team assumed that every backlog of gas event would lead to an explosive mixture. So, they call in a process engineer who points out that the Lower Explosive Limit for natural gas is around 5%, anything below this percentage is too lean to cause an explosive mixture. There is also an Upper Explosive Limit of around 15% above which the mixture is too rich to form an explosive mixture.

The kiln room is 'large' when compared to the input rate of gas so the HAZOP team concludes that on most occasions a failure to ignite event will result in a gas mixture below the Lower Explosive Limit, and decides that only one in every ten failure to ignite event results in a backlog in the explosive range. We now have a probability of one event potentially leading to an explosion once every 240 years. Returning to the risk matrix we have reduced our risk ranking from 'Very High' to 'High', but something still needs to be done.

Our target is to get to a failure frequency after instrumentation is introduced of less than one event every 1000 years.

The probability of failure under these circumstances now needs to be only in every ten years. The rating required of the instrumented control loop is now SIL1 – a considerably cheaper option.

See how quickly, with a little reconsideration, we have saved a lot of money!

The objective in the above hypothetical case is to demonstrate, using slightly different considerations, how the conclusion on what level of instrumentation to instal would be very different. In practice natural gas burners, and burner management systems, have been very well developed, and the process engineer would consult a provider of such a system and become equipped with a wide range of interlocks to protect the kiln. The above example is for demonstration only.

Illustrative Hypothetical Case – Tank Overfill Interlock Application of SIL

Let's consider again a tank being filled with a hazardous material. In our hypothetical Fuels Terminal, it could be any of the gasoline or diesel tanks,

As we noted previously, reliance on the operator filling the tank through the manual valve until the tank is 'full' as indicated by the sight glass level indicator is open to human error. So, we have put a level switch on the tank to automatically close the fill valve. Is this sufficient? See Figure 21.1 in Chapter 21.

The consequence of overfilling is judged to be 'Very Serious'. Suppose there is no other protection such as bunding of the tanks. In that case, the overflow will have serious environmental consequences extending beyond the site's boundaries into the groundwater and adjacent stream. Clean-up costs would be substantial. If vapours from the overflow find an ignition source, then a serious fire will result, which could escalate to adjacent facilities with the potential for fatalities.

A simple level switch has been shown to have a low reliability. Typically, a probability of failure on demand of such a level switch would be 1 in 10 to 1 in 100 demands. The maintenance engineer in the HAZOP team is there to point out the trouble which they have with these instruments. So we have 'Very Serious Consequences' with an 'Occasional Frequency' – a high risk which needs to be reduced. Let's target reducing from 1 in 10 failures per demand to 1 in 1000. This equates to a SIL2 which would undoubtedly require an improvement on the 'simple' level switch. Note also that we have to make the entire control loop to SIL 2 standards of reliability including the automatic valve.

However, again we may have made a mistake. Most such tanks have retaining walls known as bunds. Preferably the tank sits inside a retaining basin with an impermeable floor, but this is difficult to retrofit to an existing tank farm. A bund wall, which can be of simple, compacted earth construction with an impermeable lining, can be built. This reduces the consequences of the overflow. If we assume an earthen construction with a liner, then there would still be substantial ground remediation required, and work to avoid groundwater contamination. Nevertheless, we would reduce the consequences from 'Very Serious' to 'Serious'.

Furthermore, we have not taken any credit for the operator. A high alarm could additionally be provided, giving an increased likelihood that the operator would identify the potential to overfill before the interlock does so. The LOPA/SIL methodology allows us to take credit that the operators would fail once out of ten occurrences to prevent the overfill. We now have a minimum failure rate of 1 in 100 per demand with new level instrumentation. We need only a SIL 1 instrument.

Money saved again!

Once the SIL rated equipment has been installed, it needs to be routinely tested. Depending on the SIL rating, higher proof test frequencies will be required.

Layers of Protection Analysis is a structured way of arriving at SIL requirements. It can consider multiple initiating causes, independent protection layers, and what are known as conditional modifiers, and add the probabilities together in a robust statistical manner.

Notes from Trevor's Files: Same Process – Different SIL Outcomes

As an insurance risk engineer, I had the opportunity to visit many oil refineries.

Oil refineries vary considerably in complexity and technology. Nevertheless, I would expect that two refineries with similar complexities and process units would have a broadly similar number of Safety Instrumented Functions. In one such refinery, they had 92 SIL2's and 1 SIL 3. In contrast, another similar refinery had just 20 SIL2s and 0 SIL3s.

I remain suspicious that, whilst the LOPA/SIL process looks an exacting one, it is in practice somewhat subjective in the hands of the practitioner depending on the biases and pressures on that practitioner.

We will not reproduce the means of calculations of LOPA in this book. Textbooks, YouTube videos, and specialist training are readily available.

However, by now, there is a good chance that you are confused about this very promising approach of IEC 61508 with its requirements to carry out the calculation of Layers of Protection Analysis and Safety Integrity Level.

The HSE has published a fascinating study of LOPA as applied to the relatively simple case of filling of fuel storage tanks (HSE, 2009, A review of Layers of Protection Analysis (LOPA) analyses of overfill of fuel storage tanks – Research Report RR716). In this work, 15 LOPA studies were assessed, with considerable inconsistencies in both approach and conclusions. The first line of the report's findings reads as follows 'The majority of the LOPA studies considered in this work have areas that need significant improvement'. It has to be emphasised that this 2009 work was 'work in progress'. Perhaps LOPA practice has improved since that time.

The IChemE published a paper in 2019 (Casey R. 'Limitations and misuse of LOPA'), which suggests that there remain significant issues with the application of LOPA, which they refer to as a 'simplified form of numerical risk assessment'.

An important consideration in designing safety instrumented systems, or risk assessing existing safety instrumented systems is to identify what are called 'common mode failures'. I have an electrical cooker and a gas boiler for central heating. Therefore, I think that in the event of a utility failure, I will still have some heating and cooling options. In the event of a gas failure, we can still cook and keep warm in the kitchen with the cooker switched on. In the event of an electrical failure, we may not be able to cook, but we can keep warm. But this is not the case. The electricity failure causes a failure in the gas boiler since electricity is needed for the pilot light and the electrical controls – a common mode failure.

On a processing plant, failure of the air on which many instruments depend, causes those instruments to suddenly read either zero or 100%. Interlocks and alarms which were installed during the design phase, and which were considered during the risk assessments no longer operate. The poor recognition of common-mode failures is a frequent issue in process hazards reviews, since identifying them is quite difficult. They are not easy to spot.

SUMMARY: CAUSES OF FAILURE TO PREVENT CATASTROPHES WHEN ASSESSING RISK OUTCOMES AND ADDRESSING RISK REDUCTION

Reasons Why the Assessing Risk Using a Matrix often Fails to Protect against Catastrophes

- ☒ Consequences or probabilities are inaccurately estimated.
- ☒ Risk Matrix not linked to available external benchmarks as to what is acceptable or not acceptable.

Reasons Why Quantitative Risk Assessment often Fails to Protect against Catastrophes

☒ Not done or limited due to high cost.

☒ Can only be done for one scenario at a time. Do we know the most severe scenario?

☒ Outcome may be doubted for cultural reasons, or scepticism over the probability and consequence data/assumptions.

Reasons Why the Design of Safety Instrumented Systems Can Fail to Protect against Catastrophes

☒ Procedure of LOPA and SIF is not intuitive, needs training, and is not understood by senior management.

☒ Question marks as to whether it is legally essential, or not, and whether it is retrospective or must be applied only to new designs.

☒ Assessment can be skewed if there is pressure to avoid expenditure and retrofitting.

☒ Confusion about what needs to be SIL rated – omitting to ensure that the logic solver, instruments, and actuating device are all appropriately designed for the SIL requirement.

☒ Omitting to recognise the proof testing requirements for the particular SIL rated control loop.

Chapter 28

Consequence Analysis and Consequence Mitigation

CONSEQUENCE ANALYSIS AND DISPERSION MODELLING

During the risk assessment, the assessment team must assess the potential consequence of the failure. Sometimes this is reasonably intuitive. An aircraft control failure can result in the crash of the aeroplane, and loss of all passengers and crew can be reasonably assumed. A railway signalling failure or failure to recognise the signal can result in the collision of the train with fatal consequences. The extent of the crash might be reasonably estimated after consideration of the speed and size of the trains in use at that location.

In the process industry, sometimes the consequences are clear, sometimes not. For instance, a hydrocarbon leak into a kerbed concrete pad, as typically is found at the base of chemical plants, may result in a fire in an area defined by the kerb and result in the destruction of the equipment within the kerbed area. However, in many cases, a further calculation is needed to determine the true extent of the consequence. Even in this relatively simple example, we need to ask:

- How much hydrocarbon could leak out – total inventory of hazardous material in the particular process area?
- What is the rate of leakage – related to the pipe or vessel size, and perceived cause of leakage, plus the internal pressure forcing the hazardous material to leak out?
- At what distance are the potential ignition sources?
- What is the intensity of the radiant heat produced in the associated fire to assess the distance at which people would be injured or die due to the heat intensity? Also, to evaluate if neighbouring equipment might be damaged or suffer a fire or explosion due to the 'domino effect'.

On a process plant, releases might occur through loss of mechanical integrity such as the disintegration of a vessel. But many other sources also need to be considered such as release from pressure safety valves. The release may occur at height or below ground if, as is often the case, some equipment such as pumps are located in pits below ground. There can be two-phase flow – a mixture of liquid and gas being released, or the material may boil as it is being released. Liquified petroleum gases such as propane and butane are kept as liquids under pressure and transported at ambient temperature. As the materials are released, gas flashes off, at a rate determined by a number of factors including heat input. I use a butane-powered domestic 'flame thrower' for controlling weeds in my garden. When the device is first lit the flame is at maximum strength. But as I continue to use the device the flame starts to get weaker. On touching the butane canister, I find it has got very cold due to evaporation of the butane, and this in turn has reduced the gas emission and hence the smaller flame. This kind of phenomenon makes the prediction of surrounding concentrations of hazardous materials difficult to calculate.

Further complexities occur if trying to assess the potential for an explosion. An explosion is potentially much more destructive than a fire on its own. For an explosion to occur following a hydrocarbon leak, the hydrocarbon vapour must mix with air to the point where the hydrocarbon concentration is between the upper and lower explosive limits for the particular hydrocarbon mixture. For example, for methane (natural gas) the concentration would need to be between 5 and 15%. For a vapour cloud explosion, it is also necessary to have some congestion around the cloud of vapour in order to create a destructive pressure wave. This requires a quite complex computer modelling that can only be carried out by trained personnel. The models themselves are also subject to controversy.

Explosion modelling is of particular interest to insurance companies who need to assess the potential amount of financial damage which might result. The insurance company which I worked for uses – Marsh BLAST powered by MaxLoss[TM]. The technical details need not bother the reader of this book. It is sufficient to say that the BLAST approach is considered by Marsh, with good reason, to be superior to other techniques in its modelling of the effects of congestion and confinement, such as results from concrete floors and other obstructions to the progression of the pressure wave. Other techniques typically use an estimate of the destructive power of the chemical, based on a fraction of the destructive power of the same amount of the explosive TNT. My purpose is not to argue which is better. My intention is rather to point out that there is a choice to be made concerning the explosion model used and that this choice may give a different result!

A combination of fire or explosion events often needs to be considered (Fire Safety Journal, 2013, Dadashzadeh et al.). If a flammable gas leak occurs, a quick ignition may lead to different types of fire such as a fireball, jet fire, or flash fire. However, if there is a delay before the leak finds an ignition source, then a potentially more destructive vapour cloud explosion could occur depending on the level of congestion and confinement. In contrast, a flammable liquid leak may form a pool of liquid with vaporisation from the pool into the surrounding atmosphere. If the ignition is immediate, a pool fire will result, but if the ignition is delayed, a vapour cloud explosion may result. Different calculations exist for each of these fires or explosions. Which one to choose as the consequence?

A vapour cloud explosion at a distance from the release source is likely to generate sufficient heat load that there is an ignition at the release location with the generation of a jet fire. If the vapour cloud results from an evaporating pool, then a flash fire will result when the vapour cloud reaches an ignition source! All of which I am sure the reader will find quite confusing. The selection of which fire scenario to model in order to assess the consequences is something of an art, or maybe even guesswork. The vapour cloud explosion is almost certainly the most destructive (but not necessarily – it might depend on the direction of the jet fire and what it might impact, for example). However, a risk analysis team, doing say, a HAZOP, may see the consequence as one type of fire and not an explosion and miss the most severe consequence. The risk analysis team neither has the time, tools, nor training to model every type of fire or explosion which might result.

In insurance surveys, I often ask for the scenarios which the organisation submitted under Seveso regulations. It should be pointed out that the figures submitted under Seveso are intended primarily to judge the impact on human beings, whereas, in insurance, we are mainly concerned with property damage. Nevertheless, asking the questions helps to ensure that I have not missed something. I have seen some Seveso submissions which refer only to jet and pool fires and omit to mention vapour cloud explosions. I regret commenting that this may be deliberate since the vapour cloud explosion scenario is also likely to have the most significant impact on human beings. The submitting organisation may not want to highlight the vapour cloud explosion scenarios.

Consequence analyses are also needed for assessing the accidental release of toxic materials. The plume of toxic materials is likely to spread off-site and over neighbouring properties and habitation. Highly technical dispersion modelling is readily available, at a significant cost, for such studies.

Such models are usually done for COMAH/Seveso III sites to assess the potential fatality impact on neighbourhoods around the plant for various release scenarios.

Notes from Trevor's Files: What Height of Stack?

Whilst part of a group designing a plant for construction in Northern Ireland, a question arose about the required height of a waste incinerator stack. It is common for waste gases from a plant to be vented just as they are vented from your central heating boiler. The height is critical to ensure that the gases disperse in the air before reaching neighbouring property and causing a nuisance smell, or worse, toxic effects. Reputable engineers carried out the design using a tried and trusted consequence model to assess the dispersion. This model could also evaluate the impact on the stack plume from adjacent features such as buildings and landscape features such as hills and cliffs. The engineers had the foresight to try the model against the dispersion from an existing waste incinerator stack. To their surprise, the model did not work against the observed dispersion. The actual dispersion was much worse than predicted by the model. What to do with regards to the new stack? Conservatively, the engineers adapted their very precise model for the old stack by placing a fictitious building upwind of the stack. This created the observed phenomena with the old stack. The engineers then used a scaled fictional building to simulate the effect with the new stack. This work was shown to be successful once the new stack was put into service, but the explanation for the failure to model the existing stack dispersion was never found. One reason, given only half-jokingly, was that there was something unique about the air in Ireland. It must be something to do with the leprechauns!

I have gone into this detail to demonstrate that the assessment of consequences can be a complicated issue. A risk assessment team may underestimate the consequence. Consequence analysis calculations are often costly, requiring licenced techniques and software, and their reliability can be dubious.

Consequence analysis can involve complicated calculations for which several proprietary software programmes. DNV PHAST is a commonly used and well-respected software for consequence analysis. It requires trained personnel to run and is complicated to set up. It also requires an expensive licence.

A free alternative is ALOHA®.

The section below in the outline box describes how you can use ALOHA, and is intended to show the complexity, even for a simple application. *Non-technical readers may want to skip this section.*

Hypothetical Case: Use of ALOHA

The ALOHA® computer programme is meant primarily for emergency response and is structured accordingly. However, for the enquiring reader of this book ALOHA provides an inexpensive window into the issue of consequence modelling. It is ideal for getting an idea for the kind of extent of the consequences from any particular scenario. In Chapter 12 I used it to get an idea of the consequences of spillage from a whisky barrel.

The ALOHA® software is part of the CAMEO software suite, which is developed by the National Oceanic and Atmospheric Administration (NOAA) Office of Response and Restoration in partnership with the U.S. Environmental Protection Agency (EPA) Office of Emergency Management. The ALOHA name is a registered trademark of the U.S. government.

Open ALOHA®, and you will find that the first page gives a caution in interpreting the model's predictions, particularly under the following conditions:

- Very low wind speeds
- Very stable atmospheric condition
- Wind shifts and terrain steering effects
- Concentration patchiness, particularly near the source

This may be sound advice, but advice that somewhat reduces the model's value. We want to know the outcome in the worst credible conditions, including a still night! Indeed, a still night may create the worst conditions for a toxic gas release.

ALOHA® then goes on to state that the model does not incorporate the effects of:

- Chemical reactions
- Particulates
- Chemical mixtures
- Terrain
- Hazardous fragments

Many potential releases involve chemical reactions following accidental mixing. In Chapter 4, the consequences of dust explosions were emphasised. Most chemicals we actually come across are mixtures. Two of the most ubiquitous mixtures are gasoline and diesel! These cannot be modelled in ALOHA. For sure, the ground around us will have some terrain!

But, OK, we did say we only want to get a 'ball park' figure. I describe below sequentially the data which needs to be entered to get a result:

1. Enter the location under 'Site Data', which is confined to locations in the U.S.A., but this is for our purposes we can enter an arbitrary location.
2. The date and time should be entered. ALOHA uses time to distinguish between night-time air stability and daytime air stability (see below).

3. Select from three building types or enter the number of air changes. This Emergency Response programme assumes 'Shelter-in-Place' where all people are advised to go indoors in a release. ALOHA does not give a result for a person who stays outdoors. This is not particularly informative! What if the person ignores the advice or simply does not receive the message to shelter in place?

4. Selects from the surroundings to be building being used — either sheltered surroundings such as with trees and bushes or unsheltered surroundings.

5. Enter the chemical involved under 'Set Up', noting that only pure chemicals can be modelled. Five relatively common solutions are included, for example, aqueous ammonia.

6. Wind speed needs to be entered. Wind speed is essential since the plume of hazardous material will disperse more rapidly the higher the wind speed.

7. Wind direction should be entered for emergency response, but for the enquiring reader, any direction will do.

8. A measurement height above ground. This will be the height at which a chemical has been detected. The emergency responder needs to find the consequences at a specified height. It is important because the model assumes a wind profile aiding dispersion. For the purpose of a 'ballpark' estimate, I usually enter the height of release of the chemical for the scenario I am attempting to model.

9. Ground roughness selected from 'Open Country', 'Urban' or 'Forest', 'Open Water', or a roughness figure can be entered. Higher ground roughness will be modelled as creating more significant turbulence in the toxic plume. Turbulence leads to faster dispersion. Therefore, I usually use open country or even open water, which gives the most conservative, i.e. pessimistic result.

10. Cloud cover is to be selected from three alternatives or entered numerically. This is required because solar radiation increases the evaporation rate from a puddle of liquid.

11. Air temperature is to be entered. This is the air temperature in the vicinity of the release. Clearly, a higher air temperature increases the evaporation rate.

12. The atmospheric stability class can be entered from a selection of six. However, ALOHA will already suggest a stability class based on the entries so far for the time of day, wind speed, and cloud cover. The stability class represents the degree of turbulence and has a significant effect on ALOHA's predictions.

13. Inversion Height is entered or can be entered as No Inversion. Inversion is where an unstable layer of air near the ground lies beneath

a very stable layer above. The abrupt change between the two layers is the inversion height. ALOHAs 'Help' tell us that low ground fog is a good indicator of this type of inversion. However, it then says that an inversion causing smog typically has an inversion height thousands of feet above the ground. How do we know the difference if we are not meteorologists? Choosing inversion with a low height is conservative.

14. Humidity is to be entered from a selection of three options, or a value can be inserted. ALOHA uses humidity to estimate the evaporation rates from a puddle and 'atmospheric transmissivity'. ALOHA's 'Help' tells us that atmospheric transmissivity is a measure of how much thermal radiation from a fire is absorbed and scattered by water vapour and other atmospheric contents.

15. Data then needs to be entered about the source. A choice from four is required: 'Direct', 'Puddle', 'Tank', and 'Gas Pipeline'.

16. If 'Direct' is chosen, the user enters the amount of pollutants entering the atmosphere.

17. If 'Direct' is chosen, the user enters whether the source is continuous or instantaneous. A continuous source, for example, could be from a hole in a leaking pipe. An instantaneous source could be from the collapse of a tank or a railway tanker collision causing a severe breach in the tank.

18. If Direct is chosen, then the source height is entered. A ground-level height of 0 is the most conservative selection.

If 'Puddle' is selected, then the user enters the area of the puddle or its diameter and enters the estimated volume of the puddle.

If 'Tank' is selected, then the user selects from the type of tank and its length, diameter, and volume.

If 'Gas pipeline' is selected, the user enters the pipe's diameter, the pipe's length from the source, whether the pipe is connected to 'infinite tank source' or closed off, and whether the pipe is 'smooth' or 'rough'. This roughness of the inner surface – ALOHA kindly advises to use 'smooth' for new metal, glass or plastic pipe and to use 'rough' for older metal pipework likely to have suffered some corrosion. Then the pipe pressure needs to be entered along with its temperature and whether the hole size equals the pipe diameter.

For ALOHA, the above comprises the 18 parameters that need to be entered before coming to the result. The results are shown in the Display tab, which identifies the Threat Zones. For toxic materials, these Threat Zones are based on Levels of concern. It uses the level of exposure to a chemical that could hurt people if they breathe it in for a defined length of time. The user can choose from three levels of concern. They may be expressed as

> either AEGLs (Acute Exposure Guideline Levels), ERPGs (Emergency
> Response Planning Guidelines), or TEELs (Temporary Emergency Exposure
> Limits). There is also the opportunity to show flammable limits for highly
> flammable materials.

To show the sensitivity of dispersion models, I have entered below a
scenario for the breach of a propane road tanker of 10,000 USG capacity
(37 m^3) using:

A: Conservative parameters likely to lead to the most extensive threat
   zones
B: Optimistic, but credible, parameters likely to lead to reduced threat
   zones

I am hoping that the curious reader had the patience to do all of this. Of
course, we made many approximations and assumptions along the way. So,
it is interesting to do a sensitivity analysis using conservative and optimistic
parameters as I did for a hypothetical incident of a release from a propane
truck as indicated in Table 28.1.

Table 28.1 ALOHA® for Propane Truck – Sensitivity Analysis Climatic

ALOHA Parameter	Conservative Parameters	Optimistic Parameters	Units
Time	Midnight	Midday	
Building	Single Storied	Single Storied	
Surroundings	Unsheltered	Sheltered	
Chemical	Propane	Propane	
Wind speed	3	15	Knots
Ground Roughness	Open Water	Urban or Forest	
Cloud Cover	Clear	Complete Cover	
Stability Class	F	D	Selected by ALOHA
Inversion	Present	Not Present	
Inversion Height	10	–	m
Humidity	Dry	Wet	
Source	Direct	Direct	
Source	Instantaneous	Instantaneous	
Volume	37	37	Cubic meters
Source height	0	0	
Chemical stored as	Liquid	Liquid	
Temperature of chemical	Ambient	Ambient	

What you may find is that the distance to the Lower Explosive Limit, around 1,100 meters vary by no more than 10% whether conservative or pessimistic climatic input parameters are used!

However, the ALOHA output is vastly different depending on whether an instantaneous of continuous source for direct emission is selected. Now if there is a propane tanker collision, how do we decide over instantaneous over continuous. It is unlikely that the propane tank will be immediately destroyed giving rise to an instantaneous release of all the contents. It is more likely that the tank will be punctured, possible severely. The release of contents through the puncture will take place continuously over a period of time, but not indefinitely. Similarly, if the user tries to estimate using 'puddle' or 'tank' substantially different results will be generated. If selecting 'puddle', the user may have a difficult decision estimating the depth of the puddle resulting from the collision!

I do much appreciate the efforts of the U.S. Environmental Protection Agency in making this ALOHA® software available and I recommend the enquiring reader to use it to get some idea of consequences of whatever scenario they have in mind.

Commercial and proprietary software has an even greater number of inputs from which to select. Take care, however, that the user can gain an impression of precision which, in my opinion, exceeds that which these techniques are able to deliver, irrespective of the choice and cost of the software model used.

EXPRESSING CONSEQUENCES AS FN CURVES

In catastrophe prevention, we are primarily concerned with the consequences of an event which the potential to cause harm to many people. This is commonly called societal risk. If we take the example of a leak of toxic gas from a vessel, there are a range of possible outcomes and a range of possible frequencies at which these outcomes might occur. For example, a flange leak will be a relatively small emission but relatively high probability. Whereas a hole in the vessel wall is likely to result in a large emission but a relatively low probability. Other parameters other than the source of the emission will affect the magnitude of the consequences, such as wind direction, air stability, or turbulence, and how densely downwind property is populated.

A commonly used but somewhat complex way of expressing this range of possibilities is called an FN Curve. Creating an FN curve is a specialised task usually carried out by experienced consultants. The technique involves examining all credible scenarios, looking at each scenario under a variety of conditions, and calculating the total number of fatalities (the outcome) and the frequency of that outcome. These outcomes are then sorted with scenario having the highest number of fatalities first and accumulating the frequency per year as each scenario is added. For an excellent summary of

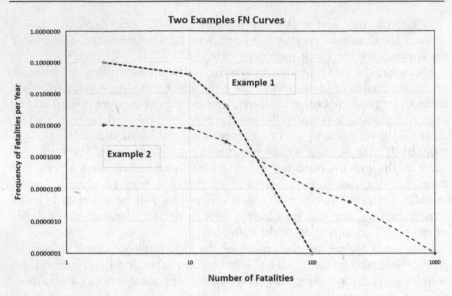

Figure 28.1 Typical FN Curve.

how this is done, see Ref 4 (Goose, 2010). Understanding does require a knowledge of exponents and logarithmic scales.

The result is a curve with a shape as shown in Figure 28.1.

The value of an FN curve is that the reader can quickly see, using this example, the risk of 5 fatalities, of 10 fatalities, of 50 fatalities, etc.

This is important in the context of Acceptable Risk see Chapter 3 on What makes Risk Intolerable or Intolerable. In Figure 28.2, the zones of

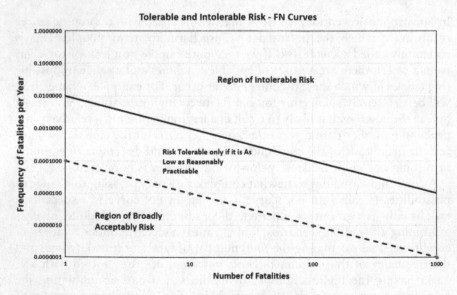

Figure 28.2 Tolerable and Intolerable Zones Shown on FN Curve.

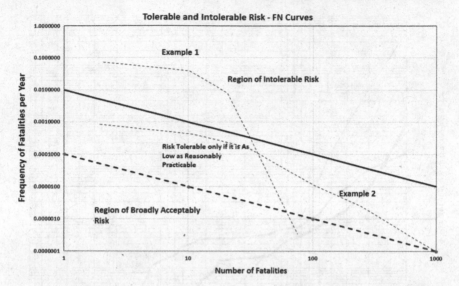

Figure 28.3 Overlay of FN Curves on Tolerable and Intolerable Zones.

Intolerable, Tolerable and ALARP are shown, broadly in accordance with R2P2 as was discussed in Chapter 3 on what makes risk tolerable or intolerable.

Now the value of FN curves starts to become apparent when they are plotted over the Tolerable, Intolerable and ALARP Zones (Figure 28.3).

As can be seen, Example 1 starts off in the Intolerable Zone for Up to around 40 fatalities, and then the potential drops rapidly. Above around 80 fatalities, the frequency per year is so low as to be in the broadly acceptable zone. Example 2, however, remains in the ALARP zone throughout.

What action does this imply? For example 1, measures must be taken to reduce the risk, for the scenarios leading to lower numbers of fatalities. For example 2, action needs to be taken to reduce the risk so long as it is reasonably practicable to do so (ALARP).

FN Curves are often produced as output from Safety Case Reports prepared under the Seveso regulations (COMAH in the UK). The question then is, do people understand them other than the specialists who prepared them?

In the HSE Report (2003) 'Transport Fatal accidents and FN-Curves', a study was made into the use of FN curves to evaluate various railway safety measures. See also Chapter 9 on Transportation. The FN Curves for Road, Rail, and Aviation are shown in Figure 28.4.

Not unexpectedly, road transport fatality frequency is higher than rail and aviation for single fatalities but falls off rapidly. In contrast, aviation has the lowest frequency for single fatalities but extends to accidents with over 100 deaths. However, we cannot use the FN curve to identify acceptable and unacceptable risk regions in this case. The acceptable risk zones

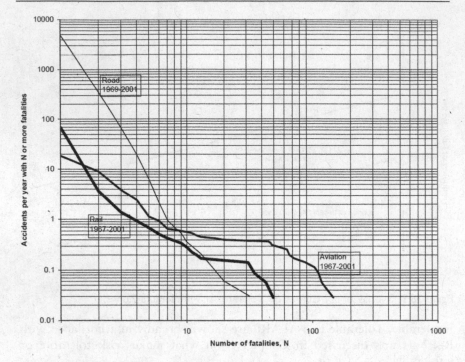

Figure 28.4 FN Curves for Transportation.

developed in R2P2 were specifically for facilities such as chemical plants, which have a single location on which the risk is centred.

CONSEQUENCE MITIGATION

Emergency Response and Readiness has been discussed in Chapter 23. Considering, in particular, the release of hazardous material, several specific mitigation barriers are available:

1. Bund walls surrounding storage tanks so that any spillage is retained. Unfortunately, these sometimes fail to provide the protection expected. Drain valves to drain the bund of rainwater can be inadvertently left open. Often constructed of rubble and earth, these bund walls are initially installed with an impermeable membrane, but this membrane degrades with time, allowing spillage to seep out through the pores of the rubble. Concrete walls are better but have joints that also provide a leakage point.
2. Steam curtains are used as a permanent installation to contain and disperse leaks of flammable gases which are heavier than air. Often, unfortunately, the automatic valves that can allow the remote

operation of such curtains from the control room or remote station are inactive. Either they are switched to local control only, or an upstream manual valve is closed. Under these circumstances, the steam curtain cannot be activated unless an operator goes into the affected zone to open the manual valve.

3. Water sprays, either fixed or mobile, can knock down vapours, especially water-soluble ones. They suffer from the same misuse as steam curtains with automatic valves rendered useless and manual valves closed, disabling remote activation.

Notes from Trevor's Files: Quantitative Risk Assessment

I was working on a project to build a Choralkali plant. The idea was to eliminate the transport of chlorine by ship to a polymer plant and instead make chlorine on the polymer plant by the electrolysis of salt. Cost savings were anticipated as well as safety benefits from eliminating the shipping. It was anticipated that some brine produced on-site from the polymer process could be used in the chloralkali plant reducing cost and total effluent. It was further anticipated that the amount of chlorine stored on-site could be reduced to below the levels required for registration under the Seveso regulations.

During this work, I was asked to participate in a quantitative risk analysis of the shipping operation. I am interested in maps, so calculating the potentially exposed populations based on large-scale maps was a fascinating job for me. For others, maybe it would be tedious. I fully immersed myself in this task.

Now there were competing interests in this study. On the one hand, the business management for the polymer plant wanted to justify the chloralkali plant, but on the other hand, not to provide evidence that might imply that the chlorine shipping operation was unsafe.

Members of the general population might think that this is a matter of putting the data into a calculation machine and coming out with an answer. In practice, there are sufficient subjective elements that self-interests can influence the whole process. For example, the concentration of chlorine at various points after a breach of the ship's chlorine tank can be calculated. I remember vividly going to see an expert in the U.S.A. who pointed out that in practice, the actual effects of known chlorine releases were much less than those predicted. His interest was in supporting the company's many operations using chlorine and finding arguments in support of that industry.

In practice, the chloralkali plant was found uneconomic, and it was never built. Perhaps we were expecting it to meet too many demanding objectives.

SUMMARY: CAUSES OF FAILURE IN CONSEQUENCE ANALYSIS AND DISPERSION MODELLING

Reasons Why Consequence Analysis and Dispersion Modelling Can Fail to Avert or Predict a Catastrophe Include:

☒ Computer modelling requires gathering and inputting a lot of data, not all of which is readily known.

☒ Some parameters may produce very different consequences, identification requires a sensitivity analysis in which each parameter is adjusted above and below the previously assumed value.

☒ Different models giving different results, or results which differ substantially from the observed event destroy credibility in the technique.

☒ The principle of FN curves can be misunderstood.

Chapter 29

Software Safety and Cyber Security

SOFTWARE SAFETY ANALYSIS

Whilst reviewing safety instrumented systems, we analysed and classified control loops according to their Safety Integrity Level (SIL). A control loop typically, but not always, involves some programmable electronics (software). Software often goes beyond simple control loops in a modern processing plant or nuclear facility. Entire start-up and shutdown sequences can be programmed. Instrumentation can detect failures in utilities and initiate a sequenced shutdown. For example, in the event of a partial loss of electricity, a complex facility can go through an automated routine called 'load shedding', where the less critical equipment is shut down first to protect and ration electricity for the most critical processes. In this case, critical usually means that the load shedding protects those parts of the process which cause the most economic loss if shutdown. They could also be parts of the process that would be the most difficult to restart if shut down suddenly.

In an oil refinery with a Fluidised Catalytic Cracker (FCC), the FCC is often a priority unit to sustain. If it suddenly shuts down, the fluidised catalyst becomes no longer fluidised. It settles in equipment and pipework from which it is difficult or maybe impossible to re-fluidise once electricity is restored.

Software that is safety-critical (part of a SIL system) should be located in a high-reliability programmable controller (logic solver). Safety-critical software should not be located in the general controls (Basic Process Control System – BPCS). If erroneously located in the BPCS, the software is more exposed to either deliberate or accidental corruption.

However, the distinction can be blurred. The load shedding example above would probably be located in the BPCS, or perhaps in Supervisory Control, even though the failure of this load shedding operation could have significant economic consequences. Supervisory control typically involves a computer that accesses set points on the BPCS to optimise the running of the process. Supervisory control uses various inputs which can change daily such as the sales value of the different products. Supervisory control is not considered safety critical.

Supervisory Control and Data Acquisitions Systems (SCADA) are sometimes used for simple control and are often used in fuel distribution. Such systems are not high-integrity devices, and they should not be used for safety-critical controls. Our hypothetical fuels terminal would have a SCADA.

In Chapter 9 on Transportation, the catastrophe of the two Boing 737 Max plane crashes was described. It is understood to have been caused by: 'Software design faults leading to conflicts with pilots' intentions. Incomplete pilot training in the software is also suspected'.

How often do we come across software faults in our daily lives? I have just finished booking a short holiday with a large hotel group. Whilst booking our room, I came across several faults in the online software. The system did not allow me to see what I had already booked. It also allowed me to double book. Whilst this is not safety-critical, one would think that it was critical for the hotel group in terms of economics and efficiency. It was in their interests to test it thoroughly.

When trying to pay for the wonderful Bow Tie software that was used to describe several incidents in this book, I found that the banking internet software's drop-down boxes are not working and the box for the IBAN is too small. One would think that software relating to financial transactions would be thoroughly tested.

I am an amateur software geek. I am proud of my hydroponic greenhouse, where the pumps and sensors are controlled wirelessly from my computer. Creating this required some considerable knowledge of the computer programming system 'Python' for the computer part, and 'Arduino IDE' for the greenhouse bit. Additionally, it was necessary to learn how to programme ZigBees for wireless communications. I have never yet written a piece of software that works the first time. Troubleshooting requires testing the software and correction of the bits that do not work. But suppose there is one aspect which I forget to test. The programme works fine for months until one day that a particular combination of inputs occurs. For example, change of the clocks in the Spring and Autumn. I could not test this at the time of my test sequence. The programme fails at the clock change, and my plants start to die!

Software programmers devising software for aeroplanes and chemical plants are, of course, much better trained and more meticulous. But I understand the general procedure is the same, write the software and then test, test, test. And as we can see, problems can still remain.

Redmill et al. (1999) describe the application of HAZOP to Software. This takes a line-by-line approach to the software, asking a number of guidewords like a HAZOP in a process operation. However, what should the software HAZOP use for its evaluation? Software is generally designed in several phases:

1. Word description (Manuscript)
2. Logic Diagrams (Diagrams)
3. Uncompiled programme (Script – Words which the software compiler can understand)
4. Compiled programme (now in machine code which the computer chips can understand)
5. Execution and Test.

Errors can occur at every step. In particular, the uncompiled script will often not work exactly as the logic diagrams require.

My wife is the proud owner of a four-wheel-drive sports utility vehicle, which is a wonderful car and totally reliable in every respect, but one. After owning the car for nearly a year, she received a safety recall notice which required an upgrade to part of the software in the vehicle. In the worst event of fire, a software bug could conceivably lead to electrical component overload and smoke. Checking the internet, I see that it is not the only brand of car requiring software recalls!

I find the issue of Software Design and Testing one of the biggest concerns in present-day Catastrophe Prevention.

Cyber Security

Considerable expense is being incurred to make cyber security more robust, yet events involving security breaches are becoming more prevalent. All of this software is open to malicious interference. Cyber security's job is to protect the software from such interference.

Incidental Case Studies: Pipeline Ransomware and Stuxnet

A company that transports refined oil products by pipeline was the target of a ransomware attack in May 2021. It was reported that the hackers began their attack by stealing about 100 gigabytes of data and threatened to do further damage if they weren't paid. This led the pipeline company to shut down the entire pipeline to stop the malicious software from spreading. While this did not result in a catastrophe, the potential is clear.

Stuxnet, a computer worm, was first uncovered in 2010. Stuxnet is believed to be responsible for causing substantial damage to Iran's nuclear programme.

Stuxnet explicitly targets programmable logic controllers (PLCs), which control machinery and industrial processes. In targeting the Iran nuclear programme, the critical machinery included the centrifuges for separating nuclear material. Stuxnet reportedly compromised Iranian PLCs, collecting information on industrial systems and causing the fast-spinning centrifuges to tear themselves apart. The worm is said to have spread beyond its original designed targets due to a software error. Targeting industrial control systems, the worm infected over 200,000 computers and caused damage to around 1,000 machines. (Wikipedia: Stuxnet)

Notes from Trevor's Files: Password Protection

Whilst surveying an oil refinery, I was amused to pass a field control station that had a weatherproofed computer above which there was a sign in large bold letters reading 'ID: ABCD, Password:1234'.

This was especially amusing to me since one of my favourite programmes is a French political puppet satire 'Les Guignols de L'Info'. Les Guignols had a sketch on a former French president who had 'gone off the rails' and was attempting to set off a nuclear attack. The Code and Password were ABCD, 1234 and were stencilled onto the bottom of the computer!

Later during the oil refinery survey, we had a session on cyber security and were told that everyone on the plant had received basic training in Cyber Security, including maintaining the confidentiality of passwords. The above observation is, of course, not as bad as it first sounds. The field computer was entirely standalone and not connected with any other systems. Nevertheless, it sets a bad example.

The oil refineries and chemical plants I visit typically describe a five-level information technology architecture:

Level 4 – Enterprise applications providing information for business planning, financials, procedures, etc.

Level 3.5 – Asset Inventories, and any remote access

Level 3 – Operations and Maintenance Management

Level 2 – Data Acquisition from the process including the Distributed Control System

Level 1 – Programmable Logic Controllers, Sensors, and Devices

Level 0 – Machines

Levels 0 to 3.5 inclusive are considered 'Operational Technology' and are protected from the Business Management system by a 'Firewall'. Many cyber-attacks, especially ransomware attacks, are targeted on the Business Management element and do not put the operating process at risk. However, an increasing number of attacks are on the Operational Technology.

Facilities in the EU are subject to a Directive on the security of Network and Information Systems (EU 2016/1148) requiring improvement and audit of cyber security.

Many facilities ban the use of 'flash sticks' or have USP ports on computers disabled.

However, access to plant operational technology is required by outsiders. For example, the organisation which provides and maintains the Distributed Control System.

Providers of complex critical equipment such as large compressors offer a service where they receive data continuously from the online vibration sensors and other condition monitoring sensors on the compressor and have their experts analyse the data remotely without visiting the site. This is a powerful service that also assists reliability as well as process safety. However, many companies now consider remote monitoring too hazardous from a cyber security perspective.

It should also be emphasised that firewalls do not provide 100% security and that hackers are finding ways to penetrate firewalls.

Cyber security will be a developing science over the next few years, and it will be a struggle to keep ahead of the hackers.

SUMMARY: CAUSES OF FAILURE IN SOFTWARE SAFETY AND CYBER SECURITY

Reasons Why Software Safety and Cyber Security Can Fail to Avert or Predict a Catastrophe Include:

- ☒ Locating the safety-critical software in a part of the IT system architecture which does not have high reliability.
- ☒ Software faults are hard to find. Test simulations may miss the combination of inputs which cause an erroneous result.
- ☒ It is a difficult job to carry out a line-by-line review (in a similar manner to HAZOP) and find the fault.
- ☒ Higher-level software versions, such as logic diagrams, may not correspond to how the software is actually programmed and responds.
- ☒ Hackers quickly identify new vulnerabilities; speed is needed to keep pace with cyber security developments.

Chapter 30

Risk Assessment Conclusions and Implementation

WHAT ARE THE CONCLUSIONS OF THE RISK ASSESSMENT?

So, in the preceding chapters, we have completed our risk assessment after considering all hazards, risk probabilities, and risk consequences and assessing the current risk's magnitude. If the current risk is unacceptable, we have proposed recommendations to improve. These have been designed. It has been shown that the risk will be acceptable once the recommendations are implemented.

Should we be happy?

Risk assessments are open to criticisms and doubts. Here are some of those listed by the HSE in the UK who consider that a risk assessment may:

- underestimate the true impact of a problem overall. A risk assessment is always undertaken for a specific purpose may therefore overlook other risks. For example, a risk assessment may be used to assess the danger to people, but not evaluate the environmental threat, cause of financial loss, etc.
- may be used is used to legitimise decisions, which have already been taken and would be hard to reverse.
- is inadequate since it often reduces a complex issue to a single number (e.g. in quantitative risk assessment) and is therefore ineffective in taking into account societal concerns or other important factors such as the degree of trust between regulators and their stakeholders (HSE (2001) Reducing Risks, Protecting People, R2P2).

Notes from Trevor's Files – HAZOP Oversight

I was asked to follow up on an incident involving a significant fire at a petrochemical plant. Fortunately, no one had been injured, but the plant suffered over EUR 100 million in damage and nine months of downtime. The particular

---

section of the plant, which was the source of the hydrocarbon release and seat of the fire, had undergone some modifications several months before, whereby a new distillation column was introduced between compression stages. These modifications had been subjected to a risk assessment (using HAZOP) before commissioning and recommendations implemented. However, the incident investigation identified that the additional inventory of hydrocarbons produced by the new column caused dynamic surging in the downstream compressor. This surging led to the rapid mechanical failure of the compressor under certain flowrate conditions. This caused the initial release of hydrocarbons.

The incident investigation also found that, whereas the original plant had emergency isolation valves on each side of the compressor, no new emergency valves were considered necessary around the new distillation column. The result was the fire could not be isolated from the flammable materials held up in the new column. These flammable liquids continued to drain out, fuelling the fire.

The failings in the risk assessment concerning the new column are reasonably subtle. The issue of dynamic surging created by the additional inventory would be difficult to foresee in a HAZOP; the HAZOP guidewords would not necessarily generate consideration of it. The site subsequently recognised that the HAZOP was too focused on the new distillation column and did not consider upstream and downstream consequences. Hence the omission of downstream compressor surging and the need for additional emergency isolation valves at the boundary of the new process.

Notes from Trevor's Files: Risk Assessment Secrecy

As a team of consultants, we were tasked with evaluating safety at an oil and gas production site in Kazakhstan, which had suffered several potentially catastrophic gas releases. The operation was a joint venture, and on my team, there were representatives of the joint venture partners. This is relevant in that it created some cultural and politically charged issues. The oil/gas field was very sour, i.e. high in contamination with hydrogen sulphide (H_2S). The facility was at the forefront of technology in extracting the H_2S and reinjecting it back into the reservoir; prior practice in the industry was to separate out the hydrogen sulphide as molten sulphur, which was then stored in yellow piles causing an environmental issue and overload of the sulphur market when companies tried to sell it.

One of the criticisms we had heard of the organisation's safety culture was that management was not open about the results of risk assessments. In particular, there had been doubts about the metallurgy of the reinjection pipework, especially in the cold (and it is very cold in winter in Kazakhstan!).

Various measures had been taken to protect the pipework, and replacement was planned. However, rumour had it that a risk assessment undertaken by external consultants, at the organisations request, had concluded that the residual risk was unacceptable.

We wanted top management to appreciate that their secrecy on this type of issue was damaging the safety culture all around the operation. The CEO got exceedingly annoyed with us raising the issue, but his annoyance prevented him from hearing the message we were trying to bring. We had no basis for criticising the technology; we wanted to help him improve the safety culture. In truth, he had not communicated the risk assessment because he and his team doubted that the external risk assessors had the correct data to come to their conclusions. In which case, please say so! I think this is a credible message to send, rather than keeping silence.

I have worked with an organisation that has received thousands of re-commendations from the HAZOP teams in carrying out HAZOP reviews of all their operations. Filtering these so that the worthiest get attention is a mammoth task. This mountain of slow-moving recommendations leaves the HAZOP team participants frustrated and leads to a feeling that the whole risk assessment task is something of a sham. My suggestion, in this situation, is to use the risk matrix described in Chapter 27 on Assessing Risk Outcomes as a filter, so that only the recommendations tackling risk towards the top right-hand side of the matrix get attention.

Often, the organisation is left without any communication about the risk assessment progress. I proposed a stage-gate system that deals with the communication process and gives higher management control over the conduct and decision-making process regarding the assessment.

Risk Assessment Stage – Gate I

This higher management meeting occurs after the Risk Identification and after the Risk Assessment Team has been provisionally named. Preliminary information has been assembled for the use of the Risk Assessment Team. Typically, this meeting includes the Operations Manager and Technical Manager plus others depending on the structure of the particular organi-sation. After this stage-gate is complete, communication is sent to the em-ployees in the part of the organisation being risk assessed. This tells the operating community of the activity. The communication explains why some of their members are being taken away in order to carry out the study, and a timeline over which the study will take place. The intention is to give the operating community value for the activity and raise awareness and expectations.

Risk Assessment Stage – Gate 2

This higher management meeting occurs after the risk assessment (e.g. HAZOP) is complete and submitted a draft report. The higher management team replies with a Management Response Letter, documenting an individual response to each recommendation, either accepting the proposal as is, suggesting modifications, or rejecting the proposal for a specified reason. This letter should be available to members of the operating community so that they are aware of the recommendations and their status. Note that the proposals have not yet undergone engineering design or evaluation of alternatives. Neither have the risks to be reduced or mitigated by these recommendations been evaluated by Layers of Protection Analysis, so the residual risk remains the informed estimate of the risk assessment team.

Risk Assessment Stage – Gate 3

This meeting of higher management takes place after Layers of Protection Analysis, evaluation of alternatives, and engineering design. The risk assessment is closed at this stage. The status of the recommendations is communicated to the operating community. At this stage, capital investment has not yet been approved, and installation of any hardware has not yet been firmly planned.

Risk Assessment Stage – Gate 4

This final meeting occurs after the recommendations have been implemented, including any hardware. Effectiveness of the recommendations should have been tested. Based on this data, the higher management meeting closes the recommendations and communicates to the operating community.

CREDIBLE SCENARIOS

In risk assessment, people sometimes ask what scenarios they should consider for two events happening simultaneously.

Notes from Trevor Files: Emergency Response Planning – Openness

A colleague of mine was very amused when watching a television programme involving a reporter visiting a chlorine plant's control room. The organisation intended to build public trust through an 'open door' policy. The plant manager was explaining to the TV interviewer the emergency response system in the event of a chlorine leak. He described how they had calculated potential exposure extents for accident scenarios. Rather ill-advisedly, he pointed to some plastic ellipses hung on the control room wall. These had been constructed to be quickly laid down on the map on a table. Knowing the wind direction, the areas potentially affected by the chlorine release could be quickly determined. The TV interviewer immediately approached the largest ellipse, which would cover several miles on the map and engulf some significant population centres. What's this one for, he asked? Stuttering slightly, the manager replied that this was for a scenario which 'was not credible' and therefore should be ignored. His explanation was not convincing!!

I expect that the large ellipse was for a catastrophic failure of one of the chlorine storage tanks. Such a failure would indeed be improbable if the tanks were being adequately inspected and so long as the chlorine was dry. Was this a credible scenario or not?

The answer can best be found by considering probability and consequence matrix again. We have Catastrophic Consequence and Very Unlikely Probability. Should we consider Catastrophic – Very Unlikely as a credible scenario? I think the experience of incidents such as the Deepwater Horizon suggests that indeed we should.

There are some exceptions. Take aircraft impact. The likelihood of an aircraft crashing into a facility is indeed extremely low. However, the consequence of such a significant event itself (i.e. the aircraft crash) would typically be more than the consequences of damage to the facility. There might be 300 people on the aeroplane, but only 50 in the hazardous facility. However, if the facility is a nuclear power plant, then a radioactive release's consequences would exceed that of the aircraft crash. This scenario should be considered, especially since the probability is increased by the potential for an act of sabotage or terrorism.

SUMMARY: CAUSES OF FAILURE TO PREVENT CATASTROPHES WHEN CONCLUDING RISK ASSESSMENTS

Reasons Why Closure of the Risk Assessment Can Fail to Avert or Predict a Catastrophe Include:

- ☒ Underestimation of the magnitude of the problem – focused on the particular risk assessment and are blind to neighbouring risks.
- ☒ Used to legitimise decisions already made.

☒ Oversimplify, for example in QRA where a risk assessment is used to come up with a single number.

☒ Organisation is overloaded with too many recommendations.

☒ Progress and status of recommendations is not communicated to the operating community associated with the operation being assessed.

☒ Not using a stage-gate process driven by higher-level management to control the process, its communication and readiness to move on to the next step.

Chapter 31

Governance of Safety

SAFETY MANAGEMENT SYSTEM ELEMENTS

So far, we have focused on 'frontline' elements of safety – on operators, contractors, and their performance, together with some consideration of their tools – the equipment they use and its maintenance. These are far from being the only elements considered essential to an organisation wanting to ensure high safety standards. We have also studied the role of techniques that an organisation might use to improve its defence against catastrophic incidents such as process safety reviews.

We now turn our attention to the role of the organisation's management. It is good practice to subdivide safety management into several different elements. This is broadly termed 'Governance of Safety'.

In Section 2, we identified eight incidents where issues with governance of safety contributed significantly to the event. These included Deepwater Horizon, Bhopal, La Porte, and Brumadinho.

Before we focus governance of safety and on management, it is appropriate to look at the structure of safety management, to provide a guide as to what management should ensure is happening.

The U.S. Occupational Safety and Health Administration (OSHA) define 14 elements (OSHA Process Safety Management 1910.119)

- Process Safety Information
- Process Hazard Analysis
- Operating Procedures
- Training
- Contractors
- Mechanical Integrity
- Hot Work
- Management of Change
- Incident Investigation
- Compliance Audits
- Pre-start-up safety review

- Emergency planning and response
- Trade Secrets
- Employee Participation

However, I suggest that this list is not holistic. Du Pont developed a list of 14 elements arranged around a wheel at the centre of which is Management Leadership and Commitment.

Within Management Commitment are considered issues such as: (Figure 31.1)

- Policies and Principles
- Goals, Objectives, and Plans
- Line Management Responsibility
- Safety Personnel

This is one of the strengths of the PSM Wheel representation. It shows the core governance issues as the driving forces behind the success of the techniques on the outer circle.

Most of the other elements in the wheel are similar to the OSHA list.

Different organisations have adopted other elements, although there are many common elements. They are not always presented in the same form or

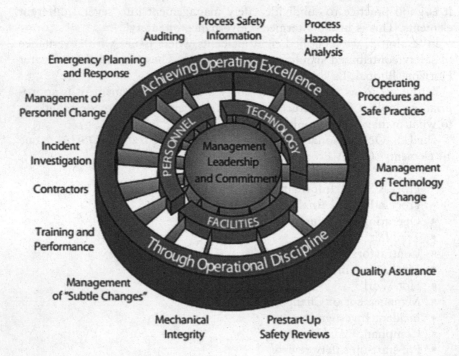

Figure 31.1 Du Pont PSM Wheel ©DuPont.

grouping. For example, Bahr (2015, p. 80) lists 5 Elements with a total of 29 sub-elements. There are reasons for the differences. The Du Pont Sustainable Solutions example is specifically targeted at process safety and the prevention of major accidents and catastrophes. The Bahr version is intended to cover all aspects of safety, including personal safety, including elements such as Medical Issues and Document Control. However, the number of variations in how a Safety Management System is subdivided are somewhat bewildering and make comparisons between organisations complex since they will probably be using different Safety Management System Elements. We will return to this in the paragraphs on Audit later in this chapter.

The challenge for senior management becomes demanding. For process safety alone, the senior manager needs to keep his/her eye on 18 or more individual elements. As a safety consultant, I have assessed many organisations against these elements of process safety. In a closing meeting lasting, typically, several hours, the results for each element are presented back to senior management. Initially, the presentation is treated with great interest. Then energy declines as each sub-element are reviewed.

This is too much for senior management to take in, amongst all their many other responsibilities. It is also often presented after an assessment as many improvements for each sub-element. Senior management, therefore, see their task as getting each of these opportunities for improvement acted upon, and then they will be in good shape. For a robust safety management system to grow strong, this all needs to 'come together', a subject we will return to in Section 4

MANAGEMENT COMMITMENT

'Safety is only as important as management wants to make it' (Bahr, 2015).

Can accompany management choose how safe it wants to be? It is most likely that an organisation will choose to be legally compliant since to do otherwise would lead to severe fines and potentially imprisonment. Organisations often point to their legal compliance as a means of demonstrating safety. Legal compliance should be regarded as a minimum.

Organisation management may choose to become safer because it makes business sense. 'If you think safety is expensive – try having an accident' is a quote often used by my consulting colleagues.

What about the moral imperative? As individuals, we develop some moral values during our childhood, and I think we would probably all agree that the moral values which a young person has when leaving school are highly varied. Moral values are then, to some extent, justified by the institutions which we join.

Notes from Trevor's Files: Safety as a Core Value

In my early days as an engineer at Du Pont, we were all told that 'Safety is a Core Value'. But that really did not mean much to me until I saw the lengths that the organisation went to achieve exemplary safety and resilience against catastrophic incidents.

It was only when I saw management's actions that Safety started to become a core value for me as a person as I grew within the Du Pont culture at the time. Such many actions included:

- Starting each meeting, even hurried ones, with a safety contact.
- Time taken to do a defensive driving course every year.
- Attention paid to audit results from safety department and corporate auditors.
- Time taken for process safety reviews including one of a plant which was imminently going to shut down permanently.

At the same time, the messages were neither perfect nor consistent. We will return to the ease with which such a safety culture can be damaged later in Chapter 33.

This author considers that Safety is only as important as society wants to make it, but the onus of responsibility for that determination lies with management. I also agree that management can decide to disagree with the social pressures and governmental pressures and operate in a more or less safe way. That is their choice within the law.

Incidental Case Study: Du Pont

Du Pont has, or had, a good safety performance driven by a strong safety culture. It also made Du Pont an excellent place to work. Some of the drivers for the management commitment which led to this culture merit consideration.

- The foundations of the company were in gunpowder manufacture in the 1800s. In the early years of the gunpowder company there were many tragic incidents. For example, in 1818 there was an explosion at the company's founding mills on the Brandywine River which killed 40 employees and injured the founder's wife. Interestingly the incident was attributed to drinking alcohol by one of the employees (Kinnane, 2002). In 1890 12 people were killed following a highly destructive explosion, also at Brandywine (www.hagley.org).

- The manufacture of Du Pont's first synthetic rubber – Neoprene – was particularly hazardous being first produced from acetylene. The process involved the manufacture of explosive intermediates. Twelve people died in an explosion at a Neoprene plant in Louisville, Kentucky, in 1965. As a young engineer, I became involved with the manufacture of Neoprene by a similar process, located in Northern Ireland. After the 1965 disaster, the process was adapted to operate with a diluent gas to reduce the potential of an explosion. I remember a 'top boss' visiting us frequently in the 1970s from the U.S. and each time walking up to the control panel to check that we were controlling the diluent properly.
- Changing the method of manufacture so that acetylene was not used and explosive intermediates made more benign was carried out at the plant for Neoprene (an example of the adoption of Inherently Safer Design). This improvement, however, also failed to prevent catastrophes with multiple fatalities resulting from an explosion in the U.S.A in 1983.

The bottom line is that the company had several shocking incidents in its history. This impacted deeply the way the company approached safety and was the root cause of the management commitment. So this brings up a question – in the absence of first or second-hand experience of a significant catastrophe, can management maintain the same commitment to safety? The answer, in this author's opinion, is that commitment declines over time. We will return to this later.

The term 'risk appetite' is often used in risk management as something an organisation's leadership must determine. Risk appetite is the level of risk that an organisation is prepared to accept in pursuit of its objectives before action is deemed necessary to reduce the risk. It represents a balance between the potential benefits of innovation and the threats that change inevitably brings. The ISO 31000 risk management standard refers to risk appetite as the 'Amount and type of risk that an organisation is prepared to pursue, retain or take'. This concept helps guide an organisation's approach to risk and risk management (Wikipedia: ISO 31000). Risk management in this context covers a broad range of risks, including financial, environmental, and reputational risks.

However, I have never seen an organisation state its risk appetite in the context of safety (and prevention of catastrophes) other than in polarised statements such as 'we will never carry out any operation unless it is completely safe'. This widely used statement only expresses an aspiration. It does not define a level of risk acceptance. So, it is often possible to say on the one hand that 'we will never carry out any operation unless it is completely safe' and, on the other hand, be operating a plant with many risks in the intolerable sectors of the risk matrix. There are such companies. I know because I have had access to many process hazards reviews, which have informed me of the current process risk.

Risk assessments themselves are open to influence by management depending on the messages they advertently or inadvertently send to their operations and technical staff carrying out the review. This theory has been used to explain why, for many new hazards, high-quality risk assessments by leaders in the field often fail to reassure people. Even using all available data and the best science and technology, many risk assessments cannot be undertaken without making many assumptions, such as the relative values of risks and benefits or even the scope of the study. Parties who do not share the judgemental values implicit in those assumptions may well see the outcome of the exercise as invalid, illegitimate, or even not pertinent to the problem (R2P2).

Discussions on acceptable levels of risk are a particular problem for government and civil servants. If I take the example of the risk to front-life staff during the Covid-19 pandemic. I ask the following questions, whether it be in the context of coronavirus or any other transmissible disease:

- Is there a risk that nurses can catch a disease from their patients? Surely the answer is Yes.
- Is this risk reduced if the nurses wear personal protection equipment? Surely the answer is Yes.
- Does the risk increase if the personal protective equipment is of poor quality or not to specification? Surely the answer is Yes.
- Would the risk to nurses during the pandemic be reduced if they had had access to more and better quality personal protective equipment? The answer can only be yes.

Yet there is considerable denial even today that nurses are at risk, and were especially at risk during the early days of the pandemic.

ACCEPTANCE OF RESIDUAL RISK

In 'Assessing Risk' in Chapter 27, a risk matrix was introduced which showed consequence ranks and probability ranks and for each combination indicated if the risk was acceptable or not.

Most organisations have such risk assessment matrix with different scales for Health and Safety Risk, Environmental Risk, and often Financial Risk.

Figure 31.2 Risk Matrix Example – Tiles of Most Interest.

Sometimes organisations also have scales for Reputational Risk and other parameters.

As part of my research for this book, I examined the risk matrices for seven medium to large companies. It should be noted that these risk matrices are not available to the general public. These were examples I had obtained through my work. I found very significant differences in how both probability and consequences are portrayed (Figure 31.2).

In the context of catastrophes, we are most interested in low probability high consequence events, so we will concentrate on the tiles to the lower right of the matrix as indicated in Figure 31.1. Tiles to the upper right are unacceptable. Typically, organisations have several categories of consequences, including health and safety, environment, and financial. Other categories can be reputational risk, quality and customers, etc. ... For our purposes, we will examine differences in definitions for the health and safety matrix.

First, let's look at how the seven organisations defined the most extreme health and safety consequence. Actual data from the sample matrices has been simplified and made more straightforward for the non-technical reader (Table 31.1).

The comparison shows some distinct differences as well as similarities. This consequence level always includes multiple deaths.

Some of the organisations studied also include health effects from which the injured are not expected to recover. Lung damage, or damage to the nervous system, due to a release of a toxic material might be an example. More significantly, some organisations do not recognise such health effects in their risk matrices.

One of the organisations in the study distinguishes injuries to the workforce and injuries to the public, with injuries to the public having a lower threshold. This is consistent with researched data on what makes risk tolerable or intolerable (see Chapter 4). The employee has knowledge of the risk and choice, whereas members of the public typically have little

Table 31.1 Risk Matrix Health and Safety Consequence Levels – Most Catastrophic

Organisation	Consequence Label	Description
A	Catastrophic	Multiple Deaths
B	Severe	Fatalities or Serious injury requiring medical treatment to members of the public
C	Catastrophic	Multiple fatalities or Significant health effects to a number of people who will not fully recover from these effects
D	Level 5	Workforce: Multiple Fatalities or More than 6 people hospitalised Public: 1 fatality or more than 6 people hospitalised
E	Severe	On-site: 1 to 10 fatalities or More than 10 lost work cases for which full recovery is expected or 1 to 10 injuries involving major health effects from which full recovery is unlikely Off-site: Fatalities or Injuries involving major health effects from which recovery is unlikely
F	Extreme	>10 fatalities, significant life-shortening effects or permanent disabilities
G	Catastrophic	Multiple deaths

understanding of the risk and little choice over their exposure. Another of the organisations makes a distinction between on-site and off-site. Note that this is not the same as workforce and public. It certainly includes members of the public but also includes members of the workforce who may leave the site on business and be injured, for example, in a road accident.

Overall, these are considerable differences, revealed by examining just one (health and safety) of the five typical consequence classes used.

Let's turn now look at how the seven organisations defined the most extreme financial consequence.

Actual data again be simplified for the non-technical reader. Currency amounts are rounded to just one significant figure, since we are interested in demonstrating how the matrices are used, not the detail (Table 31.2).

Substantial differences might be expected according to the size of the organisation. The impact of an event on a similar size would be more significant on a small rather than a large organisation. A financial comparison of the sample organisations would be unfair, and the data is not available for privately or state-owned organisations. However, an eyeball comparison, looking at available information on the number of employees and sales revenue, does not suggest any direct comparison with the chosen financial loss categories shown above.

Table 31.2 Risk Matrix Financial Consequence Labels – Most Catastrophic

Organisation	Consequence Label	Description
A	Catastrophic	> USD 10 million
B	Severe	> USD 2 million
C	Catastrophic	> USD 100 million
D	Level 5	> USD 10 million
E	Severe	No financial categories
F	Extreme	> USD 100 million
G	Catastrophic	> USD 100 million

Let's move on now to the definition of probability. We will take the probability level to the bottom of the matrix, i.e. the least probable classification (Table 31.3).

Probabilities have been converted into the likelihood of occurring in the stated number of years. Several organisations provided two-word descriptions and one quantified description, as shown. In some cases, the three probabilities are not compatible, leaving the matrix user free to choose which description most suits their need. Several of the organisations refer to 'the industry'. What is meant here? For example, a polymer plant might be considered part of the plastics industry, the petrochemical industry, or even the larger oil, gas, and petrochemical industry.

Table 31.3 Risk Matrix Probability Labels – Least Probable

Organisation	Probability Label	Description 1	Description 2	Might Happen once in xxx Years
A	Very unlikely	Happened a few times in the industry		10,000
B	Practically impossible	Practically impossible. Has happened once or not at all in the organisation	Has happened a few times or not at all in the industry	10,000
C	Very unlikely (remote)	Incident not known to occur	Very rarely <1%	100,000
D	Practically impossible	Never heard of before in the oil, gas, and chemical industry	–	–
E	Extremely unlikely	Not realistically expected to occur	about 1 in 100,000 years or less often	100,000
F	Extremely improbably	Never expected to occur	<1% in the lifetime of the plant	10,000
G	Very unlikely	Seldom heard of in the industry		10,000

Probability	Consequence				
	Very Minor	Minor	Serious	Very Serious	Catastrophic
Frequent	3	2	2	1	1
Occasional	3	3	2	2	1
Seldom	4	3	3	2	2
Unlikely	4	4	3	3	2
Very Unlikely	4	4	4	3	3

Figure 31.3 Example Matrix for Hypothetical Examples.

To demonstrate the difficulties of working with such a risk matrix, I have created in Figure 31.3 a fictitious matrix with the help of typical data from the sample organisations.

The numbers within each tile refer to the action which should be taken if a risk is identified to be in this Consequence – Probability Category. Note that four categories are used for simplicity. Some organisations use as many as six (Table 31.4).

For Probabilities we will use, for the sake of demonstration (Table 31.5):

Table 31.4 Risk Matrix Example – Actions Required

Action Category	Label	Action Required
1	Extreme Risk	Activity shall be stopped immediately, or immediate mitigating measures put in place
2	High Priority	Action plan to put in place control measures required
3	Moderate Risk	Control measures required, unless demonstrated that current measures are 'As Low As Reasonably Practicable' (ALARP)
4	Low Risk	Additional control measures not needed; risk is acceptable

Table 31.5 Risk Matrix Example – Probabilities

Probability Category	Description	Quantified Description
Very Unlikely	Seldom heard of in the industry	Less than once in 10,000 years
Unlikely	Heard of in the industry	At least once in 10,000 years
Seldom	Occurred in the industry more than once per year	At least once every 1000 years
Occasional	Happened at the location	At least once in 100 years
Frequent	Happened several times at the location	At least once in every 10 years

Table 31.6 Risk Matrix Example – Consequences

Probability Category	Health and Safety	Financial
Very Minor	First aid injury or minor illness	< USD 100,000
Minor	Lost Work Case or short-term hospitalisation	USD 100,000 to USD 1 million
Serious	Disabling injuries	USD 1 million to USD 10 million
Very Serious	One fatality or extensive injuries or work-related diseases	USD 10 million to USD 100 million
Catastrophic	Multiple fatalities	> USD 100 million

For Consequences, we will use for the sake of demonstration the following HSE and Financial Consequences (Table 31.6).

So, in these three tables plus one figure, we have described the organisation's risk tolerance.

I wonder if the Board of Directors understand what they have just done! Were they even consulted? Such matrices are, I suspect with good reason, usually put together by a safety officer and submitted for rubber stamp by the mainboard.

To demonstrate some of the practicalities, we need to ask what is 'it' – the event that is being compared and classified? Consider that we are evaluating the potential for an explosion.

Illustrative Hypothetical Case: Constructing a Risk Matrix

Let's consider our small fuels terminal. We are concerned about an explosion.

A useful reference is Marsh 100 Largest Losses in the Energy Industry, which points to approximately 20 explosions in oil refineries and a further 20 explosions in Petrochemicals over a 20-year period. So, using our example, we have a frequency that is at least 'Seldom'. However, suppose that we define 'explosion' more precisely. For example, there is a particular type of devastating explosion called a Vapour Cloud Explosion. The disaster at an oil terminal in the U.K. – Buncefield – see Chapter 13 was an example of such an explosion. What is the frequency here? Databases cannot tell you, other than we know they occur, and several are identified in Marsh 100 Largest Losses, but none of them at Terminals except for Buncefield. So, we might choose 'Unlikely' as our Probability category.

Our terminal has few personnel, but an explosion might well cause injuries at the school beyond (see map in Appendix A). Consequence modelling would likely demonstrate that the pressure wave at this location with little congestion could cause injury but not fatalities. So, the Consequences would fall in the 'Very Serious' category. The risk matrix result is shown in Figure 31.4:

Probability	Consequence				
	Very Minor	Minor	Serious	Very Serious	Catastrophic
Frequent	3	2	2	1	1
Occasional	3	3	2	2	1
Seldom	4	3	3	2	2
Unlikely	4	4	3	3	2
Very Unlikely	4	4	4	3	3

Figure 31.4 Example Matrix Results for Fuels Terminal.

The results of our analysis are either '2 – Action plan to put in place control measures required'.

Or '3 – Control measures required, unless demonstrated that current measures are "As Low As Reasonably Practicable" (ALARP)'.

Therefore, the recommendation to the Board of Directors is that it is within our risk tolerance to go ahead with the operation of the terminal, but the situation needs an action plan. What a big decision we have just made!

The actions will probably relate to making the instrumentation highly reliable see Chapter 24 on Designing a Safety Instrumented Systems.

From a financial perspective, we can do a similar analysis. Property values would probably be around USD 30 million for such a terminal. The explosion would most likely be in the tank farm area, so we could anticipate approximately USD 20 million of property damage to the terminal. Compensation would need to be paid for the repair of the school. The terminal would be out of business for, say six months, while tanks were reconstructed. So we would be looking again at a probability = consequence pairing on the matrix leading to a similar Category 2 or 3 action plan outcome.

However, there is a significant limitation when using risk matrices. A risk matrix is applied to each risk one by one. It is not used for the accumulated risk to do with an activity. To demonstrate this limitation let's assume we have an onshore oil well with just one wellhead, just by way of example, which is considered on the risk matrix to have an 'Unlikely probability' (1 in 10000 years) of 'Very Serious' consequences (One fatality or worse). This gives an 'Elevated' Risk of Category 3 for Action. However, the oil field may have 100 such wellheads. If we accumulate all the wellheads, the probability is now 1 in 100 years, which is 'Occasional' and would merit a 'High Risk' with Category 2 for Action. This is easy to see in the context of 100 similar wellheads. However often, when considering the risk to an oil refinery or chemical plant, only the risks from one particular high-risk process is given, without summing the risk from all of the processes.

DEMONSTRATING MANAGEMENT COMMITMENT

Management Commitment in Safety versus Management Commitment in Catastrophe Avoidance

Management Commitment is generally discussed in terms of the broader meaning of safety. Management concern and commitment to preventing harm to people through injuries such as those created slips, trips, and falls is important. Much has been achieved in reducing occupational injuries through strong management commitment combined with other tools. It is sometimes argued that management commitment to occupational safety translates into catastrophe prevention. 'If you look after the pennies, the pounds will look after themselves' is a common expression in the U.K., and it is in the same spirit that attention to occupational safety has broader implications in preventing more significant events.

I confess that I was a disciple of this mantra for many years, but now I am no longer convinced. Many companies with a strong management commitment to occupational safety have suffered catastrophic events. Recall that on the day of the Deepwater Horizon incident (see Chapter 6 on Oil and Gas), senior management had landed on the rig to congratulate the crew for an excellent safety performance. The focus was on occupational safety, things that might cause injury to an individual, not on how major hazard risk is managed (Hopkins, 2012, p. 129). Care for the individual does not necessarily translate into care for major accident hazards. The assumption that it does reduce catastrophic risk is, in my opinion, quite dangerous. As a consultant, I have led seminars in management commitment and left some managers with the incorrect belief that by dedicating their scarce time to occupational safety issues, they are simultaneously taking care of safety in the broader context of major accident prevention.

Having said that, I describe below some important aspects of management commitment applicable to safety management. I can only ask that practitioners continue to think about catastrophe prevention in executing such approaches, and I have included some good practices in this regard.

Rituals and Safety Culture

We will come back to the importance of organisational culture in Chapter 33 on Safety Culture and Other Social Structures. Meanwhile, there is a vital management role in guiding that culture. 'Leader Standard Work' is described as a repetitive pattern of activities carried out by a leader. The concept applies to many aspects of a leader's activity, including 'Lean' activity targeting the elimination of wasteful practices. It applies equally well to safety.

The emphasis here is on repetition – there are some rituals that a leader should practice routinely, if not every day. At Du Pont, we practised starting every meeting with a safety contact. Over the years, we extended that to a 'core value contact' so that the content might also be about issues such as ethics as an alternative to safety. Rituals such as a regular safety meeting at all levels, including shift teams, are vital not just for their content but also because all can see that they occur despite any production or other pressures exerted at the time.

Leadership Meeting

At the top of the organisation, there are typically meetings between the CEO and his/her staff to discuss performance. In larger organisations, this should and probably does always include discussion on safety performance. It is essential that safety is considered at least on par with other aspects of running a business. This should also be important for smaller businesses, but my experience is that this is rarely the case.

The same approach should also be taken at the plant level, where the plant manager or location leader would have a monthly meeting on performance, including safety.

KEY PERFORMANCE INDICATORS

Key to such meetings is the measurement of performance. With occupational safety, actual injuries can be measured. Good performing organisations will not be experiencing injuries, even minor ones, so measured results are not predictors of future performance. Regarding measures to avoid catastrophes, topics for meaningful discussion require the measurement of leading indicators that demonstrate the organisations' performance in preventing disasters. In the process industry, these are known as Process Safety Key Performance Indicators. There are analogues in other sectors, notably in air transport and rail transport.

In addition to measuring actual harmful events, it is valuable to measure events that might have led to harm in different circumstances.

So, a good list of KPIs for catastrophe prevention, in the context of process safety, might include the examples shown in Table 31.7.

Other industries will have their own KPIs.

Table 31.7 Process Safety Key Performance Indicators – Examples

Indicator	Explanation
Number of releases of any material which causes: • Injury • Community evacuation or community alert • Fire or explosion causing damage above a pre-determined value • Release of material greater than a pre-determined quantity	This is a count of actual highly undesirable events. They are called 'Tier I events according to OGP/API Guidance'. The threshold for damage is typically USD 25,000. Release quantities that would be classified as Tier I include 500 kg for an LPG or gasoline release.
The number of occurrences involving: • A recordable injury • A fire or explosion causing minor damage • Release of material greater than a pre-determined quantity	This is a count of lesser undesirable events. They are called 'Tier 2 events according to OGP/API Guidance'. The threshold for damage is typically USD 2,500. Release quantities that would be classified as Tier 2 include 250 kg for an LPG or gasoline release.
The number of excursions to: • Safe Operating Limits • Demands on Safety systems, such as a safety interlock being activated or a relief device relieving • Release of material of a lower quantity than above	Performance indicators that provide information about the strength (or lack thereof) of barriers and weaknesses in equipment and hazard control systems. Otherwise known as Tier 3 Events.
Overdue Inspections	Timely inspections to pick up corrosion, deterioration of machinery, and instrument potential malfunctions are important to catastrophe prevention.
Overdue Operating Procedure Updates	Procedures need to be regularly updated for accuracy.
Alarm Rates	Alarms can be a distraction to operators if too high.
Interlocks bypassed	Bypassing of interlocks requires close control.

The railway industry might measure measures, for example:

• 'Signals Passed at Danger' (whether or not there is a collision consequence).
• Passenger injuries
• Staff injuries
• Reported objects on tracks
• Trespassing events on railway and infrastructure

In air transport the following kinds of indicators are measured:

- Number of exceedances to noise limits
- Number of security breaches
- Bird Strikes

(List of Airport Key Performance Indicators (KPIs) (asms-pro.com))

These indicators need not just be presented at the leadership team but also discussed. The indicators are best presented graphically to identify good or bad trends. If the trend is good, it is worth determining what features of the organisation's management are helping the positive trend so that these features can be protected. An adverse trend demands action. Unfortunately, these indicators are often presented as work of the department creating them without managerial comment.

Notes from Trevor's Files: Key Performance Indicators – 100%

An organisation in the Middle East had chosen progress on Insurance Survey Recommendations as a Process Safety Performance Indicator. An excellent choice, in my opinion, and they had an unusually large number of long-standing recommendations. In preparing to help them with their next insurance survey, I noted that all the recommendations were showing as complete on the management presentation of KPIs – 100%. However, I quickly concluded that most of the recommendations were not finished in practice. The insurance surveyors came and pointed out that none of the recommendations were complete. And yet strangely, none of the indicators changed in the management meeting presentation. This is a case of the data being influenced by what the management wants to hear. It is also a case of listening only superficially to a recommendation. For example, a recommendation concerning Emergency Operating Practices was considered and reported by the surveyor as covering an inadequate range of topics. In comparison, it was interpreted by the organisation that the surveyors had concluded that the organisation did not have Emergency Operating Practices, which they did!

A further issue with Key Performance Indicators is the selection of indicators which only show the organisation as performing well. These indicators show Zero's or 100% for every month (depending on whether Zero or 100% indicates a good performance). An indicator that asks for the Number of Deviations from Legal Requirements will surely be Zero every time, otherwise, this is like 'Breaking into Gaol'. Management time is scarce – use it sparingly by choosing KPIs that merit management attention.

A few organisations following the principles of a 'High-Reliability Organisation' have a policy of 'challenging the green and embracing the red'. This slogan refers, in the first instance, to traffic light scorecards of risk indicators. Generally speaking, senior management wants to see an array of greens with as few reds as possible. Many senior managers accept that green is green, without question and press their subordinates to convert the red to green as soon as possible. Often there are ways of doing this that have nothing to do with reducing the risk – managing the measure rather than the risk (Hopkins, 2021, A Practical Guide to becoming a 'High Reliability Organisation').

HEALTH SAFETY AND ENVIRONMENT LEADERSHIP

In addition to the general Leadership Meeting, many organisations have a separate corporate-level event focused entirely on Health, Safety, and Environment, which includes prevention of catastrophes and process safety. Who should lead this meeting, at the organisation level, or for that matter, at the plant level? Should it be the CEO, or the leader of the department dedicated to Health, Safety, and Environment? As a consultant with Du Pont Sustainable Solutions, we described best practice as being that the CEO should lead this meeting. This really emphasises management commitment. Many organisations and their CEOs took this on board. But there is a counterargument. The CEO has lots on his/her plate. Does the CEO need to take the same role for other issues such as Legal, Finance and Ethics, Product Quality, etc. ... It has been widely said by Du Pont consultants that, at Du Pont, the CEO leads the Corporate Health, Safety and Environmental Committee. This was certainly true at some point in time, but it has not always been the case. I think it does have to depend on the criticality of Safety to the organisation. However, if not the CEO, then the Health, Safety, and Environmental department leader or similar? Now we send a contrary message that safety is not a line management responsibility but that of a specialist department.

Line Management is responsible for safety; this is undoubtedly true in my opinion and experience.

The role of the Health Safety and Environmental Leadership meeting does not duplicate the work of the Corporate Leadership Meeting as a whole. Instead it should focus on specific improvement areas.

Illustrative Hypothetical Case – Whisky Distillery KPIs and HSE Meeting Agenda

For the example we have been using throughout this book of the Whisky Distillery, what Key Performance Indicators might we choose in the context of safety and catastrophe prevention.

In Chapter 2, we identified a number of hazards at our Whisky Distillery. Looking specifically at the risks with catastrophic potential, we might choose to measure:

- Any releases of flammable liquids or vapours, no matter how small, with consequences or without.
- Wastewater analysis for toxic components which might conceivably come from the refinery. Probably any result above the detectable limit should be recorded as an event.
- Dust measurements will probably be required for health reasons. Excursions about a specific value could be measured as an event. There could also be a regular audit against checklist, which would count events with dust accumulations.
- We studied in Chapter 27 the need to protect the firing of the furnace beneath the barley floor with a safety instrumented system. This will be required to be tested periodically, including the functioning of the interlocks, e.g. flame out as detected by the flame scanner. Any failures of this and other safety instrumented systems in the distillery would be counted in a KPI as instrumented system failures. This would include any failures in 'real life' – failure to operate on demand, and separately failures on test.
- There have been significant catastrophic events in warehouses where whisky is held to mature. Alcohol permeating the casks accumulates in the warehouse and can form an explosive mixture with air. Frequent warehouse measurements of flammable gases should be taken, and any measurements above a certain fraction of the lower explosive limit counted as a KPI.

MANAGEMENT BY WALKING AROUND

During my consulting career, I have helped hundreds of people in the practice of Management by Walking Around. Specifically, in the context of Safety, there is the Du Pont STOP® programme (DSS Learning – STOP®). This is not only an activity for managers and supervisors but can, depending on the organisational culture, be an activity for the workforce also. There are many other behavioural safety programmes similar, and we will not attempt to compare them here.

There is a similar if not identical concept called a Gemba Walk. This is a technique promoted by consultants in lean manufacturing in which there is a workplace walkthrough that aims to observe employees, ask about their tasks, and identify productivity gains. Gemba Walk is derived from the

Japanese word 'Gemba' or 'Gembutsu', which means 'the real place', so it is often literally defined as the act of seeing where the actual work happens. A Gemba walk is a simple yet powerful lean method done by employers to promote continuous improvement (see https://safetyculture.com/topics/gemba-walk/). It is also very applicable to safety.

The key to such behavioural tours is for the leader to stop, observe, and then interact with the employees carrying out the work. The interaction is not always easy and in practice, leaders and supervisors need coaching in the conduct of such discussions to maximise the likelihood of a favourable outcome. The interaction must take place whether or not there is a problem with the work method. If all is fine, then the leader should reinforce good behaviour. This is why the stop and observe step is so essential at the start of the process. Only then can all the activity be observed. If the leader overlooks some positive or negative aspects, then they will be sending the wrong or confusing message to the workforce member.

How does this help catastrophe prevention? Convinced as I am about the value of the technique for preventing injuries to the workforce, it requires careful tuning to contribute to catastrophe prevention. Leaders often develop habits of looking for particular types of easily visible issues such as the lack of adequate personal protective equipment, incomplete barricades around a task, etc. Such issues are of limited value in influencing measures to prevent catastrophes.

Rarely will the leader go to the control room to look at control parameters or to identify the control systems which might be bypassed. To encompass some aspects of safety related to the prevention of catastrophes, the person carrying out the observations should also look at equipment isolation standards, labelling and identification of the isolation, etc. ...

MANAGEMENT OF CHANGE

Readers may be surprised to see this element here under Governance, rather than in Chapter 31 on Safety Management Techniques. Whilst Management of Change (MOC) is seen as an essential technical procedure vital to accident prevention, the way an organisation applies MOC is also vital to its success.

Failure to adequately address the consequences of technical change is to be found in a number of our example incidents. In Chapter 6 on Oil and Gas we discussed the Deepwater Horizon incident where failure in the novel nitrified cement contributed to the incident. In Piper Alpha, in the same chapter, there had been significant change in the structure of the platform. We noted that Piper Alpha was designed to produce and export oil. The requirement to export gas – with the associated separation of condensate – was an afterthought and involved extensive modification. The retrofitting went on in several phases, starting with separation of condensate and ending with production of export-quality gas.

The new facilities were located beside the control room, under the electrical power, radio room, and accommodation modules. When disaster struck, it did so with disastrous effect on the rest of the platform. The consequence of the changes had not been thoroughly considered.

There are also MOC considerations where organisations are changed, and critical skills moved, or the number of personnel involved in operating the plant reduced.

As a technical process MOC is not difficult. The initiator of the concept behind the change initiates the Management of Change Document and justifies why it is a desirable change. The change is then subject to further design and then a risk assessment or process hazards review is carried out to assess the change. If hazardous consequences are foreseen, mitigating action is taken, or the change is cancelled. See Chapter 24 on Safety Management Techniques on carrying process hazards reviews.

So, where does Governance of Safety come in, in the conduct of this technical process? Firstly, there is the recognition that the process is necessary. Large process industry organisations now all have MOC procedures, but what about smaller ones? Taking our hypothetical examples of a standalone Whisky Distillery or a standalone Fuels Terminal, it seems to me unlikely that they would have a Management of Change Process, perhaps unless the owner or manager had previous exposure to the technique.

An excellent paper describes the purpose and conduct of MOC by Marsh (Marsh: Position Paper: Management of Change). MOC is required by OSHA 1910.119 in the USA and by the European Seveso II directive, Annex III section (c) (iv), which includes the Control of Major Accident Hazards regulations in the U.K.

The number of MOCs initiated, their status, and number closed are all valuable Key Performance Indicators. Leaders can watch to see if the organisation is being overloaded. Ensuring that all the paperwork associated with a change are finalised is an essential element. This includes the revision of operating procedures.

Temporary changes, which include tests, should also be covered by MOC. Chornobyl (see Chapter 10) is a classic test not adequately controlled by MOC, with disastrous results.

Notes from Trevor's Files: Undue Delay between Risk Assessment and Implementation

I was invited to observe a technical meeting at an underground mine complex in Kazakhstan. During the meeting, an engineering leader was severely reprimanded for initiating a new spur to the mine without considering escape refuges for the miners. Clearly, a MOC issue, or maybe the engineering leader thought he would be congratulated for getting the job done ahead of schedule. Perhaps the issue was only mentioned because I was in the room.

At the same mining complex, a serious incident had occurred involving fatalities. The incident was caused by rockfall. It is sometimes said that rockfall incidents are an unavoidable risk in mining. However, any incidents where I had the opportunity to review the investigation were avoidable. The warning signs were there. Mining operations are almost always supported by specialists who listen for the movements of the rock and grade the rock stability, using instruments a little like a doctor's stethoscope.

This particular incident was located in a new mine spur, and the equivalent of a MOC had been done. This involved assessing rock stability by one of these specialists who listened to the rock movements and considered the rock stability to be satisfactory. However, this assessment had taken place some two years prior, the excavation of the new spur having been delayed. But management never thought to have the risk assessment redone on the basis that something might have changed.

AUDIT – INTERNAL AND THIRD PARTY

Local Checks

The simplest form of audit is a check against a checklist that certain things are in place. We will start with simple checks like this and build up to higher-level audits.

Suppose we decide at our hypothetical fuels terminal or whisky distillery to do a period check (simple audit) that all firefighting equipment is in place. A number of questions immediately arise:

- Who will do the check?
- With what frequency?
- What needs to be checked?
- Who will record the data?
- Who will act on any discrepancies?

We might assume that there is a list of items such as shown in the following:

- Fire Extinguishers, their type and location.
- Fire-Water Hoses and where they are located.
- Fire hydrants and Fire Water Monitors and their locations.
- Stocks of Fire Fighting Foam and where they are located.
- Etc. ...

So, it is a relatively simple matter to create a list, location by location, for someone to check that the equipment is in place.

Who will do the check? This should be done by the people who might have to use the equipment. This could be the operators of the plant or the site's fire brigade. In this way, the user doing the check becomes increasingly more fluent in the locations of the equipment.

Frequency is a more difficult question and depends on how frequently one's experience tells us things might go astray. Depending on the country and culture, theft and vandalism can be an issue. So, we might decide on a monthly check.

The person carrying out the check can, in this case, record the data.

The checklists need to go to a responsible person to correct any discrepancies in the firefighting equipment provision.

It all sounds so easy, but motivation can be an issue. Checking is to a large extent a menial task. Motivation is further impacted if any discrepancies the person checking the equipment found are not acted upon. In this author's experience, the supposed checker sometimes marks off the equipment/procedures as OK from the warmth of the tearoom.

Let's introduce a bit more complexity. Suppose we also want to check for fire prevention.

Then we have to look additionally for issues such as:

- Obstructions that might impede access to fire extinguishers, etc.
- Accumulations of paper and other flammable materials.
- Audibility of emergency alarm.
- Proper storage of goods which can go on fire, for example, new and used pallets.
- Etc. ...

The difference here is that a degree of judgement is required. What constitutes an obstruction? Something is needed to define what is acceptable and not acceptable. Pictures and examples can help. This is what is known as an audit protocol. The operator may no longer be the best person to undertake the check/audit since it is the operator himself who will have to clean up any accumulations of flammable materials and remove any obstructions. Some independence in the auditor becomes appropriate.

Local Audit

In financial control, the difference between a check and an audit is the degree of independence of the auditor(s). So far our checks comprise a local audit. So, in the context of a processing plant, the local audit would be led by an internal appointed auditor. In the context of fire safety, this might well be a fire specialist. However, this person may not be fully independent since this specialist may be responsible for the issues being audited. Indeed, the auditor needs some training both in conducting audits and on the specialised subject being audited.

My experience is that audits, especially internal ones, are a good learning opportunity. Local management do not suddenly 'know everything' the instant they are appointed.

I remember an internal audit on fire safety where it was pointed out to me that the wooden pallets stored in the warehouse were stacked too high and presented a fire hazard. The auditing specialist was very tactful and explained not just that it was wrong but also explained why there was this limitation. Anyone who has seen a pallet fire would probably understand since pallets have the suitable void coefficient to create a blazing inferno.

For a typical processing plant, it would be quite usual to have a variety of local audits such as:

- Fire safety
- Ergonomics
- Occupational health (dusts, noise, respiratory protection, etc. ...)
- Control of work, including confined space, working at height, equipment isolation for maintenance

Many checklists are available on the internet where an organisation can find a pre-prepared document and modify it to suit their circumstances. However, some of these audits are relevant to occupational safety but have little relation to catastrophe prevention.

Local Process Safety Audit

Some organisations have a local audit process dedicated to safety elements specifically relevant to catastrophe prevention. In industries other than the process industry, audits with a similar intent will have a different name. In the airline industry they might be known as ICAO compliance audits.

Such an audit would consider aspects such as:

- Control of work, work permitting, etc. ...
- Fire prevention
- Management of Change
- Safety instrumented Systems, Interlocks and Bypassing
- Risk Assessments, Process Safety Reviews including Recommendation Status
- Incident investigation including Recommendation Status
- Inspection and Inspection Results and Overdues
- Training of personnel
- Emergency planning

Etc. ...
An airline industry audit might additionally cover aspects such as:

- Low visibility operations
- Wildlife management
- Security and security breaches

Often these audits mimic the corresponding regulatory audit and are regarded as preparation for such regulatory audit. This is a mistake. We expect conformity with regulations. It is certainly right to check for it. But if we are to improve, then we need to go beyond regulatory compliance.

This will require a trained auditor, together with a team of knowledgeable people in each specific area of focus.

Corporate Audit

An organisation with several sites will typically choose to have an independent audit on behalf of the organisation. It will still expect sites to self-audit as above but will want to check on performance, at a lower frequency than would be expected by the site. Corporate Audits are usually carried out against an established protocol laid out in corporate documents. This allows for a fair audit and allows sites to be compared. However, as I mentioned above, for regulatory compliance audits, such constraints limit the value of a corporate audit in driving improvement.

One way around this is for the corporate audit to include new or recently developed standards within their audit. In my opinion, a standard is not a standard until a site has had the opportunity to implement. Until that time it is a draft standard. Otherwise, the organisation is saying that it works to a particular standard when it does not. Unfortunately, implementation of different standards takes time, and some are implemented more urgently than others. I have seen standards in organisations which have never been implemented.

Notes from Trevor's Files: Standards Implementation Process

When working on a Process Safety assessment of a large petrochemical site in the Middle East we noticed that some new standards had been well implemented and others not. We found that the well-implemented standards were ones where departments had had the opportunity to comment on the new standard, where a gap analysis of the new standard versus current practice had been done, and a plan to close those gaps prepared both at departmental and site level. Poorly implemented standards had not been subjected to the same steps and were largely issued by the safety department as a 'must do'.

> Proper implementation of a new standard requires significant effort, so only so many standards can be introduced within a given time period.

Staffing a corporate audit office in Safety, Process Safety, and Environment expertise is a costly exercise. It is also hard to attract the personnel with the right expertise, appropriate people skills and desire to do the audit role. At Du Pont, we had what I regard as an excellent practice. Firstly, there were relatively few corporate auditors. In advance of an audit, one of those core auditors was assigned to set up a team. He/she did so by calling on managers from around the region (the European region in this case) to delegate a participant to the audit team. So, for example, I would be requested to release someone from my team to join in the auditing of another plant. The lead auditor would train his assembled team of delegates. My delegate would return to his/her post with many learnings, which would then be used to help prepare for the next audit of my plant. I think this is an excellent way to use the audit process in employee development, as well as driving continuous improvement not just compliance.

Notes from Trevor's Files: Consequences of Poor Audit

During my time as a production manager at Du Pont, I experience what seems like an almost continuous stream of corporate audits of various kinds. Over auditing can cause 'deafness' to the audit process. However, in Du Pont, performance in safety-related audits was seen as a key metric of the manager's performance. I remember after one audit where my plant did not perform well. It was made clear that I would be reassigned unless I could achieve a significant improvement. Fortunately, we were offered the opportunity to be reaudited within 90 days and performed much better. Some strange things happened during the particularly poor audit. At the time, we were going through some difficult industrial relations experiences, and I was playing a prominent role in the discussions. I cannot say I was well-liked. It is possible that some of the audited events, such as the non-wearing of particular Personal Protective Equipment, were an attempt to get me removed. Or maybe I am just being defensive; I leave the reader to judge.

Note that compliance auditing by regulators is covered in the next Chapter 31 on Laws, Regulations, and Standards.

Audit Results and Actions

It is essential that all audit results are recorded and that actions required are assigned and tracked to closure.

Open action items are a useful Key Performance Indicator for review at leadership meetings.

SUMMARY: CAUSES OF FAILURES IN THE GOVERNANCE OF SAFETY WHICH MIGHT ALLOW CATASTROPHES TO OCCUR

Reasons Why the Structure of the Safety Management System Can Fail to Avert or Predict a Catastrophe Include

⊠ Too many elements, or elements being confused between organisations.

Reasons Why the Management Commitment Can Be Ineffective and Fail to Avert a Catastrophe Include

⊠ Failure to demonstrate that safety is a core value, notably at times of adversity.

⊠ Management Commitment made following first or second-hand experience of a tragic event declines with time and is weak where management does not have such an experience.

⊠ Inadvertent messages negatively influence perception of management commitment.

⊠ Commitment evident in occupational safety, but not in process safety or catastrophe prevention.

⊠ Management doesn't recognise or practice the essential elements of Leader Standard Work, such as Gemba Walks, Approach to Meetings, and Key Performance Indicator discussions.

Reasons Why Managements Acceptance of Residual Risk Can Contribute to Failure to Avert a Catastrophe Include

⊠ The risk matrix is inadequately defined or confused and does not reflect management's actual risk appetite criteria.

Reasons Why Management by Walking Around Can Fail to Avert a Catastrophe Include

⊠ Interactions are confined to occupational health hazards, such as PPE, rather than including process safety observations which could contribute to a catastrophe, such as inappropriate lockout and interlock bypassing.

☒ Correction of the unsafe behaviour is done in a 'talking down to' manner and is not open to any comments from the person being corrected.

☒ Observations result only in conversations about unsafe behaviour, and largely exclude conversations reinforcing safe behaviour.

Reasons Why Audits Can Fail to Avert a Catastrophe Include

☒ Different auditors or different audits come up with different results for the same situation. Needs audit protocols and photographs of desired versus undesired situation to align.

☒ Auditors see some issues but are blind or omit others.

☒ Assumes managers and auditors 'know everything' about good practice.

☒ An organisation or part of an organisation is exposed to too many audits of different kinds and are subject to audit overload.

☒ Audits are too few and/or too generic to be meaningful to the particular situation.

☒ Audits are to demonstrate legal compliance only.

☒ Audits are to demonstrate compliance with various ISO certifications only.

☒ Audits take too many resources and are too costly.

☒ The audit outcomes whether good or poor are not included in the managers performance appraisal.

☒ The audit actions are not traced to completion.

☒ Audit action and completion status are not indicated in Key Performance Indicators considered by higher management.

Chapter 32

Laws, Regulations, Standards, Certification

SAFETY LAWS

Laws make an important contribution to safety. In the following paragraphs, I do not seek to dismiss safety laws but comment on their limitations. Our goal should be to drive improvement. We have already discussed some aspects of safety law when we reviewed Inspection in Chapter 22.

In Section 2 some incidents were discussed where there were clear inadequacies in relation to the law, either the way in which the laws were constructed or followed. These incidents included Deepwater Horizon, Fukushima Daiichi, Chornobyl, and DPC.

In my early days as a production supervisor in the U.K. in the early 1970s I wanted to know what the law was for everything. People complained about the temperature in the warehouse – what are the legal limits? Employees complained about the weight of the tools – what were the legal limits? The production methods produced trace quantities of vinyl chloride in the atmosphere. Vinyl chloride was at the time a suspected human carcinogen (now confirmed) – what were the legal limits? How frustrating it was that often the law did not give me the answer.

I was asking these questions in 1974, the same year that the U.K. began the long journey away from prescriptive law and introduced the Health and Safety at Work Act 1974. I believe the U.K. was the first country to do so, although many other countries have since followed the same direction.

Prescriptive laws are difficult and costly for the state to maintain, especially when technologies are developing quickly. Furthermore, once having made a prescriptive law, the state is in some ways taking responsibility away from the organisation making use of the law. If an organisation follows the details of the law (and its associated regulations) then an organisation might reasonably argue that it believed what it was doing was safe.

An example here is the Grenfell Tower disaster discussed in Chapter 15. The outcome of the ongoing enquiry is not yet complete. Nevertheless, the manufacturers of the cladding which helped spread the fire made a case that the laws and regulations concerning such cladding were being met.

Apparently, the regulations required a physical test, which they argued that they did and it passed. Whether the results of the test were borderline or misinterpreted is a separate issue being considered in depth in the enquiry.

Notes from Trevor's Files: Reluctance to Set Prescriptive Laws

One of the challenges of prescriptive laws was demonstrated to me when I was responsible for a production plant in Wales in the 1990s. We wanted to expand the plant. The factory produced wastewater containing common salt – sodium chloride. The plant had a long-standing permitted daily limit on the amount of sodium chloride which we could release. The expansion would cause this limit to be exceeded. I met with representatives of the local authorities several times to try and formalise a revision to the limit. Whilst not opposing what we intended to do, it was clear that no one would sign off on approving the limit change. Whilst sodium chloride is not toxic in low concentrations, the wastewater entered a fast-flowing freshwater river. At least some knowledgeable people from the local authorities informed me that sodium chloride could be harmful to strawberries although the concentrations which might cause harm were unclear. Strawberry farms abstracted water from the river at some distance downstream. Whether our proposed expansion posed an additional risk to the strawberry farmers depended on the concentration of the sodium chloride at the point of abstraction and what other sources of sodium chloride there are along the river. The extent of dilution was vast. Nevertheless the onus was on my organisation to satisfy themselves that there would be no harm and to shoulder the responsibility, and compensation, if damage occurred.

One of the key elements of the Health and Safety at Work Act 1974 is that in carrying out any activity, a risk assessment must be carried out, and that risk assessment be 'suitable and sufficient', i.e. it should show that:

- a proper check was made
- an evaluation had been carried out to determine who might be affected
- the assessment dealt with all the obvious significant risks, taking into account the number of people who could be involved
- the precautions are reasonable, and the remaining risk is low
- workers or their representatives in the process

(Management of Health and Safety at Work Act 1999)

THE SAFETY CASE

This principle was extended into the Control of Major Accident Hazards (COMAH), first introduced in 1984 and since becoming part of the Seveso Regulations. Under these regulations, the site must have, amongst other things, a 'Safety Case' which describes how the plant is being operated in order to prevent major incidents. The safety case is reviewed and approved by the regulatory authorities, but the authorities do not detail what should be inside (although they give good guidance). The production of a safety case is a requirement for Tier 1 sites under COMAH/Seveso and the Offshore Installations (Offshore Safety Directive) (Safety Case etc.) (Regulations 2015).

There are similar requirements in some other countries. The U.K. version is particularly efficient because it covers air emissions, environment, and safety from a major accident hazard perspective. There are similar requirements for nuclear sites. However, irrespective of the legal requirements, producing a safety case is just good practice, in this author's opinion.

The safety case presents the argument that operation covered will be acceptably safe. To achieve this objective, a safety report typically includes:

- safety arrangements and organisation safety analyses
- compliance with the standards and best practice acceptance tests
- audits and inspections feedback provision
- emergency response plans

Often, they include quantitative risk analysis, but it is understood that this is not essential.

(COMAH (Control of Major Accident Hazards) HSE 2015 – L111)

The value of the safety case has been appreciated beyond the process and off-shore industries, whose incidents created the momentum to introduce safety case requirements. The nuclear industry was in fact the first to require safety cases, an initiative which followed a release from the Windscale nuclear reactor in the U.K. in 1957. Safety cases are now widely required in Railways and Military Systems, including aircraft, ships, and land-based communications and weapons systems.

Incidental Case Study: RAF Nimrod MR2 Aircraft XV230, Afghanistan (2006)

This event was included in the incident examples in Chapter 9 on the Transportation Industry.

A subsequent review (Haddon-Cave (2009), An Independent Review into the Broader Issues Surrounding the loss of the RAF Nimrod MR2 Aircraft

XV230 in Afghanistan in 2006) found serious issues relating to the safety case which had been developed for this type of aircraft. The report, which extends to 583 pages, is highly critical of those who were responsible for the preparation of the safety case. The subheading reads 'A Failure of Leadership, Culture and Priorities'. The benefit of these painstaking reviews, this one led by a Queen's Counsel, is the clues within as to how to improve in the future.

The Nimrod type of aircraft was an extensive modification of the de Havilland Comet, one of the world's first commercial jet airliners. In order to operate as a long-range reconnaissance aircraft, one of the modifications involved enabling some of the jet engines to be shut down in flight to conserve fuel. A duct was designed to enable hot air to be directed to engines on either side of the aircraft so that any combination of engines could be restarted in flight. This duct became very hot in use and passed close to some fuel lines in a compartment that had no means of fire detection or suppression. This was an issue apparently not identified in risk assessments carried out in preparation for the safety case. Why?

I have perhaps a little more sympathy for those constructing the safety case than is evident in the review. The requirement for a safety case for military aircraft was a relatively recent development at the time; the Nimrod Safety Case (NSC) commenced development in 2001. The participants had little prior experience, nor, apparently, role model safety cases on military aircraft on which to structure their work. Such experience was available in the nuclear and process industries, and the safety case would have benefited from advice from this sector, in my opinion.

Now there is a distinct difference between preparing a safety case for, say, a chemical plant and for a military aircraft. In the chemical plant instance, the owning business is seeking a licence to operate, and needs to have the safety case as part of its argument that the facility will be safe. The HSE, in the U.K., has a role in approving the safety case once developed. The cost of developing the safety case is borne by the business. For military aircraft, the cost, at least for the Nimrod example, is borne by the Ministry of Defence – the same government organisation that required the safety case development. I suspect this leads to a significantly different 'dynamic' between the parties involved.

One might expect that the aircraft manufacturer might provide a safety case as part of the total cost of the delivery. In the Nimrod instance, the safety case was being done retrospectively on an aircraft design which had already been purchased and was in service.

One of the first tasks for the aircraft manufacturer was to agree on a budget with the Ministry of Defence (M.O.D) for the development of the safety case. Along with the aircraft manufacturer, an independent adviser was appointed, together with a project team from the M.O.D. Safety cases are a source of business in the process industries also, with specialists selling their services to develop the safety case with input from the plant engineers and management. As noted by Lord Cullen when safety cases were first being introduced, 'the involvement of the company's own personnel is the best way to obtain the full benefits of the Safety Case within the company. In particular, it was desirable that the operator should deal itself with the Quantified Risk Assessment (QRA) aspects of the Safety Case rather than contract them out. Familiarity with the system was essential for good QRA and, moreover, the use of company personnel would allow expertise to be built up in-house' (Haddon-Cave 2009, p.166). The review states that one of the major reasons for the failure of the NSC was the lack of relevant operator input.

The development of the NSC took around four years before it was signed off as complete.

One of the issues the safety case team clearly struggled with is the belief that Nimrod was a safe aircraft. Indeed, Nimrod's had been flying for a long time and merited their reputation as being solid equipment. I imagine that the team largely considered it necessary to complete the safety case as something of a bureaucratic exercise. This is a danger in all safety cases, but it is especially true of retrospective safety case and risk assessment. Put yourself in the shoes of the engineer that would come up with a reason why the well-established aircraft was 'unsafe'! Safety cases need to be treated as a genuine effort to improve risk and seek out opportunities for incidents that have not yet occurred!

The end product was a safety case where 40% of the hazards were 'Open' and 30% 'Unclassified'. It seems that at handover meetings the aircraft manufacturer gave the impression that the tasks had been completed and could be signed off (Haddon-Cave 2009, p. 10), and did not disclose the scale of the hazards left 'Open' and 'Unclassified'.

During the period when the NSC was being assembled there was heavy pressure in the M.O.D. to reduce cost. No doubt when the budget for preparation of the safety case was exceeded, this had some influence on bringing the work to a close.

A particular example of the difficulties faced in preparing the safety case concerned the assessment of probability. We have come across this issue before in Risk Assessment Chapters 26, 27, and 30. Although a risk matrix was available, the accident probabilities of Improbable, Remote, Occasional, Probably, and Frequent were defined only by these adjectives. The risk matrix categorises risks as Unacceptable, Undesirable, Tolerable, and Broadly Acceptable. Examining the matrix, any probability identified as Improbable could not receive a risk classification worse than Tolerable. If the probability was listed as remote, then the worst risk outcome was undesirable, and that only for a catastrophic consequence (Haddon-Cave 2009, p. 179).

The work in preparation for the safety case included physical inspection of an aircraft, in particular inspection of each of the zone. It appears that the issue of the proximity of the cross air duct to the fuel pipes was seen, recorded, but the significance was not appreciated. Probabilities of Improbable or Remote were assigned, and therefore the risk was not highlighted. It seems that there was no facilitator involved who might have prompted this to be questioned.

A further related example concerns the probability of fuel pipe leaks at fuel connectors. Generic data available suggested that the probability of failure of a fuel pipe and associated coupling was one in one million flying hours, equating to one fuel coupling failure in approximately 66 Nimrod years. The experience of any actual Nimrod line engineer was apparently not sought. Failures of fuel couplings were not a rare occurrence (Haddon-Cave 2009, p. 179)! Imagine the scenario if a line engineer had been consulted and a more realistic failure rate included in the analysis. Would the resulting safety case have been found acceptable?

The safety case should be regularly reviewed (at least once every five years) and more often if there are significant changes.

Notes from Trevor's Files: Change to Safety Case

While carrying out some behavioural safety work on an ageing platform in the North Sea, I was coaching platform supervision in carrying out observations and interacting with employees. During one of the discussions, a group member commented that he was unhappy that he often smelled hydrocarbon when the wind was blowing in a particular direction. The reason, he described, was that the flare was no longer lit.

Earlier in the platform's life, the oil contained sufficient gas to maintain a constant flare to burn off the residual gas. Now there was insufficient gas being produced from the oil wells to fuel the flare continuously. They referred to the flare as a 'cold flare'.

I raised the issue with the Offshore Installation Manager, who confirmed that this was a deliberate change, although they had no choice.

Now to me, the 'switching off' of a flare is a very significant change to an offshore platform, and it should have been covered both by a Management of

Change Procedure and a change to the Safety Case. The employee's concern regarding the blowback of hydrocarbons onto the platform was understandable. There was a danger that the gas might meet an ignition source and explode.

This should have been considered as change to the safety case, but it was not.

SAFETY REGULATIONS

Regulations are put in place by different government agencies, including local authorities, to implement the laws. They are thus instruments of the law.

To demonstrate how difficult it can be to understand and ensure compliance with the law I will use a particular example. An important regulation in the prevention of catastrophe relates to equipment and protective systems intended for use in potentially explosive atmospheres. In the E.U. these are known as the ATEX directives (ATEX comes from ATmospheres EXplosible). Specifically, in the U.K. they are the Dangerous Substances and Explosive Atmospheres Regulations (DSEAR). These require employers to classify areas where explosive atmospheres may occur into zones. Depending on the zone, different equipment will be necessary, particularly electrical equipment, to minimise the potential for ignition of the explosive atmosphere. At first, this sounds like precise and prescriptive regulation. It was not my sphere of competence, so I decided to enrol in a course that would enable me to carry out such zoning.

In addition to a Regulation, there is an Approved Code of Practice (ACOP). The ACOP give practical advice, and if an employer follows the ACOP, the employer will be legally compliant. However, compliance with the ACOP is not mandatory, so long as the employer chooses methods that are as good as if not better than those shown in the ACOP. The course then proceeded to describe a number of different methods for Area classification:

- IEC 60079-10-2, which is dedicated to flammable dusts
- Energy Institute EI 15 for flammable liquids
- IGEM/SR/25 for Natural Gas Installations
- HSG51 concerning storage of smaller quantities of flammable liquids
- British Compressed Gases Association CGA covering pressurised cylinders
- INDG139 (HSE) covering electric storage batteries, which can emit combustible gases

So already we have a potentially bewildering set of subsidiary 'best practices' to follow.

This might be OK if the methods were fundamentally the same, but they are not.

For example, EI15 considers the occupancy level of the facility and the likelihood of a person being at the facility when the explosion occurs; IGEM does not. The methods are very complicated to follow, and further, it seems that the methods do not give the same results!

Once you have learnt how to do this in the U.K., other methods must be learnt for different countries. I have had the good fortune to work extensively in Romania, and therefore tried to read the Romanian INSEMEX rules – in Romanian! I was not successful, and fortunately, it was not critical for me to succeed!

Simply following the law and its regulations is not so easy.

COUNTRY AND INTERNATIONAL STANDARDS

Industry bodies have created many technical standards intended to indicate good practice.

Following these standards certainly helps prevent catastrophes. If it can be demonstrated that you are compliant, this will also help the legal interpretation of the situation should something go wrong.

In insurance survey work, we emphasise that technical standards compliance is the minimum expectation. We expect a location that is a good risk to go beyond these standards in appropriate areas.

There are an astounding number of bodies issuing standards.

Governmental bodies have adopted some of these standards and they now have the force of law. There are also codes that apparently differ from standards. After some time reviewing the internet to find the difference between a code and a standard, I remain confused about any substantive difference, and I will proceed with addressing both under the title of Standards.

Some Standards Bodies, especially those based in the U.S.A., are shown in Table 32.1.

A good deal of confusion arises when different bodies create standards that overlap and cover the same subject. Take, for example, the following statement concerning flanges (Ref. Engineering Toolbox). 'Although the dimensions of ASME and API flanges may sometimes be compatible for bolting – using the same bolt circle and the same number of bolts – they do not share the same pressure rating system. ASME rates the pressure of a class based on the material of construction and the design temperature. API specifies allowable materials and gives it a specific pressure rating. The difference between ASME/ANSI and API is the fabrication material and a higher rated API operating pressure'.

Table 32.1 Bodies issuing Technical Standards

Abbreviated Name	Full Name	Comment
ANSI	American National Standards Institute	ANSI facilitates the development of American National Standards by accrediting the procedures of approximately standards developing organisations, including ASME and API below. It is, therefore, more of an overseer of standards. It sometimes produces standards in its own right, primarily where different standards bodies have previously written different standards, which ANSI has then harmonised. For example, ANSI has a standard on pipework flanges.
ASME	The American Society of Mechanical Engineers	ASME issued its first standard, 'Code for the Conduct of Trials of Steam Boilers', in 1884. The ASME website lists 731 standards mostly around the construction of vessels and pipework.
API	American Petroleum Institute	API is a U.S. trade association representing all facets of the natural gas and oil industry. It has published around 700 standards relating primarily to the oil and gas industry, but also having application more broadly.
NFPA	National Fire Protection Association	The NFPA provides fire, electrical and life safety standards. There are around 300 NFPA standards. Unfortunately, there is some overlap with other standards bodies. For example, NFPA 70 E is a Standard for Electrical Safety. NFPA 20 and 25 are recognised as the minimum standard for fire water provision and fire water pumps in the insurance industry.
IEEE	Institute of Electrical and Electronics Engineers	The IEEE has some 1300 standards relating to electrical installations.
NEMA	National Electrical Manufacturers Association	NEMA publishes some 700 standards (also) relating to electrical.

The International Electrotechnical Commission (IEC) has produced over 2500 standards relating to electrical technology, including the IEC 61511 and IEC 61508 relating to Safety Instrumented Systems, much referenced in Chapter 27.

If that's not confusing enough, then let me introduce ISO standards, which are internationally agreed upon by experts. They tend to focus more on standardising the management systems by which an organisation operates. BSI (British Standards Institution) performs a similar function in the U.K. as ANSI does in the U.S. Many BSI standards have evolved and are now ISO Standards. There are harmonised European standards produced by CEN (The European Committee for Standardization), CENELEC (European) Committee for Electrotechnical Standardization), and ETSI (European Telecommunications Standards Institute). However, this does not guarantee that because a country is in the E.U. that a CEN, CENELEC, or ETSI standard will apply. Many standards specific to the country still apply, in particular with regards to Pressure Vessel construction and inspection, and on electrical equipment in potentially explosive atmospheres. In Russia, GOST refers to a set of technical standards covering more than 20 industrial branches, including the petroleum and chemical industry, power and electrical equipment, etc. GOST standards or a regional development of them are widely used in former nations of the Soviet Union such as Kazakhstan. There is also a technical standards structure in Japan, etc.

Any operation wanting to build, design, or operate something is faced with the difficult task of identifying which standards to follow. The standards are rarely freely available, and most must be purchased at a significant price – typically USD 100–200 each. A standard might be purchased only to find that it is not relevant.

Whilst many standards producing bodies are non-profit bodies, they have to cover costs, and most have entrepreneurial plans. Each has a significant number of employees whose concern is their future. Often, they are trying to diversify beyond standards production into consultancy and other fields.

Standards are not necessarily accurate, and new risks not covered by the standards can be uncovered. As an example, API Recommended Practice 941 'Steels for Hydrogen Service at Elevated Temperatures and Pressures in Petroleum Refineries and Petrochemical Plants' was found not to be sufficiently accurate – see Chapter 22 on Equipment Maintenance and Inspection where High-Temperature Hydrogen Attack is discussed.

Finally, there is an issue with the readability of a standard itself. I find most standards are soporific. They rarely make exciting reading, and it can be hard to strip away all the detail you do not need to find the detail you do need.

Sometimes the best standards to apply in a particular subject area do not arise from an obvious source. The most respected standards for oil and gas well design and operation offshore come from NORSOK, the Norwegian Petroleum Standards producing body.

The airline industry probably has the best-aligned standards. The need to unify across many countries means that the standards applied are ICAO (International Civil Aviation Authority) pretty universally across the globe.

CORPORATE OR COMPANY STANDARDS

Typically, an organisation, especially a large multi-site organisation, will develop a set of standards that are to be applied to all parts of the organisation. These are often based on practices and country standards but go beyond them. For example, Shell Design and Engineering Practices (Shell DEPS) are applied within Shell assets but are also recognised as best practices. Some other organisations have chosen to pay for licenses to use them.

CONFORMITY WITH LAWS AND STANDARDS

The application of laws, regulations, country standards, or local corporate standards can be measured by Audit. See previous Chapter 31 on Governance of Safety. From my experience in consultancy and insurance risk engineering, many gaps are to be found.

RAGAGEP (Recognised and Generally Accepted Good Engineering Practices) is a concept enveloping the concept of technical standards, but the RAGAGEP is not well defined. Rather it is the group of standards that are in the mind of the consultant advisor. For example, I recently came across a consultancy claiming to audit a site versus RAGAGEP. Beneficial though the audit was to the client, the standards applied were not defined in the audit scope document. Indeed, they were standards with which the consultant personnel were familiar, and did not seem to cover several important standards (which I consider RAGAGEP) with which the consultants were unfamiliar.

CERTIFICATION

Typically, on entering the foyer of a significant size company, the walls will proudly display certificates of conformity to:

ISO 9001 – Quality Management
ISO 14001 – Environmental Management
ISO 45001 – Health and Safety Management
ISO 50001 – Energy Management

These certifications are carried out only by approved third-party assessors. So, this must be a good, well-run company – right. Well, wrong in my opinion. ISO certification in any of these fields does require a visit by one or more assessors to the site. I have experienced several of these types of certification audits.

The assessors have a protocol of aspects they need to review, and these are almost exclusively paper-based. There is little if any field verification other than perhaps to visit the warehouse to examine labels or to the laboratory to discuss test result recording. Typically, a site will dedicate personnel to maintaining the documentation trail according to the standard.

All such ISO standards expect procedures, such as operating procedures, to be reviewed within a specified frequency and for the document to at least state the last revision date. Simple stuff! Then as an insurance risk engineer, why can I go to certified sites and find, usually within minutes, out of date documents and drawings, sometimes with glaring inaccuracies?

The answer, I fear, is that the certifying bodies are themselves businesses and clients of the site requiring certification. The certifying body has a vested interest in maintaining good relations with the client. It is rare for an ISO certification, once given, to be withdrawn. The certifier will provide recommendations at the end of the audit, which might include, for example, 'improving the updating of procedures and drawings ensuring that they are correctly dated'. This relatively trivial example is not visible to employees or clients.

Furthermore, ISO standards are, for a good reason, based on the principle of continuous improvement. ISO 90001 is one of the early ISO standards on Quality Management and emphasises improvement. However, the consequence is that an ISO standard does not say if an organisation is good at Quality Management. Certification indicates that the organisation has the structures in place to drive continuous improvement.

A key ISO concept is that of an annual review whereby progress is assessed and an improvement plan for the following year finalised. In my early days as a risk engineer, I asked, in the context of ISO 45001 on Health and Safety management, what the annual review had determined and the plan for the following year. I had to abandon this line of questioning simply because it became too embarrassing to the client. Why embarrassing? Because the list of plans typically involved issues that the site really should already have under control. Either the site was 'sand bagging' the program to ensure that issues that would be easily completed were listed or admitted that fundamental issues were not under control.

Finally, note that there is currently no ISO standard targeted explicitly at Process Safety Management or other approaches to preventing catastrophes. ISO 45001 on Health and Safety Management is concerned chiefly with occupational safety aspects but does indirectly influence Process Safety management issues.

Other Third-Party Assessments

There are a number of consultancy organisations that carry out third-party assessments at the request of the client organisation. Typically, these are requested because either:

1. The client organisation desires to periodically benchmark the organisation to measure improvement.
2. The client, usually because of a severe incident or series of incidents, wants to diagnose what is wrong and create an action plan accordingly.

Notes from Trevor's Files

Pressure for a Third-Party Auditor to Be Complementary

I spent many years in consultancy, and I believe in the value of such a service honestly carried out. There are considerable influences that can pressurise the assessment in either direction.

Recently I was involved in assessing an organisation that had received benchmark surveys over a long period. Our result did not reflect the improvement which the organisation wanted to demonstrate. We had good reasons. As we were preparing for our feedback, I was told by a senior director, in his native language. He knew I could understand if he spoke slowly. His message was that 'our future business relationship would depend on the results you are able to display'. In other words, we were expected to report back the forecast level of improvement. We remained true to the observations we had made. Our contract was not renewed. Another consultancy was given the benchmarking task. I must confess a degree of doubt about ethical values of the organisation that won the benchmarking contract.

Interestingly, the same company at a different site suffered a significant oil spill of falling within the definition of 'catastrophe'. I think this supports our critical view on the organisation's safety management.

Pressure for a Third-Party Auditor to Be Critical:

Whilst carrying out an assessment at a large site which was an amalgamation of two previous organisations, I was made aware that an earlier highly critical assessment had been made on one part of the site. My observations on that part of the site were actually quite positive. Reading the previous critical assessment, it seemed not to reflect the organisation I was seeing. Consultancies will be typically looking to expand their service beyond the assessment phase into working with the client in improving the situation through 'implementation assistance'. This provides pressure to be critical. The client leadership requesting the assessment will also expect a critical answer providing the clues on how to improve.

Third-party assessors may also be utilised to bring expertise to a particular aspect of the company's operations.

An example here is to be found in organisations in fields such as inspection and quality insurance as well as certification. Such organisations are similarly dependent on their clients for income and employment security. A particular example of the potentially catastrophic consequences of such a relationship can be seen in the tailing dam disaster at Brumadinho in 2019 (for a description of the incident, see Chapter 8 on Mining). The mining organisation, Vale, involved employed such a specialist company.

TÜV Sud, to certify the safety of such tailings dams. Some 12 months before the catastrophe, the certifier reported problems with the dam's drainage. However, the certifier's recommendations were either not implemented or only partially implemented. As a result, the dam's water level continued to rise, pressure on the dam wall increased, and the otherwise solidified sludge liquefied – until the dam broke in January 2019.

In fact, during an inspection, the certifiers employees found that the dam did not reach the necessary stability factor according to their calculation standards. Instead of refusing to issue a stability declaration, these employees looked for new calculation methods to achieve the desired result. They also consulted their head office. In the end – against its better judgement – the certifier's head office did not prevent its subsidiary from certifying the dam's stability. As a result, neither the mine operators nor the authorities initiated effective stabilisation or evacuation measures.

Several state and federal authorities in Brazil are investigating the Brumadinho dam breach. Criminal and compensation proceedings have been initiated against both the mining company and the certifiers Brazilian subsidiary. The mining company has rejected all responsibility for the failure, saying it relied on the certifiers stability declaration. The certifier says it warned Vale about safety concerns (European Center for Constitutional and Human Rights The Safety Business: TÜV SUDS role in the Brumadinho dam failure in Brazil).

It would be important to add that Vale has since done much to alleviate the situation of the displaced peoples affected by the Brumadinho dam disaster and has a far-reaching programme to reduce the risks from its tailings dams (see Chapter 8).

What is highly concerning here is not just one dreadful catastrophe. It brings into question the validity of the conclusions of third-party inspectors who are selected and remunerated by their client – the operator of a high-hazard facility.

GOVERNMENT REGULATORS

Most countries have state regulators whose role is to enforce health and safety. Their role is not confined to enforcement. Typically, they also advise government on new or improved legislation and help devise the regulations which go with them. In the event of a significant incident, they also have a crisis management role. Research is sometimes undertaken by regulatory bodies, especially in areas that are not covered by academic or impartial industry bodies. 'Reducing Risks, Protecting People' (R2P2), in which the thresholds of acceptable risks are scientifically evaluated, is a prime example of such research. Regulatory bodies are also involved with improvement programmes, especially when an industry faces new technological challenges.

Bahr (2015, p. 333) describes how the U.S. Federal Aviation Administration (FAA) recognised that the aviation industry in the U.S. was faced with a number of challenges including:

- increased demand, especially for regional jets.
- more complex technologies in aircraft.
- new technology incorporating satellite-based and digital systems to manage the national airspace.

The FAA was also faced with a staffing planning issue with many inspectors due to retire.

The FAA developed a System Approach to Safety Oversight (SASO) incorporating the following elements:

- Business process re-engineering
- Alignment of systems in support of the future processes
- Integration of technology and process across the FAA
- Change management to enable the future state to be realised.

Bahr then goes on to discuss how mistakes are commonly made in such governmental safety oversight programmes, such as the following (examples only) (Table 32.2).

As a regulator, the government body must carry out on-site inspections of hazardous facilities.

There is always a need for more inspectors and for the number of inspections to be increased. I doubt that increasing the number of checks would significantly impact safety in general and catastrophe prevention in particular.

There are two main reasons for my lack of faith in the value of inspections.

The first arises because an inspection must be fair. An inspector assessing industrial unit A should come to the same conclusions when assessing unit B, if units A and B are identical and managed in the same way. Therefore, there is a heavy emphasis on inspection protocols that restrict what an inspector might examine. Many of the protocols which the HSE follows in the U.K. are publicly available on the internet. Why is

Table 32.2 Common Mistakes in Government Oversight Programmes

Safety Practice Area	Common Mistakes
Incident Investigation	Punitive and not blame-free investigation. Undue outside influences and pressures.
Safety Regulator Executive Director	Insufficient authority or unclear mandate to manage the oversight programmes efficiently.
Crisis Management	Unclear roles. Tries to manage the crisis (e.g. the first responders) rather than support resources. Inadequate or inappropriate crisis communication to leadership or the public.
Safety Policy and Regulations	Legal framework covering regulations is overly complex and too detailed preventing future flexibility.
Audit, Compliance, and Enforcement	Escalation of enforcement actions is not clear, or not followed. Auditors are not adequately trained or do not understand the industry. Does not focus enough on evaluating the safety management system.
Safety accreditation	Accreditation focuses too much on the process and not validating actual safety.
Safety Risk Management	Incomplete definition of a safety management system. Poor tracking and trending of safety data. Industry and government oversight programs not fed back into the regulatory process.
Safety Promotion	Inadequate understanding or input from industry to appropriately design promotion campaigns.

this a weakness? Protocols prevent an inspector from delving into any particular aspect of a site's safety management in a way that might be perceived as disproportionate. As a consultant and insurance risk engineer, there were sometimes pieces of information that led me to investigate a particular area, at the expense of other areas. Apologies for the comparison, but the activity is akin to hunting dogs who have caught the smell of a prey. A good surveyor will follow the scent! Saddled with protocols to ensure evenness and fairness, the inspector cannot diverge into a new area of interest.

The second arises because inspectors are often assigned to particular sites and organisations. Familiarity occurs. After visiting a particular site on several occasions, it is difficult to suddenly come up with a new observation that could have been made in earlier visits. To do so could be seen as an admission that the inspector was not sufficiently conscientious in earlier visits. Conscious or unconscious 'blindness' can occur to specific issues. This is, in my opinion, an issue which is particularly prevalent in countries of the former Soviet Union where the Inspectors job security may be threatened if he/she 'rocks the boat'.

THE PRESSURE FOR DEREGULATION

There is significant political pressure in many countries, and it seems in the U.S.A. in particular to deregulate industry. The argument being that industry is plagued by too many regulations. This impedes competitiveness and speed of response to developing market needs. Indeed, in the previous four years of writing this book, there have been major initiatives in the U.S.A. to reduce the power of the Environmental Protection Agency and eliminate the Chemical Safety Board. The latter is, to my mind, an excellent resource from which to learn and is not bettered globally. Rationalising legislation is undoubtedly a valuable activity, but as the chapters in Section 5 will show, this author is not in favour of deregulation.

SUMMARY: CAUSES OF FAILURES IN THE APPLICATION OF LAWS, REGULATIONS, STANDARDS, AND CERTIFICATION

Reasons Why Safety Laws Can Fail to Avert a Catastrophe Include

☒ Being too prescriptive and therefore losing relevance to the particular risk being managed, especially where new technological developments are involved. Prescriptive legislation cannot keep up with the pace of technical development and is costly to devise and enforce.

☒ Being too general, placing the onus on the facility owner to devise his/her own rules as to what is acceptable and not acceptable without any guidance from the law. This is particularly an issue for smaller enterprises that are not staffed with experts who can glean from practice elsewhere.

Reasons Why Safety Cases Can Fail to Avert a Catastrophe Include

☒ Limited to a relatively small number of high-hazard industries (Seveso Tier 1), offshore oil and gas and nuclear installations.

☒ Not updated when the technology changes.

☒ Subject to regulatory approval which causes cost and delays, this is a barrier to the expansion of the safety case regime.

Reasons Why Safety Regulations Can Fail to Avert a Catastrophe Include

☒ Number of regulations addressing the same or very similar subject is confusing, and different regulations may even differ in what they measure, and in what they advise or stipulate.

Reasons Why Country and International and Technical Standards Can Fail to Avert a Catastrophe Include

- ☒ Confusing variety of standards which really require a specialist to understand.
- ☒ The cost of each standard is significant. If the safety engineer needs to understand only one standard the cost may be OK, but given the degree of overlap, he/she may have to purchase many standards.
- ☒ Many standards are very hard to read, and it can be difficult to drill down to the specific information the safety engineer needs.

Reasons Why Corporate or Company Standards Can Fail to Avert a Catastrophe Include

- ☒ Standards may be more or less stringent than the country regulations, and it is not always easy to see in which way. Clearly, the country regulations are the legislative minimum, but will this be met or exceeded by the company standard?
- ☒ A set of standards can take years to develop. This is a problem for small or young organisations. A solution is for a smaller or younger organisation to purchase standards from, say a large oil major. But this is costly – there is an annual charge – and the standards may not be found entirely relevant.

Reasons Why Certification by a Third Party (e.g. ISO Certification) Can Fail to Avert a Catastrophe Include

- ☒ The assessor is somewhat 'in the pocket' of the company being assessed since it is the company that is paying the fees for the certification.
- ☒ Certifications are based on the principle of having the essential elements of a safety management system set up and of using them to drive continuous improvement. They do not in themselves imply a minimum standard of performance.

Reasons Why Third-Party Assessments Can Fail to Avert a Catastrophe Include

- ☒ The third-party assessor is also somewhat 'in the pocket' of the company being assessed since it is the company that is paying the fees for the certification
- ☒ The third-party assessor may be biased in his/her assessment in order to find an opportunity to get further business.

Reasons Why Government Regulators Can Fail to Avert a Catastrophe Include

- ☒ A punitive role can obstruct openness and dialogue with the government regulator.
- ☒ The government regulator becomes overly familiar with the hazardous installation and its staff. He/she can become 'hostage' to the organisation and can be inhibited by suddenly raising new issues which the regulator has not seen before.
- ☒ Regulators can be inconsistent in their advice between similar facilities.

Chapter 33

Safety Culture and Other Social Structures

CULTURAL DIFFERENCES – ARE THEY IMPORTANT?

Working as a safety consultant with an international consultancy in the period 2000–2005, I had considerable success working in the U.K. In particular, I developed a good reputation for positive impact when working on oil platforms in the North Sea.

This success then led to my secondment to Russia and Kazakhstan, where I had the good fortune to be housed in Moscow and Almaty before moving on in 2013 to Saudi Arabia.

I was faced immediately with an important question. Would the same techniques and approaches which had brought me success in offshore U.K. work in these countries?

The techniques and approaches of the consultancy were based mainly on experiences in the U.S.A. and Europe. I found a considerable degree of interest but very poor willingness in implementation. Why would the barriers to implementation be so high? Does this mean we have to adjust safety and loss prevention training content, training methods, and implementation techniques and take a different approach in each country? This is a question for consultancies working across the globe in safety and other spheres.

Whilst the incidents we examined in Section 2 rarely identified poor safety culture as a root cause, it is, in my opinion, implied in many of the investigation reports. The Imperial Sugar, Piper Alpha, and some other investigations pointed to safety culture as needing considerable improvement. Safety culture is less easy to assess than other potential root causes.

Notes from Trevor's Files: One Size Does Not Fit All

While preparing for a potential project for the consultancy in Kazakhstan, I discussed the project with some of the client staff and the business development manager for the consultancy, who happened to be German.

DOI: 10.1201/9781003360759-36

> We had a long and somewhat strained discussion about incident investigation and the value of 'lessons learnt' versus 'finding the guilty'.
>
> After this interaction, the business development manager commented in exasperation, 'This will be a very expensive project!'.
>
> In practice, of course, it is rarely possible to pass on the extra cost to the client because they are 'difficult'. Like any business, revenue is critical in a consultancy, and closing the deal is paramount. To be fair, this particular consultancy has gone a long way to adapt to local cultures by recruiting and developing local nationals, putting nationals from the U.S.A. and Europe on secondment, and in other expansions of the methodology.
>
> My fundamental learning – in safety consultancy – 'One Size Does Not Fit All!'

For some other experiences in the former U.S.S.R. see Chapter 20 on Human Factors, Chapter 22 on Equipment Maintenance and Inspection, Chapter 23 on Emergency Response and Readiness, and Chapter 25 on Risk Assessment, HAZOP, and Other Techniques.

What Is Culture?

When I started working in Russia, I was doing an open learning course to obtain my Master of Science in Occupational Health and Safety Management. In preparing for the research project, I came across the work of Geert Hofstede (Hofstede (2001) Culture's Consequences: Comparing Values, Behaviours, Institutions, and Organisations Across Nations). Since cultural discussions are largely subjective, I was immediately attracted to his approach, which results from painstaking work involving the statistical analysis of over 116,000 questionnaires from 72 countries in 20 different languages (Hofstede, 2001, p. 41). I will make many references to Hofstede's work in this chapter. Hopefully, this basis in real data allows readers to accept the explanations given below and not treat this chapter as based on my own stereotypes!

Hofstede based his work on the premise that people carry 'mental programmes' developed in the family in early childhood and reinforced in schools and organisations. These mental programmes contain a component of national culture. Cultures can be considered as 'the collective programming of the mind that distinguishes the members of one group or category of people from another'. At this point, we are primarily concerned with national cultures. Later in this chapter, we will discuss cultures within organisations.

What Is the Relevance of National Culture to Catastrophe Prevention?

We are interested only in cultural traits which might help or hinder the adoption of good catastrophe prevention practices. I do not seek to conclude that one nation is more or less safe than another but rather to ensure that such techniques are introduced and implemented in a way that recognises the strengths of a particular national culture and any possible weaknesses.

Hofstede's research developed a Cultural Dimensions Theory in which various cultures were measured according to six categories, as shown in Table 33.1. His results need to be treated cautiously since this mammoth work was largely undertaken in the late 1960s and early 1970s, although Hofstede insights have been working to update it continuously with fresh data. One might argue that we need to be open to the possibility that national cultures have moved on since then. We might expect that the cultures of the countries of the former Soviet Union will have changed with the disintegration of the U.S.S.R. Similarly, with the dramatic increase in wealth of Middle Eastern Countries, one might expect that significant changes might have taken place there also. However, Hofstede produces evidence that cultures are remarkably stable, and my own anecdotal evidence suggests that that is true.

Data for any particular country can be found at the Hofstede-Insights Website – https://www.hofstede-insights.com/country-comparison/

(By courtesy of (C)Hofstede Insights, Itim International Oy.)

With the aid of Geert Hofstede's Country comparison was able to draw some conclusions as to how approaches to the implementation of good safety practices should be varied from country to country.

Two Safety Improvement Techniques as Test Cases: 'Management by Walking Around' and the 'Safety Culture Maturity Model'

In the following examination of culture, I will take two examples of the concepts believed to be very useful in the safety arena and in catastrophe prevention.

We have come across 'Management by Walking Around' previously in Chapter 31 on the Governance of Safety. It might otherwise be known as a STOP observation, Gemba Walk, Behavioural Safety Discussion, etc. These techniques have different nuances, but all involve managers getting out into the workplace, observing what is going on, and then discussing with the people working about their activity. Any unsafe behaviour or condition should be discussed, and good behaviours reinforced.

The second example is the 'Safety Culture Maturity Model'. There are various versions of this; many have similar characteristics. I will use the dss$^+$ Bradley Curve^{TM}, which came from Du Pont. See Figure 33.1.

Table 33.1 Geert Hofstede's Cultural Dimensions

Name	Low Description
Power Distance Index	This dimension expresses the degree to which the less powerful members of a society accept and expect that power is distributed unequally. The fundamental issue here is how a society handles inequalities among people. People in societies exhibiting a large degree of Power Distance accept a hierarchical order in which everybody has a place and which needs no further justification. In societies with low Power Distance, people strive to equalise the distribution of power and demand justification for inequalities of power.
Individualism vs Collectivism	The high side of this dimension, called Individualism, can be defined as a preference for a loosely-knit social framework in which individuals are expected to take care of only themselves and their immediate families. Its opposite, Collectivism, represents a preference for a tightly-knit framework in society in which individuals can expect their relatives or members of a particular ingroup to look after them in exchange for unquestioning loyalty. A society's position on this dimension is reflected in whether people's self-image is defined in terms of 'I' or 'we'.
Uncertainty Avoidance Index	The Uncertainty Avoidance dimension expresses the degree to which the members of a society feel uncomfortable with uncertainty and ambiguity. The fundamental issue here is how a society deals with the fact that the future can never be known: should we try to control the future or just let it happen? Countries exhibiting a strong Uncertainty Avoidance Index maintain rigid codes of belief and behaviour and are intolerant of unorthodox behaviour and ideas. Weak Uncertainty Avoidance Index societies maintain a more relaxed attitude in which practice counts more than principles.
Femininity vs Masculinity	The Masculinity side of this dimension represents a preference in society for achievement, heroism, assertiveness, and material rewards for success. Society at large is more competitive. Its opposite, Femininity, stands for a preference for cooperation, modesty, caring for the weak and quality of life. Society at large is more consensus oriented. In the business context, Masculinity versus Femininity is sometimes also related to as 'tough versus tender' cultures.
Long Term Orientation vs Short Term Orientation	Every society has to maintain some links with its own past while dealing with the challenges of the present and the future. Societies prioritise these two existential goals differently. Societies that score low on this dimension, for example, prefer to maintain time-honoured traditions and norms while viewing societal change with suspicion. Those with a culture that scores high, on the other hand, take a more pragmatic approach: they encourage thrift and efforts in modern education as a way to prepare for the future.
Indulgence vs Restraint	Indulgence stands for a society that allows relatively free gratification of basic and natural human drives related to enjoying life and having fun. Restraint stands for a society that suppresses gratification of needs and regulates it by means of strict social norms.

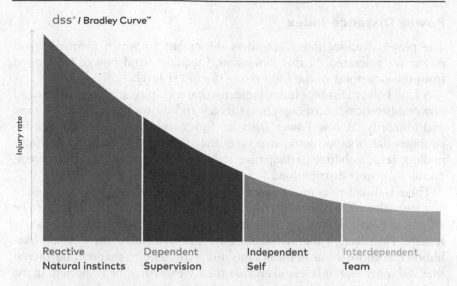

Figure 33.1 DSS+ Bradley Curve (DSS: The Bradley Curve).

The detail of the safety culture maturity model need not concern us here. The principle is that an organisation follows a progression of cultural maturity with the following four stages:

- Reactive stage – people are mainly following their natural instincts.
- Dependent stage – in this stage, there is a high degree of supervision, which achieves an improvement in injury rate over the Reactive Stage.
- Independent stage – in this stage, individual employees feel more responsible for their own conduct and safety and achieve a further reduction in injury rate over the Dependent stage.
- Interdependent stage – in this stage, employees work together well as teams looking after each other safety, thus achieving a further reduction in injury rate.

Many organisations do not, of course, progress but get stuck in a particular stage of cultural maturity. The purpose of explaining the Bradley curve is to help organisations find ways to progress to better levels of safety performance.

RELEVANCE OF HOFSTEDE'S DIMENSIONS

We will now take each of Hofstede's Dimensions and look at how a nation's culture might influence the readiness to implement the above two examples of Gemba Walk and Bradley Curve.

Power Distance Index

The power distance index considers the extent to which inequality and power are tolerated. In this dimension, inequality and power are viewed from the viewpoint of the followers – the lower level.

A high Power Distance Index indicates that a culture accepts inequity and power differences, encourages bureaucracy and shows high respect for rank and authority. A low Power Distance Index indicates that a culture encourages flat organisational structures that feature decentralised decision-making responsibility, participative style of management, and place emphasis on power distribution.

There is direct relevance to catastrophe prevention in this category. In a culture with a high Power Distance index, instructions are more likely to be strictly followed. Following instructions is generally perceived as a positive feature. However, if the instructions are defective, then there is less likelihood that those lower in the organisation would point that out. Further, if they did speak out, it is less likely that their view would be respected. In the screenplay version of Chornobyl, the nightshift control panel operators play the role of trying to communicate to the chief engineer in charge of the test. The message that they were trying to send was his instructions had the potential to lead to a massive catastrophe. They were not heard. Maybe this is an exaggeration for the screenplay, but this characteristic certainly rings true.

In Chapter 9 on Transportation, I referenced the Power Distance Index in the context of a proven relationship between aircraft incidents and cultures with a high Power Distance Index. Hofstede identifies a 1994 article written by executives in the safety section of the then Boeing Commercial Airlines Group. They demonstrated a clear positive correlation with 'power distance index' and a negative correlation with 'individualism'. If the relationship between the pilot and co-pilot is hierarchical rather than collegial (relating to or involving shared responsibility, as among a group of colleagues), this may lead to a co-pilot not correcting errors made by the pilot. Standard procedures in the cockpit assume two-way communication between pilot and co-pilot.

On the other hand, a low Power-Distance index may not be ideal in terms of catastrophe prevention. Whilst individuals may find it easier to express their views collectively to develop improved plans of action, it can take a lot of time to come to a conclusion, and sometimes the chain of command is unclear – an aspect which is critical in the event of a significant incident.

Overall, I tend to consider a low power distance index beneficial to safety and catastrophe prevention, especially considering the value of risk assessment (Figure 33.2).

As shown in Figure 33.1 the Power Distance Index is considered to be very high in former Soviet Union countries, including Russia and Kazakhstan. In contrast, the Power Distance Index for the U.K. and the

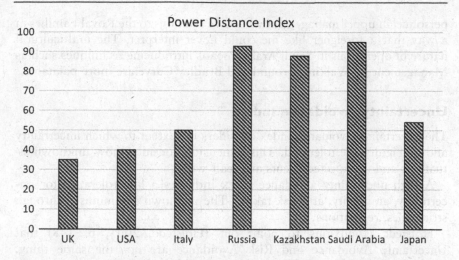

Figure 33.2 Power Distance Index for selected Countries.

(Courtesy of (C)Hofstede Insights, Itim International Oy).

U.S.A. is much lower. Also high on the Power Distance Index is Saudi Arabia. This is one of the dimensions that make the application of Gemba Walks and the Bradley Curve especially tricky.

The reader may recall how I moved from consultancy in the U.K. employing methods like Gemba Walks and the Bradley Curve to projects in Russia, Kazakhstan, and Saudi Arabia using the same tools.

Now in the standard training that accompanies introducing these and other tools, I found the managers from Russian, Kazakh, and Saudi Arabian countries very interested and attentive. Maybe a little bemused. What I found surprisingly tricky is getting the management to attempt implementation. I don't recall really seeing a manager from any of these countries carrying out a successful safety interaction whilst out on the plant – using the model for interaction which we had presented. However, things were better in Saudi Arabia, but perhaps for the wrong reasons, as described in the next paragraph.

Hofstede's data does not tackle the issue of countries with substantially mixed ethnicity, language, or identity. Thus, he alludes to differences between Netherlands and Belgium, but not between the Netherlands and the Flemish or the Walloons. In Kazakhstan, there is a mix of approximately 63% ethnic Kazakhs and 24% ethnic Russians. From my observations, the Kazakhs have some very different traits from Russians, but in Power Distance, they are similar to Russians, perhaps even stronger in Power Distance. They have a fairly strong Caste system and retain considerable pride in their ethnicity. However, in Saudi Arabia, the workforce includes a high percentage of labour, including people at management levels, from non-Arabic countries. I suspect there is also a dominance of Arabic personnel in upper management who have relations to the Royal Families in a way that a foreigner like me could never interpret. The multicultural nature of operations in Saudi Arabia means introducing techniques such as Management by Walking Around and Bradley Curve are more palatable.

Uncertainty Avoidance Index

The uncertainty avoidance index considers the extent to which uncertainty and ambiguity are tolerated. This dimension considers how unknown situations and unexpected events are dealt with.

A high uncertainty avoidance index indicates a low tolerance for uncertainty, ambiguity, and risk-taking. The unknown is minimised through strict rules, regulations, etc.

Hofstede is anxious to point out (Hofstede, 2001, p. 148) that Uncertainly Avoidance and Risk Avoidance are not the same thing. Uncertainty in this context does not have an object; risk does. 'Uncertainty has no probability attached to it. It is a situation in which anything can happen, and one has no idea what!'

Given that a high uncertainty index indicates a low tolerance for risk-taking, I suggest that a high uncertainty index is beneficial to catastrophe prevention. However, Hofstede recognises that low uncertainty index cultures stimulate creativity and flexibility to an emerging environment.

Countries with high Uncertainty Avoidance Indices have a stronger orientation towards rules and institutions which make the rules.

One of the telling questions revealing the Uncertainty Avoidance Index used in Hofstede's was 'Is it acceptable that employees break company rules if they believe this is in the company's interests?'

So, what is the relevance of the Uncertainty Avoidance Index to the Implementation of Management by Walking Around and Safety Maturity Model? In behaviour safety interactions, the emphasis, as taught by Du Pont Sustainable Solutions and others, is not to focus on the rule. If the breaking of a rule has been observed, stress the resulting risk to the individual or risk to others. Management by Walking Around is, therefore, not going to be easy to implement in countries with high Uncertainty Avoidance Indices; indeed, they might find such a conversation quite stressful since there is uncertainty in the outcome. However, there is no such ambiguity if the manager can quote the rule.

The safety maturity model or Bradley curve emphasises rule compliance in the 'Supervisory' stage but less dependence on Rules in the 'Independent' and 'Interdependent' stages. Therefore, it seems likely that countries with high Uncertainty Avoidance indices are likely to get stuck on the 'Supervisory' stage and find difficulty in moving to the right.

Overall, I tend to see a low Uncertainty Avoidance Index as beneficial to safety (Figure 33.3).

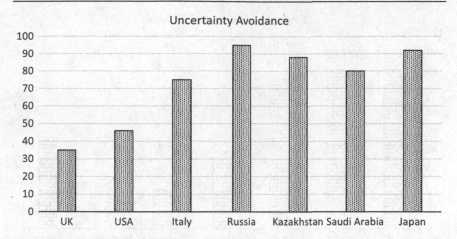

Figure 33.3 Uncertainty Avoidance Index for Selected Countries.

(Courtesy of (C)Hofstede Insights, Itim International Oy).

So, who are these high countries with high Uncertainty Avoidance Indices? Russia, Kazakhstan, and Saudi Arabia are all countries with high Uncertainty Avoidance Indices. The U.S.A. and U.K. are low in the table. There is more significant variation in Western Europe than with Power Distance Index, with France, Italy, and Spain scoring quite high. So, the difficulty I experienced implementing some of the safety methodologies, primarily born in the U.S.A., in many countries, makes sense.

The U.K. was one of the first nations to move away from prescriptive legislation with the Health and Safety at Work Act 1974. The onus was then on the employer to determine what is safe and unsafe using a suitable and sufficient risk assessment and the need for a low Uncertainty Avoidance Index. But what does this mean for other countries following suit? I suggest that countries with a high uncertainty avoidance index will struggle to deviate from prescriptive legislation. Hofstede (2001, p. 174) points out that the U.K. does not even have a written constitution!

Individualism vs Collectivism

The Individualism vs Collectivism dimension considers the degree to which societies are integrated into groups and their perceived obligations and dependence on groups. Individualism indicates that there is a greater importance placed on attaining personal goals. A person's self-image in this category is defined as 'I'. Collectivism indicates that there is a greater importance placed on the goals and well-being of the group. A person's self-image in this category is defined as 'We'.

One of the revealing questions in this category 'Are decisions made by individuals usually of a higher quality than decisions made by groups' – those agreeing with such a statement score more highly in individualism.

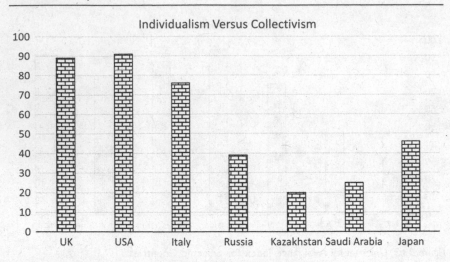

Figure 33.4 Individualism vs Collectivism for Selected Countries.

(Courtesy of (C)Hofstede Insights, Itim International Oy).

The risk assessment process is based on the concept of a team getting together. It is a team activity. Whilst collective decision-making is open to errors (see Chapter 20 on Human Factors for 'Risky Shift' and 'Group Think'), it is generally endorsed in modern safety management methods. For example, in the Safety Culture Maturity model, as expressed in the Bradley Curve, 'Interdependence' to the right is strongly favoured in terms of injury reduction (Figure 33.4).

For this category, we find a current in a contrary direction to that of Power Distance Index (PDI) and Uncertainty Avoidance Index (UAI). Countries such as Russia, Kazakhstan, Saudi Arabia score low on Individualism, i.e. they are strong on Collectivism, whereas the U.S.A. and U.K. are highly individualistic. This suggests that to achieve the team collaboration necessary to implement the Safety Culture Maturity model (e.g. Bradley Curve), a more significant hurdle should be experienced in the U.K. and U.S.A. than in Russia, Kazakhstan, and Saudi Arabia. This may demonstrate why the concepts went down well in these countries, even if PDI and UAI prevent implementation.

Long-Term Orientation vs Short-Term Orientation

The long-term orientation vs short-term orientation dimension considers the extent to which society views its time horizon. Long-term orientation shows focus on the future and involves delaying short-term success or gratification in order to achieve long-term success. Long-term orientation emphasises persistence, perseverance, and long-term growth.

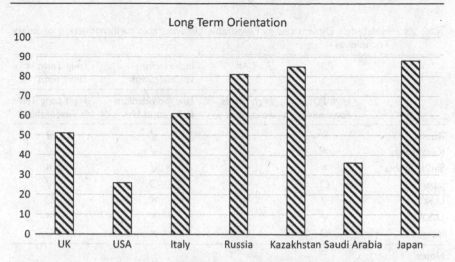

Figure 33.5 Long Term Orientation vs Short Term for Selected Countries.
(Courtesy of (C)Hofstede Insights, Itim International Oy).

Short-term orientation focuses on the near future, involves delivering short-term success or gratification, and places a stronger emphasis on the present than the future. Short-term orientation emphasises quick results and respect for tradition.

Long-term orientation I consider to be beneficial to catastrophe prevention. Work on catastrophe prevention needs looking long term. In the short term, the odds are that the big event is unlikely to happen (Figure 33.5).

So again, we have a counter-current, with Russia and Kazakhstan showing a high Long-Term Orientation, the U.K. much less so, and with the United States one of the most short-term.

Table 33.2 needs to be treated with caution. No data is available that suggests that the four dimensions are equal in terms of techniques for catastrophe prevention. The purpose of constructing the table is to show the variation. Even between U.K., Germany, and France, different approaches need to be taken to maximise strengths and recognise barriers to implementation.

Hofstede's remaining categories for:

- Masculinity vs Femininity
- Indulgence vs Restraints

These dimensions, whilst valuable in other spheres, did not seem, to me, to offer any particular insights on catastrophe prevention. Interested readers are recommended to read Hofstede's original accounts (Hofstede, 2001, Culture's Consequences: Comparing Values, Behaviours, Institutions, and Organisations Across Nations).

Table 33.2 Hofstede's Dimensions as Favourable/Unfavourable to Introduction of Safety Techniques

	PDI	UAI	Individualism vs Collectivism	Long Term vs Short term
	High PDI Is Favourable	High UAI Is Favourable	Low Individualism Is Favourable	High Long Term Is Favourable
Russia	✗	√	√	√
Kazakhstan	✗	√	√	√
Saudi Arabia	✗	√	√	✗
Japan	O	√	O	√
U.K.	√	✗	✗	O
U.S.A.	√	✗	✗	✗
Italy	O	√	✗	O

Notes:
✗ = Unfavourable
O= Neutral
√ = Favourable

Notes from Trevor's Files: Attitudes to Risk Assessment

In research done on Attitudes to Risk Assessment (Hughes, 2007) a questionnaire was sent to many of my colleagues in the UK, European, and post-communist Eurasian countries. In a statistical analysis of the responses, I found no significant regional differences regarding risk assessment understanding. However, U.K. respondents were found to be significantly more confident of risk assessment processes than from elsewhere in Western Europe and in Post-Communist Eurasian countries. The U.K. respondents generally also felt that it was more likely in their organisations that risk assessments are of value to peoples' safety at work. They also thought that both management and people, in general, were more concerned about safety in their workplaces. It was suggested that the extent to which the cultural mentality that accepts corruption may also be an influence. There will be more acceptance of influence on the risk assessment process to produce the outcome perceived to be desired by management.

Overall, the project data substantiates the view that U.K. respondents are more effective in risk assessment, perhaps because of the U.K.'s long history of implementation of risk-based legislation. However, the data suggests, contrary to the expectations, that Western European respondents (excluding the U.K.) had equal understandings and effectiveness as achieved in Post-Communist Eurasian countries, where risk assessment is a relatively new concept.

So far, we have discussed the influence of culture in the community. However, an organisation can create a culture within. I consider that Du Pont, for whom I worked for many years in different countries, was particularly adept at creating cultures that were effective in improving safety wherever in the world they chose to build a hazardous facility. I referred to techniques such as the core value contact and other techniques in Chapter 31 on the Governance of Safety.

Role of Organisational Cultures and the Development of a 'Safety Culture'

Does local culture influence safety in general and catastrophe prevention in particular?

It may be best to attempt to define culture first. 'Culture' can be defined as the way of life, general customs, and beliefs of a group of people. 'A safety culture' can be seen as the product of individual and group attitudes, perceptions, values, competencies, and patterns of behaviour with respect to workplace health and safety (IOSH Magazine Sep/Oct 2021 'Strengthen your Safety Culture'). Where does the culture come from? It is suggested that it develops from the combined experience of the people within the organisation.

The above analysis of National Cultures ignores local differences. However, Hofstede (2001, p. 93) acknowledges that different cultures can exist between home and work. Here we will concentrate on creating a different culture at work to prevent catastrophes in our own particular work organisation.

A poor safety culture is sometimes identified as a significant factor in the cause of major catastrophes. For example, Lord Cullen's report into Piper Alpha stated that 'it is essential to create a corporate atmosphere or culture in which safety is understood to be, and is accepted as, the number one priority' (Source Oliver William, 'Strengthen you safety culture', IOSH Magazine, October 2021).

The local residents living in the neighbourhood of a high-hazard facility are often dependent on that facility for employment. It contributes to the wealth of the area either directly or indirectly. Local residents are, therefore, often quite protective of the facility. Local regulatory bodies will often be under some pressure to approve permit and planning permissions because of the wealth and employment which the facility will bring.

Mining organisations in third world countries often operate with questionable standards. The challenge to improve is unlikely to come from local officials whose neighbourhood depends on the mine. Similarly it is unlikely to come through interference from foreign Non-Governmental Organisations. I have some hopes that publicly traded mining companies will, through the pressure of their shareholders and share market requirements, improve in catastrophe prevention. But what about non-traded mining companies.

It should be noted that some mining companies, such as ENRC and Kazakhmys in Kazakhstan, have gone through public listing on the London Stock Exchange, and the public scrutiny which goes with that, and then decided to go private (with a name change) after testing the regulatory pressures of a London Listing. There are, of course, many potential reasons for this unrelated to safety.

Groups of freelance miners in Columbia choose to continue to extract gold from tailings and rock deposits using mercury in full knowledge of the risks to their health and the dangers to health in the community from the discharges into local watercourses. This is their only source of income; necessity causes them to take the risky option.

Cultures differ in the way they welcome regulation or desire deregulation. For example, this can become quite political, democratically minded people in the U.S.A. want more governmental control, whilst those who are republican-minded want less government control.

SAFETY CULTURE

Much is said and written about safety culture. I, as a consultant, helped many organisations with programmes to improve their safety culture. Management's role in developing a strong safety culture is described under 'Management Commitment' in Chapter 33 on the Governance of Safety. This is particularly relevant in workplace safety, where the link between workers' attitudes and workplace incidents is intuitively clear.

In 1931 Heinrich first proposed a triangle representation which has since been updated, notably by Bird, and has become known as the Bird Triangle. The principle of the triangle is that there is a relationship between major injuries, minor injuries, near-miss incidents, and unsafe acts. A broad estimate that over 90% of injuries were caused by unsafe acts of people is commonly used. The statistics used vary, but many are based on some statistical basis. The U.K. HSE in 'The cost to Britain of workplace accident and work-related ill health in 1995/96' suggests the figures shown in Table 33.3.

Table 33.3 Accident Triangle

Fatal injuries	1
Serious or major injuries	207
Injuries causing over 3 days away from work	1402
Minor personal injuries	2754

The figures can be contested, but the principle is clear. If a person carries out an unsafe act, there is a degree of chance as to whether that unsafe act results in an injury. Suppose I am doing a piece of DIY drilling a hole in a piece of wood, and I forget to put on my safety glasses. Most of the time, nothing will go wrong until one day, a splinter flies into my eye!

We can accept the basis of safety culture as appropriate for occupational incidents. In this context, there is much excellent advice available from Du Pont Sustainable Solutions and other sources on creating a robust safety culture. See, for example, IOSH Magazine Sep/Oct 2021 'Strengthen you Safety Culture'.

However, does the Bird Triangle also apply to catastrophe prevention? At first sight, intuitively, a similar relationship should exist. If workers, pilots, train drivers, etc. ... are carrying out unsafe acts, then there is a statistical relationship with a chain of events that could lead to a catastrophe.

Where an organisation has had a catastrophe, poor safety culture is often identified as a cause. In Chapter 9, the train crashes in the U.K. in the 1990s, in particular at Ladbroke Grove (the 'Paddington' rail crash), were thought to be due to a poor safety culture existing in the railway organisation at the time.

In the energy insurance industry, we talk about losses, not fatalities. A loss comprises a significant financial loss, often born mainly by the insurers, of major accidents in the oil, gas, and petrochemical industry. A detailed study of financial losses over USD 50 million showed that over 40% of such losses were due to mechanical integrity failures such as leaks caused by pipework corrosion. Of the remaining, over 35% were attributed to the control of work (primarily maintenance work) and over 40% to operations practices and procedures. It is in these latter two sectors where unsafe acts come into play, but the situation is more complex than for occupational safety injuries (Lloyd's Market Association. 'An Analysis of Common Causes of Major Losses in the Onshore Oil, Gas and Petrochemical Industries').

Culture can be considered as the 'way we are wired'. Sorry to put it so crudely, but we are partially wired at birth, then more firmly wired with all sorts of biases and preferences as we go through our teenage years. National traits appear at this time also. When we develop our career and join an organisation, some rewiring occurs also. It certainly happened to me when I joined Du Pont in 1977 and became imbued with their obsession with safety at that time.

Religions require that the religious believe. There are hundreds of religions in the world with many millions of believers. They cannot all be right, but for sure, members of each religion believe in their religion. I chose my Catholic religion late in life. I believe in my religion; I choose the moral framework for my life which goes along with that belief.

I am always somewhat amused when a politician says that they 'believe' something. When we say we 'believe' something, we do not know or cannot prove it. If I join the Labour party, I would choose to believe some things about socialism and the worker. If I join the Conservative party, I believe some things about the free market and the power of capitalism. No one can say which is right or wrong. We probably choose one or another path because we have a particular affinity for the resulting ethos.

Sorry for my divergence into both politics and religion. The relevance is that we can develop cultures that have particular characteristics and desirable consequences. In the final chapters of this book, we will discuss the characteristics of high-reliability organisations. We will see short-termism as being the enemy of catastrophe prevention. We can 'wire out' short-termism and 'wire in' catastrophe prevention into our local organisations and even national cultures if we wish.

Is this in any way possible? I think there are examples that demonstrate that it is possible. Take waste recycling as an example. In the 1970s, if you had asked me if people would take the trouble to separate their waste and put it in different bins, I would have said that this was a non-starter. Now we all take the extra trouble to do this well. People's attitude towards sustainability and greenhouse gas emissions is developing. People are buying more expensive electrical or hybrid vehicles. We can rewire for catastrophe prevention, too, if we choose.

ROLE OF PUNISHMENT – BY AN ORGANISATION ON ITS MEMBERS

If an employee does wrong, he or she should be punished. This seems fairly straightforward. However, in determining that the employee has done wrong, words of caution are needed. Let's use the example where an employee, whether a plant operator, train driver, or airline pilot, does not follow the procedure. Before debating whether a caution, written warning, or dismissal is appropriate, pause a moment to ask some questions.

- Did the employee know there was a procedure?
- Had the employee been trained in the procedure?
- Did the employee understand the procedures?
- Were there conflicting procedures?
- Is the employee the only person not following the procedure or is there more widespread non-adherence to the procedure by others who are (unjustly) not being considered for disciplinary measures?
- Is the correct procedure practical?
- Did the employee put others in danger, or the company at risk of financial loss, by not using the correct procedure?
- Was the incorrect procedure they actually used better in any respects than the correct procedure?

- What was the employee's motivation for using the incorrect procedure?
- Was the employee encouraged to deviate from the correct procedure by others, especially any supervision or management?
- Does the employee have any past record of not following procedures or other practices which have put the employer at risk of disciplinary action?

I have come across a concept in the oil and gas industry of 'Just Culture', which uses a flowchart to navigate through the above types of questions and conclude if the incident is in order of seriousness:

- Malevolent Act
- Reckless Violation
- Negligent Error
- No Blame Error

A key test in such a process is the 'Substitution Test'. In carrying out the substitution test, you are asked to imagine whether another employee with the same competency and qualifications would have made the same error. I have found that if the concept of Just Culture is honestly applied, then 'No Blame Error' is often the outcome.

Many organisations and cultures are so content when, after an incident, they have someone to blame. Just Culture takes them right out of their comfort zone. However, there are some cultures where I have not come across the concept of 'Just Culture', and I confess to having been unsuccessful in getting it considered.

I am not suggesting that there is no role for punishment. The tool of punishment needs to exist. An employee needs to do his job. Punishment is just part of the toolbox and is not an answer to catastrophe prevention.

ROLE OF PUNISHMENT – BY THE STATE ON THE ORGANISATION AND ITS MEMBERS

If an organisation suffers an incident or misbehaves by not following state rules, it can be punished either by fines or the imprisonment of its leaders.

In the U.K., and I suspect most other countries, the criminal prosecution of one or more leaders in a large company is rarely, if ever, feasible. In a large company, responsibilities are distributed amongst the board, and any board member's wilful or gross negligence is hard to establish. Sadly this leads to a considerably injustice – leaders of smaller companies are easier to prosecute in terms of legal responsibility as individuals, and to punish accordingly.

In the U.K., fines for Health and Safety breaches can be up to GBP 10 million and GBP 20 million for manslaughter (IOSH Magazine Nov/Dec 2021). This is for criminal prosecution. A civil case involves damages, which could be larger in the context of catastrophe prevention if proven. Note that insurance often bears a significant portion of the cost of civil claims, so the organisation at fault does not feel all of the financial pain.

The reputational risk associated with prosecution is a more significant source of pain for organisations in the public eye.

Perhaps the most considerable fines to date are those imposed on BP following the Deepwater Horizon incident. See Chapter 6 on Oil and Gas. Mainly as a consequence of the environmental disaster, fines exceeded USD 20 billion, and total costs exceeded USD 65 billion. Two manslaughter charges against two well-site leaders onboard the Deepwater Horizon were subsequently dropped.

There are several instances where individuals have been prosecuted. Successful prosecutions, perhaps unfairly, tend to be on the directors of smaller companies where responsibilities are more concentrated on one or two individuals.

After the fire at Bosley Mill (2015) in the U.K. (see Chapter 12), where four workers died, a prosecution was taken against the company that operated the mill, the managing director and two other directors. Six years later, the managing director was given a 9-month prison sentence, suspended for 18 months and fined GBP 12000. The firm was fined GBP 75000 (BBC: Bosley Mill: Fine for the owner over explosion which killed four people). The managing director had been acquitted of gross negligence manslaughter. The judge described the managing director as being 'totally inadequate'. Nevertheless, it was ruled that there was not enough evidence to prove that gross negligence caused the explosion. My learned colleagues may want to correct me on a subtlety here. There may have been negligence, but it is necessary to prove that the negligence led to the accident in a criminal prosecution. The defence questioned if those in charge recognised and really understood the risks that followed from too high dust levels. I must confess I thought that ignorance of the law was no defence. See Chapter 44 on suggested changes to Government and Legal Structures for further discussion on this subject.

After the Mont Blanc Tunnel fire in 1999, in which 39 people died (see Chapter 9 on Transportation Industry – Road and Rail Tunnels), a prosecution was brought against 13 defendants. Despite the loss of life and relatively simple organisational structure, the most significant sentence was six months in jail plus a two-year suspended sentence (Wikipedia: Mont Blanc Tunnel Fire).

Successful prosecutions against individuals following catastrophes are rare. Successful prosecution seems to be more successful against offences with occupational health and safety consequences than those with catastrophic potential. Perhaps this is simply because there are many more instances of occupational health issues meriting consideration of prosecution.

For example, a manager of a car salvage company received a jail sentence although no injuries had occurred at his yard. He had ignored multiple safety enforcement notices served over a sustained period (IOSH Magazine: Nov/Dec 2021, Car Salvage Firm Boss jailed for ignoring HSE notices).

In the U.K. the costs of litigation can be significant. Prosecutions under criminal law were subject to revised sentencing guidelines in 2016, with most potential sentences being increased. Fines can be up to GBP 10 million for health and safety breaches and up to GBP 20 million for corporate manslaughter. However, in such criminal cases, the prosecution must prove the case 'beyond all reasonable doubt'. This is a higher standard of proof than that required in civil cases where the plaintiff sues for damage which are in principle unlimited. In civil cases, in the U.K. the burden of proof is 'on the balance of probabilities'. In the context of health and safety, the claimant must prove the employer owed them a duty of care, that that duty was breached, and that their injury and damage were reasonably foreseeable consequences of the breach. The duty of care is discharged if the employer does what any reasonable and prudent employer would have done in the circumstances.

Minimising media attention can be an important priority for a company potentially being prosecuted or sued following an accident. A prompt guilty plea to a criminal prosecution may bring about lower fines and reduce media attention. A civil award of damages may attract some limited publicity but is primarily a matter between the parties and their insurers (IOSH Magazine Nov/Dec 2021 'First Line of Defence').

FALSIFICATION, NEGLIGENCE, AND DISHONESTY

Some catastrophic incidents are associated with acts of dishonesty. At the Listvyazhnaya mine disaster in Russia in 2021, the President of Russia himself announced that managers had falsified methane sensor data (see Chapter 8). In Chapter 22 on equipment inspection, we discussed an incident at a nuclear power plant in which an engineer either concealed or was pressurised to conceal excessive corrosion. The incidents we have examined so far reveal, in several cases, negligent acts. For example, at the Soma mine disaster, a considerable number of methane and carbon monoxide detectors were not operational. At the Brumadinho tailings dam disaster, the inspection service found that the dam did not reach the necessary stability factor but then found alternative calculations which justified certification.

What is it that causes human beings to be either blatantly dishonest, overlook critical issues, or feel they should find alternative explanations for the data? There is no motivation for the average engineer, laboratory technician, or plant operator. They are employed to carry out a particular job according to prescribed procedures. However, management are employed to ensure that a facility achieves particular targets. In my early days of management, Peter Drucker's 'Management by Objectives' was required reading. It is a good book, and the fundamental principles are very appropriate:

- People should be set challenging yet achievable goals
- They should receive regular feedback on achieving objectives
- The focus should be on rewarding good behaviour rather than punishing failure
- Emphasis on personal growth and development

The last three bullet points can easily be forgotten at the expense of the first. And even the first can be abbreviated to 'People should be set goals', especially at higher levels.

The primary goal of any facility manager is to make the required amount of product or supply a service in accordance with the planned output. Whether the goals are achievable or not, given the constraints of a facility, is rarely, in my experience, seriously discussed.

Notes from Trevor's Files: Ethics

Whilst the following story is quality-related, not safety-related, it sadly demonstrates a time when I was drawn into ethically questionable behaviour. We had acquired a plant to make paint for automobiles. Specifically, we made special paint to apply to the plastic parts of the vehicle. Once applied, the resulting paintwork had to match the different paint which was applied to the metal portions of the car. Paint matching is a highly skilled yet somewhat subjective art. The plant was way behind schedule. Several prominent vehicle manufacturers were in danger of being brought to a halt because of the lack of paint from my plant. We had a deviation procedure. If the plant manager thought that a paint batch was sufficiently close, he could sign off the batch; I hasten to add without notifying the customer. I made use of this procedure far more often than I wished. After I was effectively fired due to poor performance, the following manager had considerably more moral backbone and stoically refused to use this procedure himself. Although it may be that others were prepared to sign off. I retain just a little pride in the fact that the week when I left this job was also the week of maximum site production. Were the goals for the facility ever achievable? I think not. The site was subsequently closed about a year later. Whatever the arguments, what I did was wrong.

If there is falsification, it is because personnel want to meet the expectations of management and fear if they do not. We will return to pressures on-site management, and consider some solutions in Chapter 43 on Changes to Leadership Mindset.

Newly appointed managers in their early 30s to early 40s are mostly in a vulnerable situation. Probably also newly married, starting a family, finding a house to suit expectations, etc. ... Under pressure at work the young manager will probably put family first. Under such pressure compromised ethical standards can occur, if permitted by the higher management applying the pressure.

ROLE OF INSURANCE – HELP OR HINDRANCE

Most companies operating high-hazard facilities are insured. The three main areas of insurance in terms of catastrophe prevention are:

- Liability – protects the business against compensation claims to a third party or damage to their property. Employers' liability is mandatory coverage protecting the company against liabilities arising from injuries to an employee.
- Natural catastrophe – protects the business against specified natural catastrophe risks such as an earthquake. Premiums depend on the natural catastrophe risk of the location.
- Property Damage – protects the business against damage to its own property. Such damage might be caused by a fire, explosion, or natural catastrophe.
- Business Interruption – replaces income lost in the event that a business is halted due to physical loss or damage, which might be fire, explosion, or natural catastrophe.

Not all of the cost is borne by insurance. First, there is always a deductible – an amount of loss up to the insured's deductible limit. Losses above this amount can be passed on to the insurer. Secondly, there is always a cap; the insured, not the insurer, bears losses above the cap. The lower the deductible and the higher the cap – the more the insured needs to pay in terms of premium.

For high-hazard facilities in the energy industry, the insurer will typically require a risk engineering report, whereby a team of risk engineers visit the facility, ask questions, make observations, and, in most cases, raise recommendations. The insurance companies use the risk engineers report to determine if they wish to participate in insuring the facility, and, if so, what the premium will be.

Liability coverage is almost always in place. Interestingly liability coverage is typically provided without risk engineering. A description of the facility and its hazards is sufficient. This is because premium income from liability on a hazardous facility is lower than income from Property Damage and Business Insurance premiums. This may sound surprising given the extent of potential liability, but liability insurance is stringently capped.

To some extent, insurance takes away the financial penalty of a major incident and could be seen as reducing an organisations motivation for protecting itself with good practices. On the other hand, insurances are capped, so a major catastrophe may still have severe financial implications on the company.

A benefit from insurance is the risk engineering surveys which contribute to the process safety effort and help transmit good practices globally in a way that an individual organisation cannot do.

However, insurance is a commodity market. Therefore, premiums are not necessarily significantly reduced for plants that 'go the extra mile' in terms of implementing good hardware and management systems to prevent catastrophes. Whilst insurers take note of the findings from risk surveys, there is no direct relationship with premiums charged; the beneficial effects can be drowned out under other commercial pressures in the insurance marketplace.

SUMMARY: WHY ATTEMPTS TO IMPROVE SAFETY CULTURE CAN FAIL TO AVERT A CATASTROPHE

Reasons Why Attempts to Improve or Introduce a Safety Culture Can Fail Include

- ☒ Approach being taken is not adaptable to different National or Local Cultures.
- ☒ National or local culture has significant characteristics which obstruct or are in disagreement with the safety culture being introduced.

Reasons Why the Introduction of 'Management by Walking Around', Gemba Walks, Safety Observations Can Fail Include

- ☒ Persons in power cannot take input from regular employees; it is too much of a break from their national culture.
- ☒ People carrying out the observations adopt narrow habits with regards to the types of observations they make and the corrective action which they adopt.
- ☒ Observers forget to compliment workers on positive aspects observed.

⊠ Observers jump into having a dialogue with the very first aspect of the work, which they see and do not take time to see the big picture. Therefore they may omit critical issues, and the worker misinterprets the absence of comments on the critical issue as being OK.

Reasons Why Organisations Can Fail to Make Progress along the Safety Culture Maturity Model Include

⊠ Excessive reliance on supervisory management – cannot let go to promote the independent phase and let interdependency grow. Legal constraints in some nations also emphasise the importance of direct supervision.

Reasons Why Organisations Are Restricted to Adopting Only Short-term Strategies Include

⊠ Expectation of being in-post is short.
⊠ Remuneration, promotion are both geared to short-term results.

Reasons Why Punishment (by an Organisation on Its Employees) Can Fail to Avert Future Catastrophes Include

⊠ Employee is only punished because his/her error was 'found out' by an incident – other employees make the same mistakes/errors without punishment.
⊠ Punishment occurs because of an incident, even if the employee had no direct involvement.
⊠ Punishment was unjust because the employee did not have a correct procedure, had not been trained, conflicting procedures, impracticality, etc.

Reasons Why Punishment (by State Authorities on an Organisation or Its Members) Can Fail to Reduce the Potential for Catastrophes Include

⊠ Responsibilities are too diffuse to punish any one individual.
⊠ The fines are too low, not impacting the minds of the management.
⊠ Organisations plead guilty, even if they are not, so as to avoid media attention and to avoid prolonging the exposure in the courts.

Reasons Why Insurers Can Fail to Influence the Prevention of Future Catastrophes Include

- ☒ Liability insurance is not usually subject to risk engineering, and therefore without examination of how the organisation is mitigating liability risk there is limited motivation for the organisation to improve.
- ☒ Whilst organisations buying business interruption and property damage insurance are subject to examination by risk engineers, the relationship between the risk quality as assessed and premiums is not direct and depends mostly on insurance market conditions.

Chapter 34

Managing Legacy Issues

WHAT IS A 'LEGACY ISSUE'

'Legacy issues' is a rather posh term used in industry to describe the issues that a business or organisation has due to its past history. It could be looked at, somewhat unfairly, as paying for the sins of the fathers!

In the incidents we examined in Section 2, a number were related to Legacy issues. The Brumadinho mine was not an active tailings facility at the time of the disaster. At the Fukushima Daiichi meltdown, the tsunami defences were based on outdated estimates of the potential tsunami magnitude and frequency. In Beirut, at the time of the warehouse explosion, there was the legacy of a shipment of unwanted ammonium nitrate with no owner.

HANDLING LEGACY ISSUES

The standards we expect today in operating a hazardous facility are much higher than they were 10, 20, 50 years ago. Many managers inherit legacy issues. One of the more prominent examples is environmental leakages.

A process plant that started operating many years ago may have leaked toxic materials into the ground. This leakage may not have been known at the time or may have been accidental. Present-day management has the unenviable task of deciding what to do with the problem that is not of its own making.

One possibility is to do nothing. Firstly, the local legislation may not actually require the management to do anything. One of my former bosses used to say, 'don't break into jail'. If the pollution is static, i.e. not putting anything or anyone at danger, it may be appropriate to 'let sleeping dogs lie'.

On the other hand, the pollution may be endangering the groundwater, it might be migrating as a plume away from the original source of leakage, driven by each rainfall, towards a vulnerable watercourse.

The manager has to decide what to do, with what budget, and with what speed.

Remote management who controls the budget will be resistant to spending money. They do not sense the danger, nor will they be the ones likely to be prosecuted.

Legacy issues may also relate to the quality of the equipment, which might be below current standards or be constructed inadequately, for example, with poor quality welds. Buildings may not be up to current building codes. In earthquake vulnerable areas, foundations may not be earthquake proofed to current standards.

In Chapter 27, we discussed Safety Instrumented Systems and IEC 61511. This is a relatively new standard, and many organisations have not yet implemented it. Furthermore, when they evaluate, they find that many instrumentation systems need to be upgraded.

Process safety studies on existing plants that have been operating for several years almost always result in the study group producing many recommendations for implementation. Either the plant was not designed appropriately, or standards and expectations have moved on since design.

I know of several operations where such recommendations are to fix risks that are in the 'Critical' and 'Very High' categories of the risk matrix. This may seem surprising. On further examination, however, I found that operator actions were considered sufficient in the original design. An indicator and alarm was provided the operator was entrusted with taking mitigating action if the alarm sounded or the indicator exceeded a specific value. To most people reading this book, doesn't this sound like a reasonable approach? Perhaps by now, the reader is more cautious. We have seen in Chapter 20 that humans are vulnerable to error. Current Layers of Protection Analysis thinking is to allow for an operator to be making a mistake 1 in 10 times. This is an example of new understandings being applied retrospectively and leading to legacy issues. So, we cannot rely on a human to avert major catastrophe and must retrospectively install instrumentation accordingly.

Asbestos and the resulting diseases from exposure to asbestos is a classic legacy issue. The material, a naturally occurring rock which can be crushed and comprises fine fibres, has been used to make many useful materials. During the Second World War many lives were saved because asbestos-based fireproofing had been applied to the ship's internals. It was a low-cost cladding for buildings. It was an essential component in vehicle braking systems. My very first job after graduating was in the asbestos industry, where my employer was grappling with trying to move away from asbestos-based products as quickly as feasible. Adverse health effects, on a mammoth scale, had been discovered. Why not earlier? It takes some time for asbestos-related diseases to appear, and when they do the disease was not immediately attributable to a person's employment or association with asbestos. Whilst the use of asbestos has been largely discontinued, some developing countries continue to use this cost-efficient material. Even in countries which no longer use asbestos, such as the U.K. asbestos continues to be present in many buildings clad with asbestos cement, and in other locations. In the U.S.A. its use continues in certain specialised applications. For a fascinating book on the discovery of the effects of asbestos see Tweedale, G. 'Magic Mineral to Killer Dust: Turner and Newall and the Asbestos Hazard'.

Asbestos materials which become cracked are a health hazard. Removal will take many years.

The chemical industry faces ongoing legacy issues. Waste dumps are found to contain the chemicals of former chemical manufacturers. New hazards are discovered about former products and wastes. Take, for example, the issue of Perfluorooctanoic acid (PFOA).

WHY LEGACY ISSUES DON'T GET ADDRESSED WITH SUFFICIENT URGENCY

By definition the current management of an organisation did not create this issue! When appointed to their current positions, their first task is unlikely to be to saddle the organisation with enormous cost. They have to 'manage their way around the issue' whilst maintaining the viability of their business.

Incidental Case Study: Perfluorooctanoic Acid

As a former employee of the Du Pont company, I feel much pain over the issue of PFOA, a precursor to PTFE, the latter being a very useful material. When PFOA was first manufactured, its toxic effects were not known. The material is used in several industrial applications, including the manufacture of carpeting, upholstery, apparel, floor wax, textiles, firefighting foam, and sealants. PFOA has been manufactured since the 1940s in industrial quantities.

Concerns began to arise in about the toxic effects of the chemical in the late 1990s.

Note that these were concerns not established issues. This is an excellent example of a legacy issue. Do organisations making PFOA now shut down on the basis of these concerns, and deprive the world of an extremely useful material?

Subsequently a study of workers living near a PTFE plant found an association between PFOA exposure and two kinds of cancer and four other diseases. A positive exposure-response trend for kidney cancer is supported by many studies. PFOA has been detected in the blood of more than 98% of the general US population in sub-parts per billion (ppb) range, and levels are higher in chemical plant employees and surrounding subpopulations. How general populations are exposed to PFOA is not entirely understood. PFOA has been detected in industrial waste, stain-resistant carpets, carpet cleaning liquids, and water.

However, we know this now, but such chronic toxicity was not known in the 1940s nor were methods available to evaluate it (Wikipedia: Perfluorooctanoic acid).

There is a fascinating screenplay relating to the legacy issues of PFOA entitled 'Dark Waters'. As for the screenplay versions of Chernobyl and Deepwater Horizon, the film writers choose hero figures and enemy figures, possibly unfairly, in order to satisfy the needs of the viewing public. Therefore, all such screenplays of historic events need to be treated with caution. In this case the hero figure is the tireless defence lawyer who pursued the complaints of animal diseases in the region of the Du Pont plant making PFOA. According to the film, he overcame mammoth obstacles before eventually securing multi-million-dollar settlements. He had to overcome objections from the law firm he was working for. The law firm wanted Du Pont as a client, not as someone they were working against. The EPA was ambivalent, at least initially, PFOA was not a regulated material. His wife questions his judgement in spending so much time on this case, which was putting his career and the family's livelihood at stake. Local residents at the plant objected to his actions – they considered the company to be a very good company, and it was providing substantial employment in the area. This employment might be put at risk by the investigations.

I felt bad after watching the film. It will be hard for the reader to accept that our knowledge advances each year. At what point does a company decide that acts of the past are no longer acceptable and put thousands of people's employment at risk. In my opinion, the balance is not as clear as viewers of this film might conclude.

OTHER EXAMPLES OF LEGACY ISSUES ON A LARGE SCALE

The nuclear industry faces vast legacy issues. Radioactivity decays slowly. Outdated reactors are being shut down but are very costly to dismantle. Whilst they remain, the potential for a catastrophe involving the release of radiation continues to exist.

The mining industry faces an ever-growing number of tailings dams that have served their purpose but still need to be inspected and maintained.

The Grenfell Tower disaster (See Chapters 15 and 44) will result in the need to review and possible replace large amounts of building cladding in the U.K. and potentially further afield.

These are just a few examples of legacy issues.

SUMMARY: WHY LEGACY ISSUES FAIL TO GET ADDRESSED BEFORE A CATASTROPHE OCCURS

- ☒ The risk was not created by current management.
- ☒ Current management have other priorities, chiefly maintaining the viability of the business.
- ☒ Many different bodies may be involved including multiple regulatory authorities.
- ☒ Deciding what to do takes time.

Vulnerability to Natural Catastrophes, Sabotage, Terrorism, and War

Chapter 35

Introduction

When I started on this book, I did not plan to cover natural catastrophes and diseases. Natural catastrophes, and pandemics such as Covid-19, are terrible events over which we have little if any control, so there are no reasons to discuss them in catastrophe prevention. The following considerations changed my mind:

1. Climate change will increase both the number and the impact of natural catastrophes.
2. Several major incidents have occurred where high-hazard facilities are impacted by a natural catastrophe. The hazard from the facility may be released by the natural catastrophe, making the total catastrophic consequence worse.
3. There are measures that high-hazard industries can take to help protect against natural catastrophes.
4. The Covid-19 pandemics may have followed from an escape of the virus from a laboratory.
5. The consequences of the pandemic caused risks in hazardous facilities to increase since skilled staffing became scarce or unavailable.
6. Dreadful diseases, such as asbestosis, can be caused by hazardous processes and materials. We need to protect ourselves from future such developments.

What natural catastrophes are worthy of consideration? In the following chapters, we will discuss:

- Hurricane and Tornado
- Flood
- Earthquake
- Tsunami
- Wildfires
- Extreme Temperatures
- Drought

DOI: 10.1201/9781003360759-39

- Diseases and Pandemics
- Chronic Diseases resulting from Industrial Exposure

We will also include consideration of some risk exposures to which the operators of a hazardous facility have very little, if any, control:

- Terrorism
- Sabotage
- War

The likelihood of a natural catastrophe is usually expressed as its return period, which is an estimated average time between events. The theoretical return period between occurrences is the inverse of the average frequency of occurrence. For example, a 10-year flood has a $1/10 = 0.1$ or 10% chance of being exceeded in any one year. A 50-year flood has a 0.02 or 2% chance of being exceeded in any one year.

This does not mean that a 100-year flood will happen regularly every 100 years or only once in 100 years, despite the connotations of the name 'return period'. In any *given* 100-year period, a 100-year event may occur once, twice, more, or not at all, and each outcome has a probability that can be computed (Wikipedia: Return period). In the USA a 100-year flood is often used as the basis for insurance rates.

What is the likelihood of a natural catastrophe where you live? Any of the above could occur, except for a tsunami, if you live well away from the sea. Natural catastrophes are more or less likely according to your location.

Insurance companies such as Swiss Re and Munich Re provide, at a cost, maps showing any location's vulnerability to a variety of natural catastrophe risks. However, free maps are often available for any particular natural catastrophe risk. For example, in the U.K. the government's environmental agency publishes flood risk maps. In the U.S.A. detailed maps are available for an earthquake.

Return frequencies are based on historical data. Therefore, it is likely that in the case of climate-related phenomena – hurricanes, tornadoes, and floods – the actual frequencies in the future are likely to be higher as a consequence of global warming.

In addition to having a direct impact on the facility, natural catastrophe events can cause power cuts and loss of other utilities such as steam supply. Whilst hazardous facilities should be protected against utility failure, for example, by being equipped with emergency generators, the loss of the utility and damage caused by the natural catastrophe event can seriously complicate the consequences.

A study by the European Commission Disaster Risk Management Knowledge Centre notes the following:

The past years set a record in the number of natural disasters accompanied by unprecedented damage to industrial facilities and other infrastructures. In addition to the Japan twin disasters in 2011 [earthquake and tsunami], recent major examples include Hurricane Sandy in 2012 that caused multiple hydrocarbon spills and releases of raw sewage, the damage to industrial parks during the Thai floods in 2011, or Hurricanes Katrina and Rita in 2005 that wreaked havoc on the offshore oil and gas infrastructure in the Gulf of Mexico. These events clearly demonstrated the potential for natural hazards to trigger fires, explosions, and toxic or radioactive releases at hazardous installations and other infrastructures that process, store, or transport dangerous substances. These technological "secondary effects" caused by natural hazards are also called "Natech" accidents. They are a recurring but often overlooked feature in many natural-disaster situations and have repeatedly had significant and long-term social, environmental, and economic impacts.

Let's look at some of these extreme weather events in more detail in the following chapters.

Chapter 36

Hurricane, Tornado and Flood

ASSESSING THE RISK: HURRICANE

The Gulf Coast of the United States and its Eastern Seaboard are particularly prone to hurricanes. The number of storms each year, and their severity, is increasing due to global warming. Hazardous facilities need to be more prepared than ever to ensure that a hurricane does not initiate damage, causing a catastrophic event. In these coastal regions of the United States, there are many high hazard facilities comprising oil and gas processing, chemicals manufacture, and nuclear facilities.

Tornados can occur in many locations, including the U.K. The most vicious and frequent tornados are to be found to the Midwest and Southeast of the U.S.A. in a region known as Tornado Alley. Such high windspeed events are usually also associated with heavy rainfall.

Weather forecasts give information in advance about potential hurricanes and tornadoes. Therefore, the management of hazardous facilities have the opportunity to decide to shut down and protect critical parts in advance of the weather event.

Oil platforms in the Gulf of Mexico are required to take the following precautions if a hurricane is forecast (National Ocean Industries Association: Hurricanes and the Offshore Oil and Natural Gas Industry, 2013)

- Evacuation of non-essential personnel
- Suspend drilling operations
- Close-in all wells

Storms in 2005, including Hurricane Katrina, destroyed 113 offshore platforms and seriously damaged 52 more offshore platforms. (Lees, 2014). It is reassuring that no significant spills occurred from any of these offshore facilities, although some oil did escape from damaged pipelines. Oil facilities built since 1988 are designed to withstand 100-year storms. The platform deck, the lowest deck on the platform, must exceed the average height of 80 feet. However, these requirements are not stipulated for oil

DOI: 10.1201/9781003360759-40

platforms outside of the Gulf region. I have seen platforms in the Caspian Sea sitting very close to the water. Whilst the Caspian Sea is not a hurricane region, high winds are not uncommon. In 2015 a storm resulted in a severe fire on a Caspian Sea platform with 32 fatalities (Entirely Safe: Oil rig fire in the Caspian Sea).

ASSESSING THE RISK: TORNADO

Most tornadoes have wind speeds of less than 110 miles per hour (180 km/h). They are typically about 250 feet (80 m) across and travel a few miles (several kilometres) before dissipating. However, extreme tornadoes can attain wind speeds of more than 300 miles per hour (480 km/h), and can be more than two miles (3 km) in diameter.

The most significant density of tornados is in "Tornado Alley", a wide stretch of inland Central and Eastern United States. This is not a dense area for hazardous industry, and perhaps for this reason, I have been unable to find any direct effects of tornados which might have led to catastrophic effects on the hazardous facility itself.

Tornado alarms are now common in the area and are initiated based on radar and other data. This allows people to take shelter. Aeroplane flights are delayed whilst the tornado risk passes.

However, at the time of writing this book, in December 2021, a large number of tornadoes associated with approximately 150 tornado warnings ripped through Kentucky, taking over 70 lives. A candle factory, a non-hazardous facility, was the location of eight fatalities, and a warehouse containing non-hazardous goods partially collapsed, causing a further six deaths.

A difficulty here was that the tornado warnings had happened some hours earlier, some tornados had passed through the area, and it was believed the danger had passed.

ASSESSING THE RISK: FLOOD

In July 2021, the German states Rhineland-Palatinate and North Rhine-Westphalia, among others, experienced some of the worst flooding in recent German history. Extremely heavy rainfall for three days, caused by cyclone Bernd, led to catastrophic flooding in multiple areas, including parts of eastern Belgium, with subsequent destruction and fatalities. Many places recorded 3-day numbers, which significantly exceeded the average for a whole month. Fatalities exceeded 100 as a direct cause of the flooding.

Floods are often attributed as the cause of hydroelectric dams and mining tailings dams failing. These dams should, of course, be designed to withstand the severe climatic conditions and be able to withstand 'overtopping'. See also Chapter 8 on Mine Waste for Tailings Dams in flood conditions and Chapter 11 for the failure of dams used for hydroelectric and irrigation in flood conditions.

Intense rainfall can cause rivers to overflow into the surrounding flood plain. Large, long rivers such as the Mississippi and Ohio in the U.S. and the Rhine and the Danube in Europe have extensive watersheds, i.e., many tributaries that feed into them. It may be possible to predict the arrival of a flood crest travelling downstream on these rivers. In some cases, flood notice of up to two weeks is possible.

Rivers that do not have extensive watersheds will rise rapidly with only a few hours notice at best. Such an event would be considered a flash flood. Once a flood plain, such as the ones seen on the Mississippi, becomes inundated, flood conditions are likely to continue for many days or even weeks. Floods from rivers with less extensive watersheds generally have a short duration. The water may recede as quickly as it rose.

In coastal regions, either flash flood or longer-notice flood conditions may exist, depending on the capacity of the land to absorb or delay the run-off from intense rainfall. Where coastal wetlands have been reduced or altered by any encroachment, the natural distribution and ability to absorb excess water has been decreased. Along flood-prone rivers, lakes and coastal areas with man-made structures on their banks, floodwalls and levees are often built to protect city streets and buildings.

Floods are catastrophes in themselves. However, a flood can cause the isolation of hazardous facilities so that relief and emergency personnel cannot access them to deal with any incident. The incident may thus cascade into a much larger catastrophic event.

Electrical equipment and instrumentation are particularly prone to flood damage. Floods can deprive a hazardous facility of equipment and instrumentation designed to deal with an emergency. The instrumentation might not fail in an obvious way but fail to give reliable results.

A particular consequence of severe winds and flooding relates to atmospheric storage tanks. They can become dislodged from their foundations and spill their contents. Many such storage tanks are located close to ports and docks on rivers and seashores. Some such oil spills resulted from Hurricane Katrina in 2005. Apparently, the recommended practice is to fill such tanks in advance of hurricanes, making them more stable and less prone to uplift (Perrow, 2007). However, this is not necessarily practical if there is no stock or supply of the stored material in order to achieve this.

INCIDENT EXAMPLES – FLOOD

Arkema Organic Peroxide Decomposition (2017)

Hurricane Harvey, a Category 4 hurricane, produced unprecedented rainfall in the southeast Texas and Southwest Louisiana region. Sixty-eight people died as a consequence of this hurricane. The hurricane's progress stalled as it hit the mainland, hence the unusual amount of rainfall in one location.

This region of the U.S.A. includes many oil, gas and chemical processing facilities. One such facility stored organic peroxides which need constant refrigeration to keep them from decomposition. The facility was located within the 100-year and 500-year floodplain. The heavy rainfall exceeded the elevations on which the equipment was designed. The site lost both main power and the back-up power available from emergency generators. The company had had the foresight to keep standby refrigerated trailers in order to keep the organic peroxide products cool in an emergency. The flooding eventually forced all the site's employees to evacuate. A Unified Command was established, the dangers having been recognised, and a 1.5-mile evacuation zone was established around the facility. Residents were transported out of this zone for their safety. The organic peroxide products stored inside one of the refrigerated trailers decomposed, causing the peroxides and the trailer to burn. Twenty-one people sought medical attention from exposure to fumes. The vapour travelled across a public highway adjacent to the plant. This road had not been closed because it served as an important route for the hurricane relief efforts. Over the next several days, a second fire occurred, and the Unified Command initiated a controlled burn over eight more trailers holding the remaining organic peroxide products.

The site did have a written hurricane preparedness plant. For the anticipated hurricane, a 'ride out' crew was established in preparation for widespread flooding, which would prevent a normal change of personnel. The level of flooding which actually occurred exceeded that for which the plan had been prepared.

A key recommendation of the investigation is improved evaluation of flood risk. It seems to this author that this recommendation lacks credibility. Arkema did have available base flood data, which questioned their flood plan but did not adopt it. An organisation is free to choose the base flood elevation it considers credible and is unlikely to extend this to cover higher flood heights at lower return frequencies.

The CSB investigation further recommended that flood data be part of process safety information to be reviewed during process hazard analyses.

(CSB, 2018, Organic Peroxide Decomposition, Release, and Fire at Arkema Crosby Following Hurricane Harvey Flooding – August 2017)

Chapter 37

Earthquake

ASSESSING THE RISK

The risk of earthquakes is concentrated in zones where the Earth's tectonic plates are moving, either colliding, moving apart or sliding over each other. Having said that, an earthquake could occur almost anywhere. What various is the frequency of occurrence?

Media will typically report an earthquake on the basis of its Richter scale. The Richter scale describes the earthquake's magnitude by measuring the seismic waves that cause the earthquake. The Mercalli scale represents the intensity of an earthquake based on its observed effects. The two scales have different applications and measurement techniques. The Mercalli scale is linear, and the Richter scale is logarithmic. i.e. on the Richter Scale a magnitude 5 earthquake is ten times as intense as a magnitude 4 earthquake.

Maps are available showing earthquake risk using the Modified Mercalli Scale intensity with an exceedance probability of 10% in 50 years, which is equivalent to one occurrence in 475 years (return period) on average, for medium sub-soil conditions.

The Modified Mercalli scale is observation-based (for example, akin to the Beaufort wind scale). Another important scale is the peak ground acceleration data for an area. The peak ground acceleration is a location-specific, calculated number, dependent on the local soil conditions and more appropriately used for design purposes. (See Table 37.1)

The intensity of a particular earthquake will depend on the distance of the observation point from the epicentre of the earthquake and also of the soil/rock conditions together with the moisture content of the soil.

Soil/surface conditions cause the effect of the earthquake to change considerably from one location to another. Certain soils/gravel can liquify under the shaking from the earthquake, causing structures to move considerably laterally and resulting in considerable damage. Subsurface conditions are now researched in geotechnical reports before building new facilities and, one hopes the results are reflected in the foundation structures built.

DOI: 10.1201/9781003360759-41

Table 37.1 Relationship between Earthquake Magnitude, Mercalli Intensity and Earthquake Effects

Magnitude	Description	Mercalli Intensity	Average Earthquake Effects	The average frequency of occurrence globally (estimated)
1.0–1.9	Micro	I	Microearthquakes, not felt, or felt rarely. Recorded by seismographs	Continual/several million per year
2.0–2.9	Minor	I to II	Felt slightly by some people. No damage to buildings	Over one million per year
3.0–3.9	Minor	III to IV	Often felt by people, but very rarely causes damage. Shaking of indoor objects can be noticeable.	Over 100,000 per year
4.0 – 4.9	Light	IV to VI	Noticeable shaking of indoor objects and rattling noises. Felt by most people in the affected area. Slightly felt outside. Generally, causes zero to minimal damage. Moderate to significant damage very unlikely. Some objects may fall off shelves or be knocked over.	10,000 to 15,000 per year
5.0–5.9	Moderate	VI to VII	Can cause damage of varying severity to poorly constructed buildings. Zero to slight damage to all other buildings. Felt by everyone.	1,000 to 1,500 per year
6.0 – 6.9	Strong	VIII to X	Damage to a moderate number of well-built structures in populated areas. Earthquake resistant structures survive with slight to moderate damage. Poorly designed structures receive moderate to severe damage. Felt in wider areas; up to hundreds of kilometres from the epicentre. Strong to violent shaking in epicentral area.	100 to 150 per year
7.0 – 7.9	Major	X or greater	Causes damage to most buildings, some to partially or completely collapse or receive severe damage. Well-designed structures are likely to receive damage. Felt across great distances with major damage mostly limited to 250 km from epicentre.	10 to 20 per year
8.0–8.9	Great	X or greater	Major damage to buildings, structures likely to be destroyed. Will cause moderate to heavy damage to sturdy or earthquake-resistant buildings. Damaging in large areas. Felt in extremely large regions.	One per year
9.0 and greater	Great	X or greater	At or near total destruction – severe damage or collapse to all buildings. Heavy damage and shaking extends to distant locations. Permanent changes in ground topography.	One per 10 to 50 years

(Wikipedia: Richter magnitude scale).

Modern cities in earthquake zones, such as Los Angeles, have building codes with appropriate earthquake protection. But what about cities where such building codes are less well developed or followed? I lived for two years in Almaty, Kazakhstan, a recognised earthquake zone.

Almaty city was destroyed by devastating earthquakes several times. In recorded history, catastrophic earthquakes in Almaty took place in 1770, 1807, 1865, 1887, 1889 and 1911. In each of these earthquakes, the city was heavily destroyed but was rebuilt every time. There are now building codes, but many existing structures were built before these codes. Further, one worries about corruption in this and other earthquake-prone nations. There is the potential for a contractor to charge for a particular construction standard but built to a lower standard.

The Almaty area is notable for being one of the few earthquake zones not associated with tectonic plate movement. However, there are many significant faults that have been formed with the uplifting of the adjacent Tian Shan mountains.

Several insurance companies publish data on earthquake return periods, and the estimated return period can be found for any location on the globe. These are available at some significant cost. OpenQuake (maps.openquake.org/ma/global-seismic-hazard-map) is freely available and expresses earthquake risk in terms of Peak Ground Acceleration (PGA)with a 10% probability of the stated PGA being exceeded in 50 years.

How does this help? If we want to avoid a catastrophe, then let's assume that we have a plant designed to good building codes. Therefore, we should consider Major earthquakes or greater – i.e. 7 Richter or greater, X MM or greater. For Tokyo, Japan, therefore, we have a return period of 400 years, and for Los Angeles, 1,800 years.

What is adequate? Returning to the issue of tolerable risk, it does depend on the consequences. If the consequences are a refinery fire, then maybe this is OK in Los Angeles, but not in Japan. If the consequences are potentially a significant radiation leak, this would be perceived as an intolerable risk in any region listed.

Earthquakes science is improving, and earthquakes can be forecast in some cases. Six Italian scientists were sentenced to six years in prison following the L'Aquila in 2009 earthquake, finding them guilty of manslaughter. The L'Aquila earthquake killed 309 people. The scientists had not predicted the earthquake despite some tremors which might have enabled them to do so. The scientists were subsequently acquitted on appeal.

An earthquake forecast would certainly enable hazardous installations to shut down. The difficulty for the scientists in this field is that they might be wrong and cause undue shutdowns and distress, or they might not issue the warning and be guilty of not alerting people to the danger.

It should also be recognised that a few earthquakes, but only a few, have been caused by manmade activities, such as dam construction.

Incident Examples – Earthquake

Earthquake at Refinery with Fire (1999)

The earthquake was approximately 7–8 on the Richter scale and had a maximum Modified Mercalli Index of IX.

A large refinery in the area experienced strong ground shaking with damage concentrated at the tank farm, the crude oil processing unit, and at the port and unloading area. The plant had more than 110 tanks of varying sizes containing hydrocarbons. Tanks tend to be more vulnerable to earthquakes since they are large and mostly atmospheric tanks. In contrast, process equipment is more robust and typically includes pressure vessels with much thicker walls. Ground shaking vertically displaced the floating roof of one tank, creating sparks that ignited escaping oil, and the fire spread to destroy a nearby wooden cooling tower completely. Floating roofed tanks are common in the oil industry. The advantage of a floating roof is that there is no vapour space, reducing the chance of explosion, which can happen inside a tank with a vapour space. Floating roof tanks also reduce the amount of hydrocarbon vapours displaced during the emptying and filling of the tank. Shell buckling at another tank base and resulting oil leakage may have contributed to the fire's spread.

A 115 m (380 foot) tall stack collapsed onto a boiler and crude oil processing unit, significantly damaging the processing unit, boiler, pipe alley and surrounding facilities. Fuel released from piping systems caused a second fire that destroyed the adjacent crude oil processing unit.

The refinery's cooling towers also sustained heavy damage, including one collapse. Ground settlement also damaged the port and loading area, a common problem at most port facilities.

Extensive damage to the pump stations and pipelines cut water supply into the site. Stored water at the plant was insufficient to deal with the two major fire outbreaks. The crude unit fire was quickly brought under control, and the available resources were then concentrated on the fire at the tank farm, including aerial water bombardment. The fire took three days to control and three more to extinguish. After initial clean-up and repair of minor damage, the plant resumed operations at about half capacity using the undamaged production facilities.

There are a couple of points worth emphasising here. The damage was focused initially only on the vulnerable tank farm. Much of the processing equipment suffered only minor damage due to the earthquake. It was one fire at one storage tank, which then spread to create significant damage and fire elsewhere in the refinery.

In 2011 the U.S. Nuclear Regulatory Commission reviewed the earthquake risk to the 104 nuclear power plants operating in the U.S.A. at that time. The review was considered necessary because of upgraded earthquake risk information and consideration of the age and construction of the reactors. The study was quickly published following the Fukushima

Daiichi catastrophe in Japan. Surprisingly the reactors built within earthquake zones were not found to present the most significant risk. The explanation given was that these reactors were made with the earthquake risk in mind. The Nuclear Regulatory Commission calculated that there is a 1 in 74,156 chance of an earthquake strong enough to damage the core of a reactor each year and seems to be OK with that result. I am not as convinced as the Nuclear Regulatory Commission. The U.K. HSE in 'The tolerability of risk from nuclear power stations' suggests that 1 in 100,000 is more appropriate, but they are both in the same 'ball park' number. However, the Nuclear Regulatory Commission analysed the risk at each of the reactors and found that the highest reactor at Indian Point not far from New York showed the highest risk at 1 in 10,000. Indian Point and a number of the more risky reactors from an earthquake perspective have subsequently been closed. The highest risk from the 2011 study was Limerick 1 & 2 in Pennsylvania with a risk of 1 in 18,868, which I find in the unacceptable zone. Risks are additive; this is the earthquake risk alone (MBC News: What are the odds? U.S. nuke plants ranked by quake risk).

Japan (2011)

The 2011 earthquake off the Pacific coast of Tōhoku was 9.0 Richter scale and was an undersea megathrust earthquake off the coast of Japan with the epicentre approximately 70 km (43 miles) east of the mainland. It occurred with an underwater depth of about 30 km (19 miles). It was the most powerful known earthquake ever to have hit Japan, and the fifth most powerful earthquake in the world since modern record-keeping began in 1900.

The effect of the earthquake and tsunami on the Fukushima Daiichi nuclear plant were covered in Chapter 10. Here we will discuss the broader earthquake damage.

The main earthquake was preceded by several large foreshocks, with hundreds of aftershocks reported. The first major foreshock was a 7.2 Mw event on 9 March, approximately 40 km (25 miles) from the epicentre of the 11 March earthquake, with another three on the same day in excess of 6 Richter. Following the main earthquake on 11 March, over eight hundred aftershocks of magnitude 4.5 Mw or greater have occurred since the initial quake.

The earthquake moved Honshu (the main island of Japan) 2.4 m (8 ft) east and shifted the Earth on its axis by estimates of between 10 cm (4 in) and 25 cm (10 in). Soil liquefaction was evident in areas of reclaimed land around Tokyo.

In addition to the damage to Fukushima Daiichi and other nuclear power plant the earthquake itself cause a fire at a refinery which was only extinguished after 10 days.

Ichihara 2011

A medium-sized oil refinery east of Tokyo had a fire in its LPG storage area due to an earthquake in 2011. The fire was extinguished after ten days, injuring six people, and destroying 17 LPG sphere tanks plus damage to asphalt tanks. This resulted in asphalt running out into Tokyo Bay as well as damage to some processing plants and several jetties and offices.

The initial tank fires and subsequent vessel explosions were caused by the mechanical failure of a sphere being hydro tested at the time of the earthquake, which subsequently collapsed, impacting adjacent spheres and causing leaks.

Whittier Narrows, U.S.A (1987)

An earthquake displaced a one-ton chlorine tank that was being filled, releasing a half ton toxic cloud. The power failure disabled the company's warning siren, and malfunctioning telephones made it impossible to warn the authorities. Whilst there were no injuries, in this case, it is included here to demonstrate the potential. This type of incident could happen to a 90 ton tank car! This case also indicates that emergency plans are quickly rendered unworkable due to associated communications and power failures.

Chapter 38

Tsunami and Other Natural Catastrophes

ASSESSING THE RISK: TSUNAMI

A CEO of a utility company with operating facilities on the Mediterranean coast asked me to estimate the potential damage to the operating facilities from a tsunami. I did so according to the methodology established by the insurance company for which I was working at the time. On receiving my result the CEO angrily exclaimed that my conclusions were 'impossible'. He added, 'In that case, the whole of Athens would be underwater'. Well, I guess there is a good reason that there is a Hellenic National Tsunami Warning Centre (NTWC)! The Hellenic NTWC was established officially by Greek Law in September 2010 and is centred at the National Observatory of Athens.

In fact, I had been careful not to exaggerate my estimate, which was based on a tsunami of 3 metres. The particular facility was somewhat protected against the full force of a tsunami by some islands. The facility was protected by cliffs of 3 m elevation on the seaward side.

There have been some tsunamis in the recent past in the eastern Mediterranean. An earthquake at Cyclades Island group in 1956 caused a 20 m tidal wave, and in 2020 there was an earthquake that caused a 3.4 m wave in the area. (World Data Info: Tsunamis in Greece).

Tsunamis follow earthquakes and are caused by movements of the seabed. The wave is small whilst at sea but grows as it approaches land. A gradually sloping seashore, or narrows between land, can substantially increase wave height due to a phenomenon called 'run up'. There are Tsunami Warning Systems in the Pacific, Atlantic, and Indian Oceans, together with the Mediterranean Sea. Tsunami warnings should precipitate the immediate shutdown of hazardous facilities. The delay between the earthquake and the tsunami depends on several geophysical aspects and maybe from just a few minutes to several hours.

Some forecasts of potential tsunamis make excellent sensationalistic news. But there is usually some credibility behind the story. The Storegga slide, which inundated the Shetland islands with waves over 20 metres high, is

DOI: 10.1201/9781003360759-42

believed to have occurred around 8,000 years ago. In case this seems remote from our scope, it is worth reminding readers that there is considerable oil extraction in the North Sea, together with oil terminal facilities at Sullom Voe.

There are also incidents of 'rogue waves', exceptionally large waves not caused by earthquakes of volcanoes, probably caused by constructive interference of the overall sea wave pattern in the area. Whilst these are rare, and quickly disappear, they can crash into the shore.

Incident Examples - Tsunami

Tohoku Earthquake and Tsunami (2011)

The tsunami which caused the Fukushima Daiichi nuclear incident provides a realistic case study concerning tsunami preparedness. It actually happened in the recent past!

The tsunami hit in just under one hour following the earthquake. The massive earthquake, the fourth largest in recorded world history, caused the automatic shutdown of the reactors. See Chapter 10 for detail on the reasons why this became a nuclear catastrophe. In this chapter, we will discuss only the tsunami-related aspects.

The Fukushima Daiichi site was first commissioned in 1971. At that time the plant was designed based on a maximum wave height of 3.12 m, being the largest wave height observed at Onahama port in Japan following an earthquake in Chili in 1960. Tsunami prediction technology raised the predicted maximum tsunami heights to 5.7 m in 2002 and 6.1 m in 2009. A further study in 2008 suggested that facilities should be designed to withstand a wave height of 10.3 metres. In practice, the largest tsunami wave in 2011 was 13–14 metres. What is most significant to me here is:

- The original wave height of 3.12 m is clearly woefully inadequate. This isn't convenient déjà vu. The propensity for earthquakes off the coast of Japan was well known. Many tsunamis were known, producing higher wave heights, albeit not in Japan.
- There may be some sympathy for the earthquake's magnitude and tsunami size. However, a significantly lesser event would still have caused substantial flooding and may also have flooded the emergency generators, perhaps with a similarly catastrophic effect.

The sea wall is understood to have been 5.7 m above sea level, and the ground level of the reactor facilities at 10 m.

The Fukushima Daiichi plant was, of course, not the only hazardous facility damaged by the tsunami, but the disaster which resulted far overshadows the other facilities. Power generation facilities at Fukushima Daini, Onagawa, and Tokai Daini were all inundated by the tsunami and damaged to some degree but there was no loss of control of the nuclear reactors. (Atomic Energy Society of Japan (2014). The Fukushima Daiichi Nuclear Accident).

It should not be forgotten that the earthquake and tsunami itself caused over 19,000 deaths.

Indian Ocean Tsunami (2004)

Like the 2004 Indian Ocean earthquake and tsunami, the damage by surging water, was particularly destructive. The epicentre of the earthquake was off the west coast of northern Sumatra, Indonesia. Communities along the coats of the Indian Ocean were devastated, and the tsunamis killed an estimated 227,898 people.

The tsunami was detected at great distances, including a town in South Africa 8,500 km (5,300 miles) away, Japan and Antarctica.

Although there was a delay of several hours between the earthquake and the tsunami, most of the victims were taken by surprise. At that time there was no tsunami warning system in the Indian Ocean, and tsunami awareness was low, unlike the Pacific Ocean there is little history of tsunamis in the area.

Other Natural Catastrophe Threats to Hazardous Facilities

A number of other types of natural catastrophes are worthy of consideration:

Wildfires and Bushfires

Wildfires seem to be increasing in recent years, with major outbreaks in the USA, Australia and Russia. There is the potential for a wildfire to overrun facilities containing hazardous materials. The smoke can impact operations and could precipitate a catastrophe. Consider aeroplane operations, for example.

Volcano

Facilities carrying out hazardous operations would not normally be constructed in areas where volcanoes are active. Volcanoes can be dormant for many years before bursting into activity. Furthermore, some activities benefit from the same geology, which leads to volcanoes. This can be mineral extraction or geothermal power generation.

The islands of the Hawaiian chain in the Pacific Ocean are formed from volcanoes, with active vulcanicity on the eastern and largest island known as the Big Island. Beneath the surface, there are hot rocks resulting from the cooling magma, which is being exploited for electricity generation. Water is pumped down into the fractured rock, and steam is emitted and used to drive turbines that generate electricity. In 2018 during an eruption of one of the local volcanic systems, the facility had to undergo an emergency shutdown as molten lava encroached on the plant.

Volcanic ash can also lead to hazards to aeroplanes since it can obstruct the jet engines and limit visibility.

Lightning

Lightning can cause damage and fires to facilities where flammable materials are processed. In general, I think we have become smarter in installing lightning protection. However, sometimes I have seen such protection broken due to poor connections or, in some cases, never properly installed.

In 1807 300 people died when lightning hit a gunpowder factory in Luxembourg.

Subsidence and Landslip

Subsidence can lead to the destruction of hazardous facilities on a large scale.

Notes from Trevor's Files: A sliding oil operation

I was asked to review the potential insurance needs of an oil extraction operation that was under construction. The whole project had to be postponed for several years when shortly after construction commenced, one of the construction areas on the top of a hill started to slide downwards! Fortunately, the fault was identified and managed before the oil facility started operation.

Temperature Extremes

Temperature extremes, both hot and cold, would not usually be considered potentially causing a catastrophe. Certainly, severe disruption can be caused when plants in areas not equipped for cold temperatures are exposed to freezing weather.

Particular points of vulnerability are remote facilities for extracting oil, gas and minerals.

I remember, on many occasions when visiting sites in remote desert parts of Kazakhstan and Uzbekistan and asking about the water supply. Water is often piped over long distances, and there is usually only one pipe!

Thawing of the permafrost

The permafrost is thawing due to global warming in northern parts around the Arctic Circle. This can have a severe effect on the stability of the ground.

In May 2020, an atmospheric storage tank located above the Arctic Circle containing diesel fuel leaked 21 kt of diesel fuel. The bunded area of the tank overflowed, and the spill overcame a designated collection pit, and into the neighbouring streams. It is understood that the melting of the permafrost may have contributed to destabilising and rupturing the tank. Corrosion of the tank bottom and inadequate inspection and maintenance may also have contributed to this incident. (Norilsk Nickel Sustainability Report 2020)

Thawing of the permafrost could lead to a mammoth legacy issue (see chapter 34) in polar and subpolar regions.

Chapter 39

Diseases

OCCUPATIONAL DISEASES

The whole concept of an 'occupation' brings man into contact with hazards, and with those hazards comes the risk of occupational disease. You are probably reading this book sitting in a chair. If your posture is not good, or the chair does not support you ergonomically there is a risk that over time you may develop signs of ill health such as a bad back. People using computers can develop problems with their arms and wrists through repetitive strain injuries. One of the earliest recognised occupational health diseases was byssinosis, a lung disease found in cotton workers. The diseases take a long period of exposure to develop, and so do not fall readily into the scope of catastrophes targeted in this book. Nevertheless, they are man-made and kill or injure a very large number of people. Therefore, I will briefly cover a couple of examples of occupational diseases, and other diseases, in this chapter.

Asbestosis – an occupational disease catastrophe

Asbestos is a naturally occurring material. It is a rock which can be dug out of the ground in locations where there are asbestos deposits. In the early 19th century, large scale mining of asbestos began. There was a demand by manufacturers and builders for its unique physical properties. Asbestos is fire resistant and was used to create much valued fire-resistant coatings. It was widely used for this purpose in ships, and undoubtedly, this saved many seamen's lives during the world wars. It is an excellent electrical insulator. Mixed with cement, it can be formed into inexpensive sheets for the roofing and cladding of buildings.

Being naturally occurring, no one thought much about any potential harmful effects from this dream material. The tiny fibres which give asbestos its desirable properties are also the reason for its downfall, which took many lives in its turn. There are six types of asbestos, all of which are composed of long and thin fibrous crystals. Each fibre is composed of many

DOI: 10.1201/9781003360759-43

microscopic fibrils. These fibrils can be released into the atmosphere by abrasion. Inhalation of asbestos fibres can lead to various lung diseases, including mesothelioma, asbestosis, and lung cancer. At least 100,000 people are thought to die each year from diseases related to asbestos exposure (Wikipedia: Asbestos).

Whilst the use of asbestos is banned in many countries for construction and fireproofing, there are still specialist materials that continue to incorporate asbestos. Developing countries continue to use the material. Russia is one of the top producers, with a reported output of 790,000 tonnes in 2020.

Asbestos remains in the materials of construction of many older buildings and equipment. See also Chapter 34 on Legacy Issues.

Notes from Trevor's Files: Asbestos

In the first job of my career, I was assigned to a factory that produced goods incorporating asbestos. The company involved was engaged in a serious fight for its life, being heavily dependent on asbestos-based products. Daily I passed the plant where asbestos fibres were broken up, mixed with other components, bagged and sold as spray fibre insulation. For the experience, I spent a day operating a machine making asbestos cement. The company was rushing to bring new products to the same market without using asbestos. I supervised a plastic mouldings shop making rainwater goods and worked on the replacement of asbestos cement sheets with materials impregnated with glass fibres.

The ethical dilemma here was that, as the dangers of asbestos became progressively discovered. it was still essential to continue manufacture whilst the asbestos material was phased out. Alternatives took time to research and develop.

Fortunately, I suffered no ill effects, although I have kept papers as evidence of my employment. Related lung diseases can appear years after exposure since the fibres can enter the lungs and remain there indefinitely whilst scarring the lung tissue.

Asbestosis is the best example of an occupational disease catastrophe. See Tweedale,G.: Magic Mineral to Killer Dust: Turner and Newall and the Asbestos Hazard.

Other occupational disease catastrophes

Fortunately, there have been no other occupational health diseases identified on the scale of asbestos related disease which linked to a specific hazardous material. There may well be some unidentified one. The extent of damage from some exposures will never be known.

A further example concerns the use of tetra ethyl lead. In 1921 engineers discovered that tetraethyl lead could make internal combustion engines run more smoothly. For almost 100 years, leaded gasoline was manufactured, and its combustion in the vehicle engine widely spread this material and its combustion products into the atmosphere. This exposed workers and the general public to lead compounds. Concern gradually developed about the toxic effects of tetra ethyl lead. Indeed, some argue that its toxic effects were known or suspected from the day it was first introduced. The lead in vehicle exhaust may well have accumulated within the soil adjacent to busy road. There is continuing concern about harm to children, including damage to the brain and nervous system, learning and behaviour problems and slowed development.

There are other occupational health issues such as exposure to cancer-producing chemicals, lung irritants etc... These are chronic diseases that develop over time and with ongoing exposure to the chemical.

Viruses, Bacteria and Pandemics

Whilst I was writing this book, the COVID-19 pandemic spread across the globe. I had no intention of including this disease when I started the manuscript. However, there is no question that the virus has caused a massive catastrophe. As of February 2022, it has led to over 5.5 million deaths. Many industries and businesses have suffered, and some will not recover. There is, of course, much speculation as to the source of the virus. I am not a medical man, so I have limited knowledge to contribute to the discussion. What does seem clear to me is that there is potential for dreadful viruses to appear at any moment. The question, therefore, arises as to why governments were so ill-prepared when COVID-19 started to spread.

Notes from Trevor's Files: MERS

In early 2013 I was working at an oil refinery on the West Coast of Saudi Arabia. In June 2012, a new virus causing Middle East Respiratory Syndrome (MERS) had been identified. The first identified patient was from Jeddah. My organisation became increasingly concerned to avoid the exposure of any of its personnel to MERS and made an edict that all travel to Saudi Arabia temporarily cease. Now here is an ethical dilemma. We were under contract to provide a service to the oil refinery. Our office was expected to be constantly manned by at least two consultants. Not only would the travel ban cause loss of revenue, it would also seriously damage our relationships with the oil refinery management. At what degree of risk is it really necessary to discontinue travel? Eventually, it was resolved that journeys could continue, but only if Jeddah was avoided. I recall arriving into Medina at 03:00 in the morning, with no means of travelling from the airport to the oil refinery. No one could understand why I. a non-Muslim, had arrived in Medina and not used Jeddah airport. Fortunately, the airport is outside of the Muslim only area. Eventually, with the production of adequate cash, I was able to secure a taxi for the 3-hour journey to the refinery.

While there remain occasional infections by MERS, it has thankfully never created a pandemic like COVID-19.

When Covid-19 first became a public concern, my occupational health and safety training told me to get a good mask. Fortunately, I had in my PPE drawer some FFP2 masks (KN95), and I immediately ordered more. Whether simple cloth face-covering ever really did any good is a matter for considerable debate. At best, they might reduce the airflow, and both reduce the risk to others from your breath or the risk to yourself from the breath of others. An FFP2 mask or higher provides a light seal around the masked area. The simple surgical mask provides no seal. Probably the reason FFP2 masks or better were not recommended is simply that there would not have been sufficient.

During the pandemic, I had to go for a hospital appointment concerning an issue unrelated to COVID-19. The specialist in charge of my case wore an FFP3 mask, whereas the nurses carrying out the tests wore a simple surgical mask. They were at much more risk than the specialist since they were much closer to me, the patient!

My advice, order FFP2 masks now. Hopefully, you will never need them!

Whether or not COVID-19 initially escaped from a laboratory, or not, raises an issue of concern. Many laboratories in the world carry out research on viruses. There are four standards for biosafety, with category 4 being the most tightly regulated. There are more than 3,000 Category 3 laboratories in the world. The laboratory in Wuhan, China, is a Category 4 facility, as is Porton Down in the U.K.. Note that staff at Wuhan were trained in biosafety techniques to category 4 level in France, Australia, Canada, and the United States. (Wikipedia: Wuhan Institute of Virology).

One of the differences between a Category 3 and a Category 4 facility is that in a Category 4 lab, all work on infection material must take place in a high-grade biosafety cabinet by personnel wearing a positive pressure suit. To exit the Category-4 laboratory, personnel must pass through a chemical shower for decontamination, then a room for removing the positive-pressure suit, followed by a personal shower (Wikipedia: Biosafety Level).

It will be interesting to see how governments prepare for future pandemics in the future, once the threat of COVID-19 has receded.

In Chapter 3, we discussed what makes risk tolerable or intolerable. It is known that people will accept a higher risk when they have some control but are demand a lower risk if that risk is beyond their control and includes elements of fear and dread. It has been interesting to see people's attitudes change during the course of the pandemic. At the start, so little was known. People respected lockdown because there was indeed considerable fear and dread. There was little people could do to control the risk other than stay at home. With the mutation to Omicron, suddenly, there was less fear. There was a greater feeling of personal choice as to whether to mingle with others or not. The risk level was suddenly far more tolerable, and compliance with health recommendations such as social distancing and the wearing of masks deteriorated

Chapter 40

Terrorism, Sabotage and War

ASSESSING THE RISK TO HAZARDOUS FACILITIES

Acts of Terrorism and Sabotage may also initiate a chain of catastrophic events.

Terrorism can be defined as the use of violence and intimidation, especially against civilians, in the pursuit of political aims. The terrorist act is intended to influence a government or governments. Terrorism cannot sometimes be differentiated between acts of war, and we will not attempt to distinguish this here. For example, the attack on the Saudi Oil Facilities at Abqaiq and Khurais in 2019 could be regarded as war or terrorism. These attacks were probably connected with the situation in the Yemen where Saudi forces are supporting one side in a difficult civil war (Congressional Research Service, October 2019' Attacks on Saudi Oil Facilities: Effects and Responses).

Sabotage is an act carried out to deliberately destroy, damage or obstruct something. The saboteur may be doing so for political or military advantage (in which case it is indistinguishable from terrorism), or it might be carried out by people who are disgruntled at the organisation which is the target of the sabotage. Reasons might be an employee who feels poorly treated, or it could be sabotaged by members of the local community unhappy with the company's environmental performance. It could be by an organised group hostile to the organisation on ecological grounds. Sabotage can also be by a person who has a psychological problem. Vandalism is also a form of sabotage, but in this case, there may be no specific intent; the act is considered 'mindless'.

Site security is one aspect helping to prevent terrorist and sabotage acts. As well as traditional fences and security cameras, elaborate intrusion protection is now increasingly used. Intrusion protection can detect a person coming close to the fence.

Notes from Trevor's files: To fence or not to fence ...

When surveying a gas field in the U.S.A., I noticed that many of the wellheads and valve stations were not fenced. I raised a recommendation that all such facilities be fenced to prevent against sabotage and vandalism. I felt my case was reinforced by the fact that this same organisation was engaged in fracking and that people opposed to fracking who may want to carry out some kind of attack.

Some of these valve stations were close to villages and small towns in locations where young people could easily access them and opening the potential for vandalism. My recommendation was further reinforced by a report by the CSB (U.S. Chemical Safety and Hazard Investigation Board, 2011, Investigative Study – Public Safety at Oil and Gas Storage Facilities), which had identified 44 fatalities in the period 1983–2010 resulting from public interference with oil facilities. In particular, in 2009, two teenagers were killed when a petroleum storage tank exploded in a rural oil field in Mississippi. In the same year, a group of youths were exploring a similar tank in Oklahoma when an explosion and fire fatally injured one of the group. Also, in the same year, a man and woman (in their 20's) were on top of an oil tank in Texas when the tank exploded, killing one and injuring the other. These incidents all involved rural unmanned oil and gas storage sites that lacked fencing and signs warning of the hazards. Despite this background the organisation was opposed to my recommendation, given the cost of implementation over so many remote installations.

Illustrative Case Study: Pilot Suicide

In 2015 a commercial airline flying with 144 passengers and six crew from Barcelona to Dusseldorf crashed in the French Alps. It was concluded that the crash was caused by the co-pilot who had been previously treated for suicidal tendencies. The co-pilot kept this information from the airline. The co-pilot locked the cockpit door when the captain was out of the cockpit. New regulations were put in place to require two authorised personnel in the cockpit at all times, but this has not been maintained by all airlines for practical reasons.

The actual cause of the Bhopal catastrophe (see Chapter 7 on the Chemical Industry) will never be known. One interpretation, favoured by the parent company, was that the catastrophe was caused by worker sabotage. It is known that a significant amount of water entered a tank of methyl isocyanate, causing a chemical reaction and the release of the highly toxic gas. This point of view argues that it was not physically possible for the water to enter the tank of methyl isocyanate without concerted human effort. The water, it has been argued, therefore must have entered the tank when a rogue individual employee hooked a water hose directly to an open valve on the side of the tank. Given the awful consequences, even if this interpretation is true, there should have been considerable additional barriers to prevent such a possibility. This also leads to the question, given that methyl isocyanate and materials with similar chemical properties are still produced and stored, what precautions now exist against employee sabotage?

I fear that there are no precautions against sabotage by employees. The airline industry does not have an answer for disgruntled pilots and co-pilots, neither does the chemical industry for rogue employees.

The extensive training required for people working in the offshore oil industry probably provides some protection, at least in the North Sea and Norway. It may have the incidental consequence of reducing the potential for psychologically weakened personnel to work offshore.

Some dreadful acts of terrorism have occurred on railway transport, notably on underground transport.

In 2004 some 193 people died during a coordinated attack on the metro system of Madrid involving 13 explosive devices planted to explode at the peak of the rush hour (Wikipedia: 2004 Madrid Train Bombings). In 2005 52 people died in a coordinated attack involving 3 separate suicide bombers on London underground July 2005 London Bombings). During my work in the City of London, I passed the memorial at Aldgate station many times. Apart from increased surveillance, railway transport does not have an answer either.

Road and rail tunnels (see Chapter 9 on Transportation) have not so far been targeted by terrorists, so far as I could find in my research. It may surprise you that dangerous goods are allowed to be taken through tunnels. Rail tunnels at least have some knowledge of the contents since there are controls during loading of the train, and documentation needs to be completed. For road tunnels, there is no control, and it seems to me this remains a point of vulnerability.

The London metro bombings were carried out using improvised explosive devices made with triacetone triperoxide. This explosive can be produced using the readily available materials of acetone and hydrogen peroxide. Triacetone triperoxide was used in a number of other terrorist bombings worldwide, including the 2016 bombing in Brussels, both at the airport and on the metro. Homemade manufacture is extremely dangerous, but the suicide bomber has already calculated that he would lose his life.

The objective of a terrorist is to create a shocking event either by an attack causing many fatalities or news headlining stories involving photographs of fires and destruction. It would be inappropriate for me to give too many details here, and I limit what is written to fairly obvious public domain facts.

Attacks on oil and gas facilities may create an impactful picture, but it is actually quite difficult to cause massive damage. Refinery equipment is well laid out and usually secure. An attack on an oil storage tank creates a lot of smoke, which is not particularly hazardous. All smoke is hazardous to the extent of being a lung irritant, but smoke from a burning oil refinery tank farm is not particularly dangerous.

Chemical plants are a different matter since there is the potential to release significant quantities of highly toxic material. These plants are normally relatively well secured, and a successful attack would probably require a degree of local knowledge. Facilities handling highly toxic materials such as ammonia and chlorine usually have remotely operated isolation valves that can be closed from the control room, limiting the extent of the release. Puncturing a storage tank would, of course, be less easy to mitigate, but these tanks are often double-walled, or are thick steel pressure vessels requiring a missile to penetrate.

Pipelines are vulnerable places. Sometimes it is not terrorism, but a desire to get free fuel, which leads people in third world countries to drill through oil pipes and drain off some material. This is, of course, highly hazardous because of the flammable gases also emitted. Some toxic gas pipelines cross country. I have seen chlorine pipelines alongside popular footpaths, even in France. A terrorist event on a pipeline could release the entire contents of the pipe between valve stations. A pipeline will have valve stations along their length, but these are not always capable of remote operation. Various attempts have been made at pipeline bombings, but pipelines can be relatively quickly repaired and are unlikely to result in shock worthy fatalities and fires.

Nuclear facilities have a clear potential for a high-profile terrorist attack. Probably they are well defended, and an attack would have to be well planned and perhaps include some inside knowledge.

Whilst the nuclear reactor itself is well enclosed against a terrorist bomb, the spent fuel pools which accompany many nuclear installations are more vulnerable. Disabling the cooling and water supply will lead to exposure for the fuel rods and radiation release within days. If the fuel pool could be punctured by terrorist bomb or sabotage act, a radiation release of substantial proportions could result (Perrow, 2007, p140). Aircraft impact is not routinely regarded as a credible event, but after 9/11, German regulators and industry focused on defining, and then addressing, the threat of a severe accident caused by terrorists aiming a passenger jet into a nuclear reactor.

The potential for bombing a road tanker containing toxic materials is, for me, a continued area of concern. A chemical tanker provides to the opportunity to be hijacked and moved to a location of dense population before an explosive device could release the contents. In some countries,

I expect the routing of tankers of toxic material to be tracked by GPS, but would an errant truck be spotted in time?

Biochemical attacks or attacks with biological agents such as anthrax and ricin have already occurred in Japan and the U.S.A. The potential is perhaps not fully recognised.

Regarding secured facilities, a terrorist has other possibilities other than penetrating the fence or gate. A point of vulnerability, if the plant is sited adjacent to the sea or lake, is to access by boat. This access point is hard to fence and often hard to protect with extensive surveillance cameras. Cliffs can conveniently act as cover for illegal entrants, and this part of the site will often be less often visited than elsewhere. Even where there are facilities close to the water, these will often be water reservoirs and wastewater treatment facilities that are infrequently manned. Jetties are more frequently manned but also provide another source of an intruder – a person on the ship!.

Intelligent drones provide a means of access from the air, and if the attacker is sophisticated, drones can be weaponised in order to penetrate the shells of tankage and vessels.

There is a grey area separating the definition of war from terrorism. Since this book was conceived before 2022 I had no intention of covering war, but with the sad military conflict ongoing in the Ukraine at the time of submitting this book for publication, I fear I have to include a mention.

In general, there is not much we could reasonably expect a prudent operator of a hazardous facility to do to defend against hazards resulting from war. As we have seen, fuels terminals are an important target at times of war, since, without fuel, tanks and other military vehicles cannot operate and logistical operations are interrupted. With the outbreak of war however, we would reasonably rely on military operators to put measures in place to attempt to protect the fuels terminal.

War potentially enables the enemy to use hazardous facilities as a weapon. Storage tanks of toxic chemicals might be accidentally or deliberately damaged by explosive devices including shells releasing the materials with the potential to cause widespread injury and even death to the community in the vicinity. A particular example are spent fuel pools associated with many nuclear reactors. Whereas the reactors themselves are typically within containment vessels which are likely to repel errant shells, spent fuel pools could be more readily damaged. A major consideration for an enemy carrying out such acts would be the potential harm to the enemy's own troops, and to members of the community who may be otherwise in favour of the enemies intentions. It seems that not all enemies would worry too much about that!

The issue of emergency response to a major catastrophe, including one caused by terrorism and sabotage, is raised in Chapter 23.

Section V

Preventing Catastrophes – Potential Solutions

Chapter 41

Is There a Need for Change?

APPETITE AND REVULSION

In the aftermath of a major catastrophe, there appears to be sufficient revulsion at the event that there is a considerable societal demand for change. However, this urgency fades as time progresses, especially in locations unaffected by the catastrophe. There is also a strong tendency to attribute the event to someone, or something, which distances or isolates the threat, thus making the chance of recurrence feel less likely. This, I contend, is a mistake.

Thus, Deep Water Horizon led to massive pollution in the Gulf of Mexico on an appalling scale. We surely cannot live with incidents like this. Yet the pollution along the Gulf has been cleaned up. Not so much is known about the damage on the seabed. People typically attribute the incident to a 'rogue' British company intent on cost-cutting at the expense of safety, which I consider to be unfair. My feelings are beside the point. More important is whether there is any evidence of change to the deep-water drilling and exploration and production, which would suggest that the risks of such an event happening again are significantly reduced. If it did happen again, it would result in substantially less damage? There are words about tougher safety standards and better response plans, but I cannot identify anything specific. Automated well control is being developed and could result in a significant reduction in the number of blowouts. I doubt whether automated well control would have prevented the Deep Water Horizon incident, given the extent of misunderstandings about the condition of the well at the time. As a regular reader of some oil and gas journals, together with internet searches on technical developments in deep water drilling, blow out preventers, etc. ... I do not find anything significant to convince me that the chances of such are catastrophe are significantly reduced.

Almost. But I have to give some credit to a development in the U.K. whereby a new well capping device has been developed by the Oil Spill Prevention and Response Advisory Group (OSPRAG). There is an excellent YouTube video that describes how it works. This is a device that, in the event of a blowout, can be lowered onto the top of the Blow Out Preventer

DOI: 10.1201/9781003360759-46

after the drill pipes are severed and will progressively stop the flow of oil. Key to its abilities is the provision of chemical injection points to prevent hydrate accumulation – a solid product when oil meets water. A device for this purpose deployed at the time of Deepwater Horizon failed because of hydrate formation, which plugged the device. So far, I have not found evidence of similar devices being available elsewhere in the world.

The Bhopal tragedy occurred over 35 years ago (see Chapter 7 – Chemical Industry). For contemporary young and mid-career engineers and managers, this is an incident from the dark ages well before their time. Its recurrence seems inconceivable. But why? The causes of the incident have never been finally resolved. What is known for sure is that a large amount of water entered a tank of methyl isocyanate, resulting in an exothermic reaction and the release of large amounts of highly toxic gas. Two conflicting points of view are typically voiced with regards to the cause. One is that the plant suffered from chronic under maintenance, another that sabotage was involved. The parent company Union Carbide distanced themselves from the incident (see Jackson Browning Report, Union Carbide Corporation 1993) because of the suspected sabotage and because the plant was part of an Indian company operated by Union Carbide India Limited (UCIL). Just over 50% of the company was owned directly by Union Carbide Corporation. Now joint ventures formed overseas are quite common, in some countries encouraged. Does this mean that the international company can absolve themselves of responsibility over joint ventures? I think it is especially questionable when the international company would be thought of as the technically more competent partner. A joint venture has board meetings in which the joint venture partners are represented and can voice their opinions.

Methyl isocyanate is a particularly hazardous material. It is still produced and used in industry. Even if it were not, many other toxic materials can readily release toxic gases, especially if they are mixed with other materials, including water.

Maintenance costs are a significant issue in most industries. What is to suggest that the issues at Bhopal were particularly poor compared to other locations where costs are an issue?

A further aspect concerns the proximity of poor-quality housing to the site of the Bhopal tragedy. Whilst one might be reassured that the planning permission process might prevent the construction of housing close to a toxic facility, planning permissions in many parts of the world are not thorough. Nor are the hazards of a facility recognised by the planning authority, who may not have asked or know of the dangers within. In the case of Seveso III regulated sites, then planning authorities would be aware. Still, Seveso only applies in Europe and then only to operations storing above a specific threshold of hazardous material. I can think several Seveso regulated sites even in Western Europe and the U.K. where residential housing is close to the site boundary.

Further afield only needs to look at the warehouse explosion in Beirut (see Chapter 13 Process Industry Infrastructure) in 2020, which resulted in over 200 deaths and left 300,000 people homeless, to see an example of inhabited structures, including poor quality housing, being built close to hazardous facilities. At the disaster in West, Texas, in 2013 (see also Chapter 13 on Process Industry Infrastructure), a school and a nursing home were sufficiently close to be severely damaged by the explosion.

For four years, I managed a small chemical factory in the U.K. that had been acquired from a large international organisation. I managed the factory, which was within the boundary of the larger chemical site. One day they had a very significant leak of hydrogen sulphide. This was an old chemical site that dated back to around 1901. Housing had built up around the industrialised area over the next two centuries. The nearest domestic housing was just 100 metres from the operating plants. Examination of aerial photos of several other chemical plants in the U.K. and other developed countries indicates similar problems elsewhere. It is not easy to solve such legacy issues. Land planning can only help forward-looking plans.

At the time of writing this book, I am doing some risk engineering work on the construction of a renewable diesel fuel plant in the U.S.A. The development reuses the site of an old oil refinery. The new plant is surrounded by residential and retail properties together with a mobile home park, which is just 100 feet from the nearest pieces of process equipment.

Some countries, notably the Netherlands and Australia, have regulations prohibiting new vulnerable installations such as kindergartens or hospitals from being within the risk zone. A risk zone represents the risk of 1 fatality in 1 million years and for less vulnerable facilities such as offices not to be within the individual fatality risk zone of 1 fatality every 100,000 years. But not that the requirement is in respect of new facilities. Existing facilities are not required to relocate.

The Chornobyl catastrophe (see Chapter 10 on Nuclear Power Generation) is attributed, unfortunately, in people's minds to be specifically a problem likely only in the former Soviet Union. Indeed, there have been improvements in reactor design, including those of the RBMK design still existing in Russia. But such designs do not eliminate the potential. In 2001 we had the Fukushima Daiichi nuclear catastrophe. Is there evidence here that there are improvements in the protection of high-hazard facilities against some of the aspects we identified in Chapter 10 where the Fukushima Daiichi incident was discussed?

- Underestimation of potential effects of natural catastrophes?
- Poor recognition of common-mode failures?
- Inadequate major emergency response planning?
- Improved operator training in responding to extraordinary 'out of design' events?

- Provision of Uninterruptable Power Supply/Batteries to maintain critical valve operations and maintaining control in a blackout situation?
- Etc. ...

As described in Chapter 9 on the Transport Industry, major software flaws are believed to have caused the Boing 737 Max plane crashes in 2018/2019. Software is extremely complex to follow, and difficult to test given all the different inputs and outputs that a piece of software handles. I am fortunate to have learnt some simple Python and Arduino IDE software for some hobby activities at home. After every few lines of programme, I test it to see if it runs and what errors occur. I use it to control the conditions in my greenhouse. Different unforeseen errors occur despite months of running. What happens when communication is lost between the computer and the microprocessor in the greenhouse for example. Lettuces and beans which have dying of dehydration are hardly catastrophic, but the loss to me is considerable! I find little assurance that latent software faults in commercial operations can be totally eliminated even in commercial software (see Chapter 29 on Software Safety and Cyber Security). Whilst personnel writing such programmes are considerably more competent and experienced, these programmes can be very complicated. A sudden combination of inputs might occur which have not previously been tested. This can have disastrous consequences in any sphere of activity, including aerospace, oil and gas, chemical manufacture, nuclear, etc. ...

NORMAL ACCIDENTS AND EXECUTIVE WRONGDOING

Perrow (1984) argues that complex, tightly coupled technologies will continue, inevitably, to have accidents. He calls such events 'normal accidents'. Different parts of tightly coupled systems have prompt and significant impacts on each other. The nuclear industry is therefore tightly coupled. Perrow questions whether such industries should continue to operate since, in his view and supported by his research, catastrophic incidents will continue in such complex, tightly coupled systems.

'Complexity' relates to the uncertainty that if a change occurs, a given outcome will result. Complex systems may have several potential outcomes. 'Tightly coupled' is a measure of the interconnectedness between system components and the speed with which one element of the system affects other components. In this context people are also regarded as components.

Perrow followed up with a publication in 2007 entitled 'The Next Catastrophe'. Here he emphasises the role played by Executive Malfeasance (Wrongdoing) in creating catastrophes and illustrates this by a number of catastrophes including the Economic Meltdown of 2008. For example, in the context of the 2008 financial meltdown, Perrow points out that the financial institutions received many warnings but considers that these institutions were profiting greatly at the time and were deaf to hearing anything that would damage their prosperity. He then suggested that one justification for this behaviour is that 'everyone was doing it'. A financial firm that pulled back from financial sector work subsequently seen as risky would at the time have seen its stock value go down and top traders leave. Catastrophes which are seen as having occurred primarily due to cost cutting such as the Prudhoe Bay pipeline leaks (see Chapter 13) also fall into this category.

So according to Perrow's analysis we have two primary causes of accidents:

Two major types of fundamental cause of failure are identified by Perrow:

1. 'Natural Accidents'. Since nothing is perfect, there are bound to be failures. In complex and tightly coupled systems, those failures almost inevitably sometimes lead to accidents. These are unwitting failures.
2. 'Executive Malfeasance', or executive wrongdoing. In its broadest context, members of management deviate from good practice to achieve a particular goal, typically a financial goal. Such failures can at first be considered wilful. In my opinion, Management gets swept along during their periods of tenure, perhaps recognising their sins but believing that at the next opportunity, they can 'right the ship' from within.

In Chapter 47 'In pursuit of High Reliability Organisations' we will discuss a potentially approach to avert both 'Normal Accidents' and 'Executive Wrongdoing' kinds of catastrophes.

However, at the time of writing this report, this author has to accept that there is only a weak social appetite for any major change to reduce further the risks of such incidents. We are pre-occupied with two other types of catastrophes:

- Covid-19 pandemic (briefly discussed in Chapter 39 on diseases).
- Climate change (briefly discussed in Chapter 46).

Social appetite may not be there just at this moment but after the next major catastrophe.

IF YOU CAN'T AFFORD IT – STOP IT

Cost-cutting is often portrayed as a cause of an incident. Indeed, it is vital for organisations to manage cost, whether they be in the public or private sector.

BP's incidents at Texas City, Prudhoe Bay, and Deepwater Horizon disasters are all considered in the context of the cost-cutting drive led by the CEO (Perrow, 2007). See Chapter 6 on Oil and Gas and Chapter 9 on Process Industry Infrastructure on these incidents. I have previously defended BP in Chapter 6. The BP described by Perrow is not the BP which I knew in the U.K. offshore in the North Sea, which was, in my opinion, highly focused on safety. Perhaps I am in denial.

Perrow offers the argument that maybe BP was a rogue outlier, but perhaps typical of oil and gas companies of the time but was unlucky to suffer three catastrophes in a row. Exxon is held out as an example of a safer oil company. Following my own interactions with Exxon, I highly regard their expertise. However, Perrow points out that Exxon is a highly profitable company and so can perhaps afford the level of safety attention that it demonstrates. If true, then this does not bode well for those companies which are less well financially established.

Cost-cutting, or at least 'putting production first', has also been linked to many other catastrophes. Strangely the same companies have safety policies that often use words similar to the all too commonly used phrase 'if we cannot do a job safely, then we will not do it'.

We need to find a way to ensure that organisations are true to their promise. If they cannot do so, then the operation needs to stop. In the following few chapters, I propose some means to do this.

I appreciate that some of these thoughts will be considered by many readers as impractical or even outrageous. My position is that failure to achieve such revolutionary changes leaves us open to the next series of catastrophes.

POLITICS

For some reason, most of the general public scoff at politics and politicians. Respect for our political process and those who make it their career is profoundly lacking. Imagine we were emerging from a period under the dictatorship of an authoritarian regime. Democratic politics were being installed. May I suggest that in such circumstances, our respect for politicians would undoubtedly be greater, at least for a while?

The intent of a political process is to form a government, and a government is indeed what we all need. In a democracy or even semi-democracies such as those typical of the former Soviet Union, the politicians need the support of the people. To get our support, they tell us what their policies are and tell us what is wrong with the policies of the opposition.

Most of my proposals, which follow in this section of the book, would involve our democratic institutions getting involved in the implementation and, therefore, require involvement, support, and leadership from our politicians. Now 'here is the rub'. Politicians will only be successful and interested if they think any particular topic can win your support.

There is a lot of criticism of politicians as being primarily concerned about their re-election. But isn't that generally true of us all in the context of our careers. Our priority for ourselves and our families is to maintain secure employment and if to advance our careers.

Politicians are often criticised for not having the capability for to implement systemic change. Yet there are some examples where political leadership has been successful. I select the 1974 Health and Safety at Work act as an example. This, as written elsewhere in this book, was a turning point between ever more prescriptive legislation, and placing the onus of good risk management on the organisation. Here risk assessment was largely born. The leader of the Committee on Health and Safety at Work, Lord Robens was appointed in May 1970 by Barbara Castle, Secretary of State for Employment and Productivity. This was an extraordinarily insightful appointment by Barbara Castle. Lord Robens, apart from his leadership qualities, was still smarting after handling the aftermath of the Aberfan disaster (see Chapter 8). Barbara Castle was an MP for 34 years, holding the record for the longest-serving female in the British Parliament.

A further barrier in the context of the political world is the subject of governmental organisation. Perrow (2007) points out that, in response to many incidents where the government departments are not perceived to have performed well, the government response is to restructure. I have seen similar actions by the U.K. government. Reorganising, I suggest, is often not the right answer. Political leaders can then point to something they have done to prevent future issues. They are rarely in-post to answer to the result.

Commercial interests are often embroiled in politics. Disasters are funding opportunities for government departments and opportunities for commercial ventures to push their products and services. As a management consultant specialising in safety, I cannot deny that we always looked to any major incident as an opportunity to promote our consultancy help at an appropriate cost. Shamefully it has been shown that many apparently well-meaning companies used the Covid-19 pandemic as an opportunity to overcharge, or supply equipment which was below specification.

You may well conclude that the proposed solutions in this section are just too crazy or too long term to merit consideration by a politician. Politicians are interested in short-term wins. They are interested in short-term wins because you, the general public, are interested by and large in short-term wins.

There are some indications that this is changing. Politicians are now well engaged in the subject of climate change. Whether this is real or lip service remains to be seen. But because the general public are interested, the politicians are too. The general public are interested because of the impact on their children and grandchildren. Impact on our loved ones draws our attention.

Therefore, in completing this section on potential solutions I realise that politics is a barrier. But you can change this, keep your grandchildren in mind.

(Apologies to our direct children, but for them, it is already too late.)

WHERE WE HAVE FOUND NO ANSWER

In some of the sections of this book we found answers to some of the past catastrophes. In many others we have found no credible answer to prevent future catastrophes.

In particular, we found no answer being realistically implemented for the following types of catastrophes:

- Oil well blowout in deep water or shallow water drilling
- Sabotage in chemical operations
- Tailings dams collapse in mining
- Nuclear reactor core meltdown
- Hurricane and flood damage to storage tanks of hazardous materials
- Fires in industrial estates
- Terrorism from destruction of Road and Rail transport carrying hazardous materials
- Fires and terrorism in tunnels
- Suicide by aeroplane pilots

Let's just see if we can explore some imaginative answers in what remains of this book.

Some readers will, understandably, consider the ideas proposed in Chapters 42–44 a little outlandish. Therefore at the end of Chapter 44 I have included the likely conclusions resulting from the Grenfell Tower catastrophe (see Chapter 15) for comparison.

Chapter 42

Improvements to Risk Management Methodology

Reasons why many risk management methods fail to prevent catastrophes were discussed in Section 3. In this chapter, some of these reasons will be revisited in order to indicate what could be done to improve each piece of methodology. Why only some? In Section 3, a number of issues were identified with the various aspects of safety management, and the path forward is simply to identify, emphasise, and attempt to correct those faults. In this section, I will only focus on areas where I think I have potentially new insights or areas of particular emphasis.

REDUCING OPERATOR FAILURES AND ISSUES WITH OPERATING PROCEDURES (REFER CHAPTER 19)

Consider this in three parts:

- Is the procedure correct – the most efficient and robust in reducing risk?
- Does the operator/pilot/train driver, etc., fully understand the procedure?
- Does the operator/pilot/train driver, etc., do their job following the procedure?

Many organisations have periodic reviews on operating procedures so that each procedure has to be reviewed and brought up to date, say once per year. Whilst this is undoubtedly good practice, it can lead to 'light touch' review to stay in line with the schedule. What is needed is some depth of enquiry, probably involving both an operator and a supervisor/engineer.

Can the operator think of a better procedure in terms of efficiency and safety? Would this procedure involve any additional risks or reduce them? Is the operator tempted to deviate from the procedure, even perhaps by a small change in sequence, because this is more logical to their workflow or maybe because they can complete the task more quickly? Does this insight lead to a better procedure or a more risky one?

DOI: 10.1201/9781003360759-47

One can see that such searching requires an open culture for discussion and not a punitive one. Each procedural review should be seen as an opportunity to improve.

MAKING TRAINING MORE 'WATERTIGHT' (REFER CHAPTER 19)

Does the training match with the way operators actually carry out the task? As previously pointed out, if the operator is new or new to this particular equipment finds out after training that people do something different in practice, then the value of the training is effectively destroyed. This match between what is taught and what is carried out in practice is the single most valuable improvement I can foresee in training.

Ensure the training is complete, covering all aspects, including abnormal operations such as loss of electricity. For the aeronautical industry this would include abnormal situations such as flying in bad weather.

Simulators using computers and real-life control screens are ideal, so long as the programming of the simulator is such that the simulator behaves like the real-life equipment. Beware 'generic' simulators not modelled for the particular plant. They are not likely to behave in the same way in real time. Simulators can be very expensive, although I have the impression that they are becoming better and more efficient.

NEUTRALISING ERRORS DUE TO HUMAN FACTORS (REFER CHAPTER 20)

It would be highly beneficial if there were general recognition that human beings make mistakes. In the chapter on Human Factors, we discussed how often humans make mistakes depending on the stress levels of the situation. This can be as high as one error in every ten operations, or people working under stress. Therefore, in our work, to prevent catastrophes, we need to allow for this kind of level of human error. We cannot rely on some magic that will eliminate them. At the time of finalising this book a dreadful incident had occurred in Aqaba, Jordan. A pressurised container of 25 tonnes chlorine was being loaded onto a vessel when there was a crane malfunction and the container dropped releasing the contents. At least 12 people died. It is not known at this stage whether the cables being used for hoisting failed, or there was a failure in the operation of the crane or in the mechanism of lifting. However, dropped loads during port operations occur all too frequently. This looks like being some kind of human failure, and if such an operation can produce such dreadful consequences one questions whether it should be carried out at all. At least one solution to reduce the risk is to ship smaller cylinders of chlorine, considerably reducing the consequences of dropped loads.

The Normalisation of Deviance is a theme that has recurred repeatedly as a contributor to the incidents described in Section 2 Those inside the organisation where the deviations occur are mainly oblivious to the issue. It requires an external pair of eyes from someone independent who is motivated to see and point out the deviation and explain why this is important. Reduction in the normalisation of deviance is, in my opinion, the single most significant potential contributor to the prevention of catastrophes.

SELECTING AND DEVELOPING ERROR-FREE CONTRACTORS (REFER CHAPTER 20)

Looking back on the six-step process outlined, contractor evaluation is the link that is likely to best improve the quality of the contractor. Someone in the organisation employing the contractor – the contractor sponsor – should be responsible for gathering all data on contractor performance, including that obtained by their audits and the audits of others. The Contractor Sponsor is not responsible typically for all contractors and is rarely someone from Purchasing. I first came across the Contractor Sponsor concept when working as a consultant for a major oil company operating in Kazakhstan. I found the process extremely valuable, especially when connected with contractor re-selection.

RAISING THE BAR ON EQUIPMENT SELECTION AND DESIGN (REFER CHAPTER 21)

At this stage, there is the strongest opportunity to consider Inherently Safer Design. If there is no alternative to a design involving hazardous materials or operations, is there an opportunity to minimise the inventory of the hazardous material? This consideration requires creativity and in-depth thinking rather than replicating the design used elsewhere.

ELIMINATING ERRORS IN MATERIALS OF CONSTRUCTION (REFER CHAPTER 21)

Consider two parts:

1. Specifying the correct materials of construction.
2. Ensuring that the specified materials have been provided.

Keep an eye on the 'Procurement department' to ensure that the specification does not change to something less expensive during the procurement process, to something which might not be as resilient. Thoroughly understand the corrosion mechanisms which might occur in the process, not just in normal operation but also in abnormal conditions, including start-up.

For example, plants making concentrated sulphuric acid work just fine once the sulphuric acid has reached a particular concentration. Below that concentration, the acid eats through the carbon steel.

TAKING CARE OF AUTOMATION AND INTERLOCKS (REFER CHAPTERS 21, 27, 29)

It is essential that all industries step up to the principles of IEC61508, together with the use of Layers of Protection Analysis (LOPA) and that of Safety Instrumented Systems (SIS). We are not there yet. Even when an organisation says they apply these principles, a little questioning soon reveals that they do not.

Carrying out proof testing at the required time is, of course, essential in addition to the correct classification and correct equipment for the Safety Instrumented Level determined by the LOPA.

There are a lot of Legacy issues here. Plants designed in the past will not, in general, have gone through this process. Retrospective application is slow to occur, and this remains a critical issue in catastrophe prevention.

IMPROVING THE EFFECTIVENESS OF MAINTENANCE ACTIVITIES (REFER CHAPTER 22)

Reliability is closely related to catastrophe prevention. Many excellent reliability programmes are being implemented to improve plant uptime and improve process safety. These are often based on a 'bad actor' approach. A 'bad actor' is a piece of equipment or a particular operation that is regularly causing problems. Typically, the top three bad actors are selected for focused attention and creative thinking among the reliability team members, including operations, maintenance, and process engineering personnel. I have seen this approach working successfully in many organisations.

Interestingly, I have a colleague in Russia who disagrees with the bad actor approach and does not think this reliability approach will ever be accepted in his country. 'You can't put focus only on selected pieces of equipment; you must manage all the pieces of equipment!'

The path forward here is, I think, to speed up the bad actor and reliability team approach, perhaps by having more of them and ensuring that process safety bad actors are also included.

ELIMINATION ERRORS IN THE CONTROL OF WORK (REFER CHAPTER 22)

Errors in the control of work and work permitting remain a significant contributor to many of the disasters described in Section 2. The importance of the control of work is, in my opinion, well understood. However, the number of work permits in any typical day, for example, in an oil refinery, is enormous, so some do not get the attention they deserve.

I don't have any particular insight to offer here over the many excellent practices I see in industry, other than the energetic application of work permit audit. In a work permit audit, a selection of permits is examined and marked. In any bunch of permits, I will guarantee that several errors can be found, not necessarily critical errors but opportunities for considerable improvement.

IMPROVE THE ROBUSTNESS OF INSPECTION AND BUSINESS SIGN-OFF ON CORROSION RISKS (REFER CHAPTER 22)

The evaluation of the results of inspection is a critical step. As I pointed out in Chapter 22 on this subject, the interpretation is often left to the Inspector.

My proposal here is that all inspection results be reviewed by the Inspector and a Process engineer together. The risks must then be put in a report to business management for them to sign off on collectively.

CREDIBLE AND PROVEN EMERGENCY RESPONSE PLANS (REFER CHAPTER 23)

Emergency response plans need to cover a credible number of scenarios and expand each scenario in sufficient depth.

The issue as to what is a credible scenario is a contentious one. In Deep Water Horizon (see Chapter 6), the emergency response to a deep water blowout resulting in a sizeable environmental spill was clearly not considered sufficiently credible for either the operators or the environmental authorities along the Gulf Coast to deal with in any depth.

We saw a similar issue with the emergency response plans relating to the Fukushima Daiichi incident. A meltdown was never envisaged as credible, so the emergency plan for that size of scenario does not seem to have been seriously reviewed and practised.

There should be a direct relationship between the outcomes of a risk assessment, say a HAZOP and the emergency response plans. Testing this relationship could be a productive path forward, and it should be seamless. All outcomes in the Risk Matrix, which are in the Critical, very High, and

Elevated zones, should have an Emergency Response plan; and that Emergency Response Plan should be practised.

MAKING HAZOP OR OTHER RISK ASSESSMENT TECHNIQUES EFFICIENT (REFER CHAPTER 26)

HAZOP stands for Hazard and Operability Study. Scrap the Operability part; this overloads the HAZOP team and is a distraction. For many hazardous facilities priority is to manage the 'Red Risks'.

Human factors should always be considered as part of the HAZOP. The failure of an operator to see an alarm or respond to it, etc. should be part of the same work, and guidewords, if necessary, should be added to include the human HAZOP. To divide them causes additional work and may hide the vulnerability.

IDENTIFYING COMMON MODE FAILURES (REFER CHAPTER 27)

In Chapter 27 on 'Assessing Risk Outcomes and Risk Reduction', we discussed common mode failures. In several of the incident examples in Section 2, common-mode failures were a contributory cause. This was most notable in Fukushima Daiichi in Chapter 10 on Nuclear Power, where flooding was the common-mode cause of failure of both emergency generators on Unit 1, DC power failure was the common-mode cause of failure of most instruments, etc.

Common-mode failures are not easy for a HAZOP team to identify, so I put the section mentioning common mode into the part addressing Layers of Protection Analysis (LOPA), where the necessary skills should be available.

I suggest that a separate section to the Risk Assessment of facilities above a particular hazard rating be dedicated to identifying common-modes failures, reviewing the LOPAs recognising this, and deciding if further action needs to be taken to separate the common-modes.

Notes from Trevor's Files: Hiding Common-mode Failures

Whilst working as a consultant at a large oil refinery and petrochemical plant, we were tasked with reducing the number of serious incidents. This was a joint venture between organisations of two different national cultures, which perhaps contributed to some political overtones in achieving open discussion. One day the technical director passed by our offices and commented in hushed tones that common-mode failures were not well recognised in the risk

---

assessment procedures used to design the plant. Whilst we appreciated his candour, I was powerless to do anything with the information. It would have taken a mammoth effort to conduct an independent risk assessment of a portion of the plant to uncover common-mode failures in a disciplined way. Even if we had succeeded, I doubt if our motives would have been appreciated.

IMPROVEMENTS IN THE MANAGEMENT OF RISK ASSESSMENT CONCLUSIONS (REFER CHAPTER 30)

In organisations where the number of recommendations being produced by HAZOP or other techniques is high, it should be acceptable to prioritise these based on LOPA and tackle only the highest risk first. Government and insurance companies have a role to play in enabling this.

An excellent practice is to allocate a budget each year for minor projects resulting from HAZOP recommendations. Any recommendations requiring investment above a given threshold will become major projects with a separate funding request.

Innovative suggestions relating to Governance of Safety, Laws, and Culture are contained in the subsequent chapters of this Section 5.

IMPROVEMENTS IN INCIDENT INVESTIGATION (REFER CHAPTER 24)

In the chapter on Incident Investigation, it was suggested that a valuable approach to incident investigation is to find out how people's assessments and actions made sense at the time, given the circumstances that surrounded the event (Kernick's (2022) Catastrophe and Systemic Change – Learnings for the Grenfell Tower Fire and Other Disasters). I could not find an incident investigation that followed this method, although, in some incident investigation outcomes, it was implicit that the investigation team was taking this approach.

In order to test this method, I 'self-investigated' a major incident in which I was involved. This was the incident I described in Chapter 7 on the Chemical Industry 'Notes from Trevor's Files: One of My Mistakes – Latex Spill'.

Notes from Trevor's Files: Latex Spill – An Improved Investigation

In the latex manufacturing process, which I described in Chapter 7, I had, at different times, been the process engineer, process hazards leader, operations supervisor, and plant manager. Over the years, therefore, I had plenty of opportunities to foresee the event which occurred – a significant environmental release – and at least attempt to initiate some engineering and procedural changes to reduce the risk. Why did the actions of myself and many others associated with the process make sense at the time?

Process Engineering

As a process engineer, I was tasked with the opportunity to find ways to improve 'yield', i.e. to reduce losses and therefore improve manufacturing economics. In retrospect, I was looking at only normal operations. I was blind to thinking about losses during abnormal operation. Had I made more 'tea room' enquiries at the time, I may well have found out more about abnormal situations which might have caused the type of discharge which resulted in the incident.

It was also clear that my work was not expected to result in major engineering proposal. A second stage effluent water stripper would have been a significant and costly project.

Plant Supervision

Past practice had been that the Shift Supervision reported into a dayshift Superintendent, who had been previously a Shift Supervisor. The Superintendent then reported to the Plant Manager and the Process Engineers. The former were all people who were employed from the local community with no particular qualifications, whereas the latter were all graduates, mostly recruited from outside of the region. To some extent the Superintendent protected his Shift Supervision from decisions where the Plant Management and Process Engineers might be critical. Shift Supervision were proud of their ability to get things done. I did spend a few days on night shift, and I should probably have spent more, and I was aware that some activities were not in line with procedures, although not the particular activity which caused the discharge.

Process Hazards Analysis

Following a major incident in the U.S.A., unrelated to this specific event, the whole process was subjected to a process hazards review. I was fortunate to be assigned to play a major role in this, and I remember presenting for a whole day in New Orleans on my part of the process. Why did this team not see the potential for the event? All the attention of the reviews were on the highly reactive monomer component. In the portion of the process which caused the discharge, only minor amounts of the monomer were normally present. It was very much the least interesting part of the process. Again abnormal operations did not get enough consideration.

Plant Management

We did not accept the effluent results reported on the day of the incident. They were just unbelievable. Occasionally rogue results had occurred in the past due to contamination of the sampler with one or more droplets of monomer. However, an engineer was assigned to investigate, and he broke the news to me some three days later that a significant spill had indeed occurred.

Hopefully, the reader can see that this retrospective analysis shows at least three general issues at the time:

- Lack of consideration of abnormal operation, both in the process hazards analysis and in the process engineering. How could things go wrong, how could the shift operators be tempted to do something inappropriate? No doubt the shift operators did not conceive of the consequences, they thought they were doing the right thing.
- Poor trust and communication between the workforce and local supervision and better-educated engineers and plant supervision.
- Normalisation of deviance over laboratory results.

If the incident had been analysed in this way at the time, a broader set of corrective actions may have been possible.

Chapter 43

Changes to Leadership Mindset

DEFINE WHO IS ULTIMATELY RESPONSIBLE AT THE CORPORATE LEVEL

Who's in charge? Who is responsible if things go wrong?

The topic of who is ultimately in charge needs to be refocused. Over recent years, in my experience, this has become more diffuse. At the corporate level, the CEO is responsible. Of course, he or she must delegate, and the CEO will demand a certain performance from the Chief Operating Officer and the Director of Health, Safety and Environment in this regard. The CEO will also expect the Chief Financial Officer as the controller of budget and expenditure to feel their responsibility also. The fact that a CEO often does not have an operations/manufacturing background should not be an excuse for not protecting the company from catastrophes. One of the CEOs first tasks should be to become thoroughly familiar with risks that might lead to catastrophes and consider his or her role in preventing them.

This does not mean that the CEO needs to become suddenly familiar with the techniques we discussed in Chapters 25–27 on Safety Management Techniques such as HAZOP/LOPA and SIL. Clearly not. But the CEO does need to know the something about strengths and weaknesses of such techniques so that the CEO can take full responsibility for ultimate decision-making.

Matrix management has become a common approach in many organisations, in which different departments work together at lower levels in the organisation. An individual may report to more than one supervisor. There are cross-functional, cross-business groups. Effective matrix management has a significant advantage of enabling decision-making at lower levels, perhaps amongst people with better knowledge of the issues at that level, without passing issues back up the organisation to the place where they 'join together'. By the time an issue reaches this level, the importance and the understanding can, and often does, become lost. However, what must not be lost is the ultimate responsibility for the decisions taken. My view is that ultimate responsibility must be borne by one or other of the managers in the matrix who depend on inputs and outputs of the associated matrix manager. It cannot in some way be diffused amongst the matrix.

I believe these points of ultimate responsibility need to be spelt out in corporate policy and made public. It should also be a requirement of government – an issue to which we will return in Chapter 44.

This is not a unique concept. In the fields of finance and procurement, in many countries, there are persons with particular responsibility who, by law, must be named.

DEFINE WHO IS ULTIMATELY RESPONSIBLE AT THE OPERATING/EXECUTION LEVEL

Moving on from the Corporate Level, who is responsible at the 'front end' closest to where the catastrophe might occur?

On an aeroplane, it is the pilot. Despite all the complex and multiple organisations whose members are safely on the ground below, it is the pilot with the controls who makes the ultimate decisions. Ultimately the pilot goes down with his plane.

On a ship, it is the captain – or is it? The captain does not necessarily go down with his ship (see Chapter 9 – Costa Concordia). Marine incidents are rarely fatal to the captain. The captains' decisions are influenced by a range of instructions that play out in the event of a normally slow-moving incident. These instructions come from the ship's owners, other ships, port and maritime authorities.

On a chemical plant, oil and gas installation, nuclear plant, etc., it is surely the Site Manager who is in charge, or is it anymore? In the early days of the Du Pont Company, the Gunpower Mill Manager, who at least at the start was a member of the Du Pont family, lived at the gunpowder mill. In the event of an incident, he and his family were likely to perish along with the mill workers. This was long perceived as one of the reasons why Du Pont had such a strong emphasis on safety, and this author supports that view.

The Site Manager of today has many pressures to improve. The diversity of these pressures goes beyond the pressures some 20 years ago, in my opinion. Sustainability and diversity are just two. The site manager has to balance these. No external specialist can dictate the correct balance, but they can have valuable input.

Notes from Trevor's Files: Unnecessary Expenditure?

Being the recently appointed site manager for a plant under enormous cost and production pressure, I had a distracting problem. The scheduling office was located above part of the production facilities where strong solvents were used. Inevitably some fumes permeated up and into the office. Measurements were made, and from a regulatory perspective the levels of solvents were acceptable. To some of the office workers the associated smells gave them problems. An industrial hygienist had suggested that some plants arranged around the office might provide some improvement. I duly arranged for the plants, along with a small service contract to maintain the vegetation. When the business manager found out about this unnecessary expense, he was aggressive and critical that I undertake such expenditure when my focus should be elsewhere.

Even more seriously, there had been a fire, and very sadly a fatality, some weeks previously. Emergency response vehicles found it difficult to access the site due to employees parked vehicles. I arranged for additional parking at an adjacent factory, at a cost, and applied for capital expenditure to instal a simple turnstile to allow employees to directly access the overflow car parking. My application was denied, partially because it was known to senior management, including myself, but not other employees, that the plant would close within a year. I spent the money anyway. The overflow park was in use. On this occasion, I did the right thing!

One of the responses to incidents in the UK sector of the North Sea was the adoption of an Offshore Installation Manager (OIM). Given that there can be several lines of authority and contract organisations on an offshore platform or rig, it was clear from incidents such as Sea Gem (1965) and Piper Alpha (1975) that there needs to be only one person carrying the ultimate decision-making responsibility. The OIM principle has been widely accepted internationally.

The importance of knowing who is in charge goes beyond emergency response. The person in charge needs to have control of the means to prevent the incident or stop the operation altogether if his/her assessment is that the risk is too high.

Today's site manager rarely has such power or even respect. This is where things need to change. Typically, the site manager has a matrix of people reporting to him or her with specific responsibilities for production, process development, capital projects, IT, maintenance, inspection, health, safety, and environment. Each of these functions has reporting lines directly to corporate or localised functional heads, which are located off-site. Reporting to the site manager is sometimes a 'dotted line' relationship. The site manager often has little say in the performance reviews of these functional heads or their recruitment or succession planning. Business pressures determine the budgets by which the site operates. Typically, the site manager's tenure is temporary, either moving on within a few years or being found wanting and moving on.

What I have just described may at first seem contrary to the Technical Authority concept supported by Hopkins (2009). Following the BP Texas City disaster, BP implemented reforms which strengthened the role of engineering authority. Under the new structure the refinery manager had a dual reporting line both into the Vice President for refining (as before) and also into the Group Engineering Director. Engineering Authorities at the refinery reported to the refinery manager, as before, but with a dotted line through the engineering management levels to the Group Engineering Manager. Under this revised organisational design any engineering authority would have the power to raise any issues of concern about what the refinery or refinery plant manager is doing with the refinery business engineering authority (Hopkins, 2009, p. 102). It is further suggested that if this structure had been put in place in the Oil Exploration and Production business, the Deepwater Horizon disaster might have been avoided. The argument being that the shore-based engineers would have a direct reporting line to a director solely responsible for maintaining engineering standards. Clearly, there will always need to be a balance, and the argument for and against a matrix organisation could be swung in either direction.

To me, it comes down to the acceptance that there is always some residual risk. Someone has to be answerable for that risk. The Technical Authority route can easily lead to a situation where the engineering organisation protects itself by not agreeing to any risk. This leaves the site management in handcuffs and unable to get anything done or to continue in ignorance of what the engineering organisation protectively required. Indeed, in the author's opinion and experience, the site manager can become the 'fall guy' when things go wrong even though he has little control over the functions which can eliminate such failures.

By now I hope that readers of this book will agree that zero risk means zero activity. Risk is an uncomfortable business.

There is substantial value in the matrix structure, and Perrow (1984) points to this in praising the aerospace and airline industries over marine. The key is that the nodes in the matrix need to have real responsibility too.

Tests for real responsibility are, in this author's view:

i. Is the person with real responsibility a specific named person/title instead of a group or organisation?
ii. In the event of an incident, who gets woken up at night?
iii. Who will the press want to interview?

iv. Who will the authorities want to interview?
v. Who runs the risk of criminal liability with the potential for a custodial sentence? Crudely – who's going to jail?

In the airline industry, the responsibilities of the ground controllers, airport management, etc. ... are generally clear and defined.

There are other kinds of continuous improvement matrices that this author supports. Multidisciplinary groups working on reliability and operability have been shown to be very effective.

Typically, such groups include representatives from operations, maintenance, and technical. They work in accordance with well-proven behavioural techniques such as Plan-Do-Check-Act, Quality Circles, etc. ... They may be inter-plant or even international. But they should work on behalf of the site manager, who is the person, or should be the person who decides whether to invest resources in the activity or not.

So, this author offers some potential solutions:

1. Giving the Site Leader the true responsibility and accountability needed.
2. Getting rid of the dotted lines to the Site Leader.
3. Extending Site Leadership tenure to at least five years.
4. Respect for the Site Leader's career in Site Leadership.

In the context of other industries, the operator/execution level person responsible may be less easy to define.

Actually, for an airline, it is probably not the pilot but the manager of a group of pilots operating a particular segment or part of the business. Separately the airport manager would be responsible for the airport itself. For the ship, it could indeed be the ship captain, or maybe a shore-based manager.

For hospitals, it would be the named hospital manager, not just 'administrator'.

MINIMUM ESSENTIAL CERTIFICATION AND TRAINING FOR THE CARRYING ULTIMATE RESPONSIBILITY

Those named as ultimately responsible at both the Corporate and Operator/Execution Level need to have some minimum essential training. It is feasible that this certification could be based on prior learning or experience. What is critical is to ensure that those responsible cannot use ignorance as an excuse. We need to exclude the possibility, for example, that people responsible for a wood mill cannot argue that they do not know of the hazards of dust explosions (see Chapter 12 on Agriculture, Food, and Beverages). I am not asking that they know about how a dust explosion flame propagates or to be able to calculate the speed – just to know that it can. Sample incidents can be used to make the learning process more interesting, with the danger that people will always see their own operating process as different. The certification would include recognition of:

- Legal responsibilities
- Basic hazards
- Risk assessment basics
- Emergency response

It would be a condensed and cut-down NEBOSH (National General Certificate in Occupational Health and Safety) in the UK. Other countries have similar qualifications. New training content need not be devised!

Notes from Trevor's Files: Priorities for Emergency Responders

In Chapter 20 on Human Factors, I described 'A Dreadful Contractor Fatality'. Along with one of the top managers from the company, we went to see the family at their lawyers' offices. During the emotional discussion, a member of the family criticised the actions of the people present at the time of the incident as 'running away' from their young man, who was encompassed in a fire. The top manager present did not understand this and subsequently challenged me on it.

It is fundamental of emergency response practice that the responders first collect the equipment necessary to deal with the emergency and to protect themselves against the risk. Emergency responders are not expected to carry out heroic acts of potential self-sacrifice. I was astounded by this lack of understanding by a very senior person.

Chapter 44

Changes to Government and Legal Structures

REGULATIONS WHICH CHALLENGE ORGANISATIONS TO IMPROVE

Laws and Regulations need more challenge to ensure organisations to continuously improve. This may be difficult to enact, but it is an important and imaginative future step. Whilst the U.K. Health and Safety at Work Act 1974 imaginatively introduced risk assessments and abandoned prescriptive safety legislation this has its limitations in motivating organisations to improve. Current regulations can only issue enforcement notice against 'the sins of the past'. Any infringement must be unambiguous if prosecutions or improvement notices are to follow. For the authorities to do otherwise leads to potential accusations of bias. Regulatory audits are carried out against strict protocols which have taken years to develop and are thus 'cast in stone'. An inspector who deviates is likely to be considered biased and unfair.

The Grenfell tower catastrophe was, it seems, a hidden risk just waiting to be found, unfortunately in the most horrible way. How do we encourage regulatory regimes to look for new hazards without the individuals involved being considered biased or vindictive towards the organisation being inspected? I make the following suggestions:

1. Overcome unhealthy familiarity with a particular site and risk. Familiarity is a constraint on inspectors. If the inspector said nothing about situation 'X' last month, how can the same inspector decide to say something about it this month?
2. Identifying continuous improvement could become an expectation of the inspector. Finding 'new risks' should be encouraged.
3. Avoid inspectors becoming hostage to the organisation they are inspecting. This is especially an issue, in this author's opinion, in current or former communist countries, for example, Russia, Kazakhstan, and Ukraine. I propose cross-fertilisation and rotation

of inspectors to some degree, again with a continuous improvement expectation.

4. Improve compliance auditing. For the U.K., the law demands that a risk assessment be 'suitable and sufficient'. I think we may confidently say that if the risk assessment had been suitable and sufficient, no incident would have occurred! In the U.S.A., OSHA's PSM Rule is somewhat more prescriptive but nevertheless tends to describe what should be in place, rather than a framework for identifying inadequacies. Yet we cannot go back to prescriptive legislation, which goes out of date too soon and requires massive state resources to develop and maintain. Audit frameworks are needed so that either state inspectors or third-party auditors fairly and consistency assess compliance in a way that allows hidden risks to be exposed and mitigating measures enforced.

EXTEND THE REQUIREMENT FOR SAFETY CASES

In much of Europe, safety cases are a requirement under the Seveso legislation (COMAH in the U.K.). They are required for Upper Tier establishments. I suggest an extension of the safety case to all Tier 1 and Tier 2 establishments with a possible extension to a broader range of organisations handling potentially hazardous materials. As an example, manufacturing facilities where flammable dusts are a potentially catastrophic risk. In Chapter 12 – Agriculture, Food, and Beverages, the incident examples included a tragic fire at a wood processing mill and at a sugar factory, neither of which would be required to register under COMAH/Seveso since the materials are not in themselves particularly hazardous unless finely dispersed as dust.

Countries not benefitting from Seveso legislation should also adopt the Safety Case approach. Proponents of deregulation will protest about the cost involved. It should not be so difficult for an organisation to describe what it already does to protect itself and the public! The cost is in the approval process, which requires trained inspectors.

So let me propose a radical deviation. Criteria for acceptance of the new Tier 2 and beyond safety cases would be limited to the contents they should contain, not to the integrity of the entire document as an adequate means of protection. The document would be lodged as a promise on how the company will be operated.

Other countries have somewhat different requirements. In the U.S.A., the 1990 Clean Air Act, EPCRA, and the OHSA PSM rules combine to require something very similar to a safety case.

Other industries such as aeronautical, railways, and nuclear have similar legislation requiring something similar to a safety case.

MAKE SAFETY CASES PUBLIC DOMAIN DOCUMENTS

Safety Cases are not currently public domain documents. In my opinion this should not be the case. It is appreciated that, in some instances, the safety case may contain data that might be valuable to a competitor. My experience suggests that the amount of such sensitive material is limited, and the limited exclusion of confidential information would be acceptable. In Chapter 16, we examined what could be found out about an atomic weapons manufacturing facility in the U.K. Much more on the technology is readily available. Therefore, a redacted version of the safety case would, in my opinion, typically require less than 10% to be removed for commercial or secrecy reasons. Neither would it, I propose, take much longer to add a redacted version.

Audits can then occur against the safety case, and any incidents can refer to what the safety case says. The safety case is, amongst other things, a set of promises as to how the facility is operated. Those promises should be open to all to review.

In Chapter 32 on Laws, Regulations, Standards, and Certification, the value of safety cases was discussed, along with an explanation as to how they can be ineffective. A particular example of a Safety Case for the RAF Nimrod Aircraft was described following an independent review into the loss of a Nimrod due to fire. The review (Haddon-Cave (2009), An Independent Review Into the Broader Issues Surrounding the loss of the RAF Nimrod MR2 Aircraft XV230 in Afghanistan in 2006) was deeply critical of the leadership and some of the engineers who carried developed the safety case. The Queens Counsel leading the review went so far as to 'name, and criticise, key organisations and individuals who bear a share of responsibility for the loss of the (Nimrod). I name individuals whose conduct, in my view, fell well below the standards which might reasonably be expected of them at the time, given their rank, roles and responsibilities, such that, in my view, they should be held personally to account'. His motivation for doing so is worth some consideration. I suggest, although I cannot prove, that his motive was to send a warning to all of us who participate in the preparation of safety cases and other forms of risk assessment. That warning is that safety case development is to be treated very seriously and that if the quality falls short those participating are open to being 'named and shamed'.

However, this only works if a serious incident occurs meriting public enquiry. I suggest that the maintenance of safety cases as public documents adds considerably to the motivation of those preparing the case to be diligent. For example, using the Nimrod case, the issue of the actual frequency of fuel pipe connectors would be open to consideration by those employees who maintain the aircraft, and the optimistic figures used in the safety case would be called into question.

Of course, using the example of a military aircraft may not be ideal. A safety case may be too explicit to be able to share with a potential enemy!

SHRINKING THE TARGETS

An idea promoted by Perrow (2007) is that of shrinking the targets. Given that accidents will inevitably occur, the concept is to limit the size and avoid concentration. The concept applies to process industries handling hazardous materials, nuclear operations, as well as concentrations of energy, such as dams. Perrow also applies this concept to concentrations of populations and concentrations of power. Risk reduction attained from shrinking the targets applies to many threats, including terrorism, operator error, sabotage, etc. ...

In the airline industry context, this would mean limitations on the size of both aircraft and airports, with less concentration on the much-favoured hub and spoke system. I am a regular user of London's Heathrow airport, but I cannot help agree – an incident at this airport brings widespread chaos for days. This is unfortunately witnessed whenever Heathrow is impacted by snow. A greater spread of airport facilities would reduce the vulnerabilities at some customer inconvenience. Instead, we have elected for a third runway at Heathrow, but maybe the Covid-19 epidemic and global warming are causing the government and the airport owner to have second thoughts about this.

Perrow (2007, p. 67) has some creative proposals on having a government directorate responsible not only for emergency management, including terrorism and natural catastrophes, but also for settlement density, escape routes, building codes and their enforcement, hazardous materials, and vulnerable sites. Such a government institution might have the vision to see the importance of shrinking the targets.

This is not necessarily a government issue, but it is one that government can encourage. Perrow (2007, p. 208) points to several successes that the chemical industry has implemented in reducing the inventory of hazardous materials. For example, Du Pont reduced the need to store methyl isocyanate. This same material was involved in the Bhopal tragedy. The Du Pont improvement involved confining this hazardous intermediate to a closed loop with a maximum inventory of just 2 pounds. In other examples, a pharmaceutical company eliminated the use of phosgene by switching to a less toxic material. Aqueous ammonia solutions have been introduced to replace the more toxic anhydrous ammonia.

NATIONAL AND GLOBAL POLICIES OF VULNERABILITY REDUCTION

In the context of natural disasters in the U.S.A., Perrow (2007) suggests a national policy of vulnerability reduction. In the U.S.A., there has been a massive movement of population to areas vulnerable to natural catastrophes – earthquakes in California, hurricanes on the Gulf Coast and Florida. Wealthy people enjoy the climate, whilst less wealthy people gain from the employment needed. Natural catastrophes involving the general population are not within the scope of this book. However there are many hazardous facilities operating in these states. A policy of vulnerability reduction seems very appropriate in the context of man-made incidents. This theme will recur in Chapter 46 on common approaches to potentially resolving global warming and sustainability catastrophes. It is an issue present in many countries in the world. London has oil terminals on the Thames vulnerable to flooding, to take one example.

Notes from Trevor's Files: Small Can Be as Good as Big

As a management consultant, I was a regular traveller to offshore oil platforms during the period 2000–2010. During this time, the two largest operators in the North Sea, BP and Shell, sold many of their assets to smaller companies. I was convinced that disasters in the North Sea would increase without the support of these two large companies with departments having many core staff engineers, extensive company standards, and safety processes. This has not been the case. There is no adverse evidence of a significant trend in the types of incidents that cause catastrophes in the sector.

The Covid-19 pandemic is an extreme result of the concentration of populations combined with the interlinking between populations created by the travel of people and by trade. Concentrations of agriculture, particularly livestock, create the potential for extreme catastrophes. In the U.K. we have had and continue to have outbreaks of Avian flu, brucellosis, bovine spongiform encephalopathy (otherwise known as Mad Cow Disease), foot and mouth disease, to mention just a few. Not only do such diseases have an impact and can be transmissible to human beings, they have the potential to destroy our food supply line. Covid-19 seems to have taken the world by surprise. Hindsight is always a great thing, but I think the evidence is that we should have been better prepared. The science was there to indicate the risk. So, we should have taken notice. 'We' is not just about government. Governments are concerned with those things their societies are most concerned about, and it seems that the risk perception was not raised to a high enough level to create awareness.

CHANGING THE LEGAL PROOF OF CRIMINAL RESPONSIBILITY

In Chapter 33 on social structures and culture, some of the difficulties in bringing successful prosecutions for violations of safety laws were identified. Although this analysis was made in the context of the U.K. legal system, similar issues are believed to occur in other countries.

In the U.K., a typical defence to a health and safety prosecution is that the relationship between the violation and the incident was not proven. This, I propose, needs to change. If the violations significantly contributed to the incident, as opposed to necessarily being the root cause, this should be sufficient. If we take the Bosley Mill case (see Chapter 12 and Chapter 32), the operation of a wood mill with such copious wood dust levels is clearly negligent. Because the wood dust was not necessarily the root cause of the explosion, the Managing Director and two other directors were acquitted. Whereas the relationship between the extremely poor housekeeping and the explosion is so logically interlinked, the responsibility is clear to us all. The law should judge similarly.

IGNORANCE IS NOT AN EXCUSE

In some instances, members of top management plead ignorance of the risks as a means to escape prosecution. If a top manager does not understand the dangers or does not take the time out to understand them, then the manager should not be a leader in the organisation.

COMPARISONS WITH LIKELY OUTCOMES FROM THE GRENFELL TOWER ENQUIRY

The Grenfell Tower disaster occurred in June 2017 (see Chapter 15). At the time of writing this book in 2022 phase 2 of the enquiry was still ongoing. Nevertheless, considerable actions have already been taken, or there are specific recommendations widely anticipated. Compare these to the proposals I made in the last three chapters, put in the context of this particular hazard of fire in high-rise buildings:

- A new regulator (Building Safety Regulator) to act as the sole building control body for all higher-risk buildings.
- The Building Safety Regulator will have powers to prosecute duty holders for ignoring compliance notices or providing false information.
- New approved code of practice for fire risk assessors will emphasise the need for assessors to be accredited, and to have a thorough understanding of the characteristics of particular types of building before taking on assessment work for that type of building.
- It is expected that the legislation will be clearly framed so that it would be easier for the regulator to 'point the finger' at individuals.

- There will be emphasis on achieving and maintaining independence between certification and building control bodies and the companies whose products and projects they authorise.
- For all buildings there will be a new duty holder – the accountable person – representing the building's owner, whose responsibility is to satisfy the Building Safety Regulator that the building is safe for occupation.
- The duty holder will have to gain a safety certificate by submitting a 'safety case' report.

Recently, staying at my favourite low-cost hotel in London I noticed at reception a notice concerning the renovations that were in progress at the hotel. The fire risk assessment which had been completed ensuring satisfactory precautions during the modifications had a specific named person responsible for the assessment.

For further discussion on Grenfell Tower and its aftermath I recommend Kernick's (2022) Catastrophe and Systemic Change – Learnings for the Grenfell Tower Fire and Other Disasters and IOSH Magazine May/June 2022 'Never Again? – Grenfell Tower'.

Chapter 45

Societal Changes, including Public Disclosure and Education

POTENTIAL CHANGES WITHIN THE EDUCATIONAL SYSTEM

The concept of risk is poorly developed in general in our society. Smokers dismiss the risks of lung cancer with remarks that they know of granny Smith who never smoked in her life but died of lung cancer. Mobile phones are used ubiquitously and without questioning the risk of radiation, never mind progressive eye strain issues. The media bombards us with climate change and sustainability issues.

But try going to hospital. Even for a wisdom tooth extraction requiring total anaesthetic, it was part of the surgeon's procedure to warn me about the risks, especially since I had a minor cold at the time. I seem to remember something like 1 in 2000 because of the cold. I read that the normal risk of death or other serious consequences is more like 1 in 10,000 where there are no complications. This was about 30 years ago. On reflection, I should have asked more about this statistic since it now sounds like an unacceptable risk. However, much depends on the benefits. My wisdom teeth would have eventually caused quite serious problems without attention. Much later in life, when going to a bowel cancer operation, the surgeon explained many risk statistics. However, the higher risk numbers were now much more tolerable to me as an individual, given the risks of cancer spread, pain and death.

Therefore, I start with education as a place where the reshaping of societal thinking should start.

Some researchers in educational approaches value the teaching of risk in the context of illnesses and diseases, including dietary as well as occupational causes of ill health (Radakovic N, Towards the Pedagogy of risk: Teaching and Learning Risk in the Context of Secondary Mathematics). Radakovic specifically promotes the concept of teaching risk from the age of 14. I would like to start at an even earlier age.

DOI: 10.1201/9781003360759-50

Primary Education (children of age up to 11)

I can hear people scoff – trying to teach children in the 6–11 age group about safety and risk? Sex and relationships, and religious education are already taught at this age, so why not risk? I see this as being additive to the current studies for this age group. The need for safety is experienced by the child and parent every day. Even at this age, the child may hear about drugs and be exposed to risky activities such as smoking. The school may be in the vicinity of high hazard premises. It is not too early, and something could be taught, perhaps along with issues such as sustainable development, without cramming the young mind with too much scary information.

Children of this age love watching videos. I confess to watching Tractor Ted along with my grandchildren of 2 years old. Tractor Ted is a wonderful programme which shows lots of advanced agricultural technology. No question in my mind about the capacity of the young person to absorb information that shapes the mind.

Reviewing the school curriculum in the U.K. for this age group, I found that the subject of probability is not mentioned, although some tests involve dice and 'spinners'. At the same time children in this age range learn about lines of symmetry, basic algebra and the concept of experiment. The mathematics that is in the curriculum at this stage is quite adequate to introduce probability. What child of this age is not playing games using dice or cards?

Secondary Education (Children of age 12–16)

By the age of 11–16, children are starting to make choices, if not about what they want to do, about what they want to study. In the U.K. a foundation (compulsory) subject is 'citizenship'. Surely there is room here for teaching some of the key issues of the day, including safety, especially regarding catastrophic events.

I reviewed the U.K. school curriculum for this age group and found that probability is introduced. For example, the pupil is expected to 'explore what can and cannot be inferred in statistical and probabilistic settings'. There is no mention of how to apply this learning.

Towards the end of this period the child will be expected to learn quadratic equations. How often do we come across quadratic equations in normal life?

Earlier in this book I introduced the concept of Bow Tie as an excellent way of understanding risk. I propose that Bow Tie is included in the school curriculum for all secondary school pupils.

Some major industrial incidents should be included in both the history syllabus, and the chemistry syllabus.

Mathematics has probably changed little since my day completing school. Children still come out of school ill-equipped to handle practicalities such as calculating their mortgage or taxes. Solving a pair of simultaneous equations will undoubtedly fill them with joy or horror, depending on their 'bent' towards mathematics. Statistics are taught, but more as a support for other sciences, not in the context of probabilities of catastrophic events. The subject of risk and probability also has enormous relevance here to the subject of health, and the risks of harm through use of tobacco, alcohol and drugs.

Tertiary (University and College) Education

Whilst university education is now prevalent, student choice of subject specialisation means we have lost influence on a large part of the young community that is not pursuing subjects where safety and catastrophe avoidance would be of interest. I can hardly see its introduction into courses on art, or English history, except possibly as part of a first-year foundation course. Generally, foundation courses are not well-liked.

But what about engineering subjects. I did a little research on my own profession – chemical engineering. To my surprise, the curricula look very similar to the one I studied 50 years ago. I can see greater choice and can see the inclusion of sustainability. But loss prevention and safety are not to be found (with one exception out of eight university curricula studied).

This is particularly surprising to me since one of our lecturers back in 1977, who taught us distillation, as I recall, published a book during this period on "Loss Prevention" and later set up his own consultancy.

Why could this be? Because teachers beget teachers and lecturers beget lecturers. There is big inertia in the educational system to enter into a new field, at least with any enthusiasm.

I propose that it is now time to change something dramatically in the educational system. Bring in people with experience outside of the educational system back in and create new thinking, not just in the delivery of lessons and lectures, but also in developing new content relevant to the priorities of the modern world.

There are some early signs that the issues of catastrophe prevention may be receiving closer attention. I note an article in September 2021, 'The Chemical Engineer' – "Engineering X outlines strategy to lead on safety in complex systems". Engineering X is a partnership between the Royal Academy of Engineering and Lloyd's Register in the U.K.. Engineering X has called for changes to engineering education and professional development, and new tools and a shared language to help disciplines better manage safety. Engineering X has a GBP 5 million five year mission called Safer Complex Systems.

Education for and at Work

Apprenticeship schemes and other types of training in preparation for work clearly have a role to play. There should be specific training modules on risk assessment and incident investigations, all in preparation to meet these issues once at work.

Adult Education

Considering this outside of the training which takes place at work, or in the performance of a skill, or carrying out of a profession, adult education is, of course, voluntary. With the internet and streamed services becoming an increasingly popular source of knowledge, there is a considerable opportunity for adults to become more aware. Almost too many! Just try on YouTube searching for risk assessment.

Personally, I welcome the screenplays on Catastrophes such as Deepwater Horizon, Dark Waters, and Chernobyl (2019). Whilst the story and the characters may be dramatised, and inaccuracies introduced for effect, on the whole, I think these are positive in helping the public understand major incidents. Unfortunately, such entertainment requires heroes and villains, and this need can distort the learning opportunity.

RISK DISCLOSURE TO THE PUBLIC

A key improvement would be in broader distribution and understanding of the company's risk register. In some reports and accounts there is quite expensive coverage of risk. Unfortunately, this author has the impression that this feature tends to be most developed by companies who have suffered significant catastrophes, such as BP and Vale (with references to Macondo and Brumadinho in section 2).

Exposing risks can be done well and in a way valued by shareholders. Highly oil-dependent companies like BP, Neste and OMV have managed the message that "our future is not in oil".

The same transparency on catastrophic risk should be expected for privately owned, publicly owned or state-owned companies, but perhaps surprisingly, not so easily exposed to the public eye. Norilsk Nickel had a storage tank corrosion problem. It certainly has a major problem with permafrost melting. Yet today, the risk seems to not be mentioned in Norilsk Nickel's sustainability report except in the specific case of the May 2020 oil tank leak. What is the generic risk not just for Norilsk but for all organisations operating atmospheric storage tanks around the Arctic Circle? This is a highly political area and one maybe where politics is indeed the solution. Politicians need to be ready to ask questions and constructively expose these risks.

The concept of improved risk disclosure goes along with improved public education. Specifically in the context of Safety Cases this idea was introduced in the preceding Chapter 44 on changes to Government and Legal Structures, and is pursued further in the next chapter in the context of alignment with an organisations disclosures with regards to sustainability and the environment.

Chapter 46

Common Factors with Other 'Catastrophes' such as Climate Change

OTHER CATASTROPHES

Current and pending catastrophes of different types surround us. At the time of writing this book in 2021/2022, we have:

- The Covid-19 pandemic and its variants.
- Climate Change and Global Warming.
- Depletion of the world's resources/sustainable development.
- Wars, in particular in Yemen and Ukraine.
- Migrants in large numbers are trying to illegally pass borders between France and the U.K., Belarus and Poland, Mexico and U.S.A.
- Threats from nuclear proliferation, in particular with nations developing nuclear weapons such as Iran and North Korea.

I don't detect the same urgency level for catastrophe prevention as for the other challenges listed above. Perhaps this is because the danger from an industrial catastrophe is not so obviously 'at our front door'. Whereas when it comes to man-made catastrophes of the type that has been this book's focus, we have to go back to 2011 at Fukushima Daiichi for the last really major disaster, a year after the Deepwater Horizon and the Gulf of Mexico oil spill. Over ten years of relative freedom from major industrial tragedies, one could develop the impression that all is under control. I suspect people regard the Beirut ammonium nitrate warehouse explosion in 2020 as an 'outlier' – something to do specifically to do with the poor situation in Lebanon. Similarly, the Brumadinho tailing dam failure in 2019 is considered to be from some faraway place not relevant to people in general.

To test this view, I developed a questionnaire to assess people's comfort level with the status quo (see Appendix D). I found that people generally feel that there is a high likelihood of future catastrophes in each of the industrial sectors we reviewed in Section 2 of this book. This included other major nuclear power plant incidents, massive oil spills, and large tailings dam failures. Such was the unanimity of view I discontinued the sending out

DOI: 10.1201/9781003360759-51

Table 46.1 Questionnaire Results

	Not Possible	Unlikely	Possible	Likely	Very Likely	Not Heard
Gulf of Mexico Oil Spill/ Deepwater Horizon	0%	5%	18%	15%	62%	0%
Grenfell Tower Fire	0%	0%	29%	18%	53%	0%
Bhopal Toxic Release	0%	3%	21%	32%	32%	12%
Cost Concordia Sinking	0%	18%	24%	26%	32%	0%
Beirut Warehouse Explosion	0%	9%	26%	35%	27%	3%
Chernobyl Nuclear Disaster	0%	24%	26%	26%	21%	3%
Paddington Rail Crash	0%	12%	35%	26%	24%	3%

the questionnaire. I found that there are so many questionnaire requests on the internet that there is considerable 'questionnaire fatigue'. The version used and shown in Appendix D is a cut down version of the original and was largely completed in my presence to urge respondence on!

As a safety professional I feel, and I hope my colleagues feel that this is serious public criticism of our collective work. I can only argue, with some justification, that without us things would be worse. (Table 46.1)

Some explanations which I heard for these opinions was as follows:

- People have not heard of any measures actually taken to prevent a Gulf of Mexico type disaster, and deep water and shallow water drilling goes on. Those that felt that there was less likelihood in the future were taking into account to reduce dependency of the world on oil.
- This was a U.K. sample. Grenfell Tower was considered with a high likelihood of recurrence because people are aware at the very slow progress in upgrade where the cladding of a high-rise building is suspect.
- Views on Bhopal were mixed. It was a long time ago to most respondents, who felt that we must have learnt something by now.
- There were mixed opinions on the Concordia sinking. Many recollect the antics of the Ship's Captain, and discount recurrence on the basis that it was unlikely that others would behave similarly. Others felt that it was more likely on the basis of the continued growth of the cruise industry.
- Beirut was considered a problem of a third world country.
- On Chernobyl, people want to think that we have learnt something.
- The U.K. respondents have seen the safety record of the U.K. rail industry improving, hence the greater confidence that a similar catastrophe would be less likely.

Is it a question of choosing which catastrophes we need to tackle as a priority? This choice depends not just on the consequences of the catastrophe but also on the tools we have to tackle that catastrophe. Global warming is set to change the globe in terms of weather patterns and sea level. We seem, at last, to be developing the tools to deal with greenhouse gas emissions. It may already be too late. Nevertheless, we need to do what we can to avoid it getting even worse, and we also need to concentrate on handling the consequences of global warming.

The majority of the world's population today has not experienced the horrors of war. Experience is limited to some Vietnam war veterans and some more recent regional wars in the Ukraine and Yemen. The consequences of nuclear war have long been forgotten.

No doubt, virology will make some big strides in the wake of the pandemic. Given that escape from a laboratory is considered a credible initiator of the Covid-19 pandemic, much needs to be done to make virus laboratories more virus secure. Similarly, progress will be made, I expect, on identifying viruses in animals that have the potential to transfer to humans.

In the face of so many catastrophes competing for attention, it is right to look for common factors which might in helping prevent not one, but a number of these catastrophes.

Process Safety, and System Safety Engineering is Drowning!

In Section 3 of this book, we reviewed many techniques and approaches for reducing the risk of catastrophe. For the process industries including oil, gas, chemicals and nuclear these techniques and approaches are broadly call 'Process Safety'. There are strongly related approaches for other industries, and the industry wide approach is better termed System Safety Engineering.

Given the many facets of Sustainable Development, it seems that Process Safety is getting lost?

Most large enterprises accept that it is their responsibility to report, not only on the company's financial performance but also on the social and environmental outcomes of their activities. The Global Reporting Initiative provides detailed guidance on how to do this through some 37 separate standards, each with either a general or topic-specific role. So, I started investigating these to find evidence that Process Safety is considered.

Now I am open to Process Safety aspects having different names, as a subset of the whole:

- Major Accident Hazards
- Asset Integrity
- Spills and other Loss of Process Containment
- Emergency Preparedness and Response
- etc ...

I found that in the 37 standards of the GRI, such Process Safety Issues are scarcely mentioned. An exception is the standard on Effluents and Waste, where Significant Spills is a disclosure requirement. The occupational safety and health standard is confined to exactly that – of worker safety, not on catastrophe prevention.

This omission can hardly be on the basis of materiality. Materiality is defined by the threshold at which any particular aspect becomes sufficiently important that they should be reported. For oil, gas and petrochemical, and mining industries, Process Safety issues are undoubtedly high in materiality.

So, what about the reporting of sustainable development issues by enterprises themselves. I researched a variety of reports from major oil and gas, petrochemical, and mining companies. The reports from Vale and BP stand out as having Process Safety/Major Accident Prevention as prominent issues on which to report. This should not be unexpected, given the consequences to Vale from Brumadinho and BP from Deepwater Horizon. What is concerning is that other reports from companies who thankfully have not suffered the same nightmares do not give process safety the same prominence. Perhaps they or their stakeholders do not feel the same vulnerability – and that must surely be a mistake.

Most of the other reports I reviewed do include Process Safety/Control of Major Accident Hazards to some degree. But this is mostly in aspirational policy statements, together with occasional trailing indicators. Seldom is there a mention of any leading indicators; contrast the declarations on climate change, ethics and diversity where measurements of progress and details for projects are quite detailed.

Sustainability reports are often supported by data verification and sometimes scoring from organisations such as MSCI, FTSE4Good, Dow Jones Sustainability Index etc.. None of these appears to recognise Process Safety/Control of Major Accident Hazards from the available methodology.

May I suggest that they all consider the excellent term adopted by Glencore: Catastrophic Hazard Management.

So, does it matter? "What gets measured gets done" is an old but solid adage. Unless there is attention from stakeholders, then Process Safety will not get the emphasis that it needs. Maybe it is those same stakeholders: shareholders, employees, local communities who themselves need to demand more.

Worse, given the increasing demands on top management to generate improvements in many environmental and sustainability issues, Process Safety could get left behind – with potentially catastrophic consequences.

There are other societal changes demanded of today's leaders such as:

- Equal opportunity, inclusion, diversity
- Modern day slavery
- Ethics and integrity

On reflection, I now see some common factors which could be used to leverage all of these societal change issues together.

COMMON FACTORS

One factor worth consideration is the change in technologies if we are successful in phasing out fossil fuels.

Dangers from Oil and Gas, especially deep-water oil and gas exploration and production should decline. There is, of course, a threat hidden in this scenario. As investment in oil and gas exploration and production declines, pressure on the costs of running the existing assets will increase. There is a danger that measures to prevent catastrophes may be curtailed because of cost. Assets may deteriorate due to corrosion and lack of maintenance, with the potential for a catastrophe as a result.

The chemical industry is devising new technologies to recycle plastics and waste oils. The recycling processes, some of which include the oil refineries, will neither increase nor decrease the catastrophic potential. In Chapter 17 on New and Emerging Technologies, we discussed the potential consequences of increased use of, and transportation of, hydrogen and ammonia.

I expect the dependence on nuclear power to increase as an alternative to fossil fuels. With the increased proliferation of nuclear reactors, the threat of further catastrophe from them will also increase.

Mining will continue. Whilst many coal mines will be shut down, mines working ores for copper and other exotic metals used in batteries and solar panels will increase, and so will the size of the tailings dams associated with these workings.

As I read Bill Gates book "How to Avoid a Climate Disaster" the issue of climate change seemed both different and demands more attention than catastrophe prevention. As I read on, I realised that the goal is the same, and in some respects, the path to getting to that goal is very similar.

Shareholder Value vs Stakeholder Value

Compare the following statements:

1. Enterprises exist to maximise value to the shareholders.
2. Enterprises exist to create value for the stakeholders – including shareholders, employees, customers, suppliers, local communities.

This discussion often precedes a decision to adopt corporate social responsibility and sustainable development in particular (Bellucci et al., 2019).

However, I don't see the two as being mutually exclusive. The first statement does not specifically mention profit or a common, more precise definition - EBITDA (earnings before interest, taxes, depreciation and amortisation). The profit in any one year does not reflect shareholder value. I am fortunate to have a small shareholding in several companies, and I look to the long-term prospects of the company to decide whether to buy or sell.

Failure to adopt good sustainability practices is likely to affect profitability in the long term, especially if we can forecast that there will be a widespread societal backlash against organisations that do not adopt good sustainable development practices.

However, in the case of catastrophe prevention, the argument is simpler and more powerful. A catastrophe will affect shareholder value directly. Catastrophes cause the following types of financial loss:

- Damage to equipment which will need to be replaced.
- Loss of production or loss of provision of service.
- Dissatisfied customers who cannot get product.
- Reputational damage which can affect customer buying patterns.
- Loss of trust which negatively affects ability to get permits from local authorities for current and future operations.
- Loss of confidence in the organisation which leads shareholders and other investors to 'flee from the stock' and take their investment elsewhere.

The only aspect which is up for discussion is the subject of risk. An organisation's operating facilities with identified risks may be fortunate and not have a catastrophe, or it may be lucky and experience one.

Suppose the probability of a catastrophe could be calculated with confidence. We have doubts about this (see Chapter 25–27, 30). If, however, we were successful, this risk exposure could be mathematically expressed in terms of risk to future profitability and directly be computed within the shareholder value calculation. This, however, is a vague science, but one effectively practised by insurance companies who have to decide whether to insure an organisation or not. As an insurance company risk engineer, I have had the good fortune to survey organisations in the energy sector and advise insurers accordingly. Therefore, in Appendix E, I include the types of disclosures an organisation might make if it were to declare them as part of its sustainable development report.

Global Reporting Initiative

Efforts have been made to enable and encourage companies to report their 'off balance sheet' efforts in all aspects of sustainability. A company's annual reports are mostly focussed on financial report, and mostly in short term financial reporting covering the year of the report.

Before I attempt to explain I should warn the reader that this field of sustainability reporting is full of acronyms. The big accountancy firms are heavily involved in this subject, I suspect as a potential lucrative revenue stream to complement their current major accounts. As well as acronyms the accountants have 'intellectualised' a subject which should be easily digestible to people at large. As you next door neighbour what is the meaning of 'materiality'!

The Global Reporting Initiative (GRI) is an international independent standards organisation intended to help businesses communicate their impacts on issues such as climate change, human rights and corruption (Wikipedia: Global Reporting Initiative). These are broadly the same issues that come under the banner of Environmental, Social, and Governance (ESG). The GRI publishes a framework of standards which it proposes should be used for businesses to declare their responsible behaviour, including:

- Economic performance
- Market Presence
- Indirect Economic Impacts
- Procurement Practices
- Anti-Corruption
- Tax
- Energy
- Effluents and Waste
- Supplier Environmental Assessment
- Employment
- Labour/Management Relations
- Diversity and Equal Opportunity
- Training and Education
- Child Labour
- Security Practices

And ... Occupational Safety and Health

There is nothing on Catastrophe Prevention. The Occupational Health and Safety standard covers the protection of workers in the context of daily work but makes no mention of the risk to either workers or the general public.

A GRI standard on management Catastrophic Risk, probably more politely called Process Safety or similar, is an essential addition. If it can be done for the factors shown above, it can be done for Catastrophic Risk.

I felt so strongly about this that I have put together a framework for such a standard on Catastrophic Risk in Appendix E. Of course, such standards are developed through discussion with a variety of stakeholders and organisations with expertise. My proposal is a framework to be used as a starter.

The GRI Foundation Standards suggest that if an organisation determines a particular topic to be material, and there is no current standard for that topic, the organisation should report its own relevant disclosures. In other words, it seems to me that an organisation is free to develop its own standard until or unless GRI develops one. Hence, I suggest that Appendix E could be of immediate value.

Some organisations will struggle to disclose current residual risks and say that they have none. If so, then they should disclose it accordingly. I think that reality will soon come to bear.

Incidentally, I have not so far come across one single facility in my long career which did not have some current residual risks of concern. Why not be transparent with your stakeholders on this?

Some readers will immediately protest that such disclosures might cause undue concern amongst stakeholders. In the organisation's annual report, and in particular SEC 10-F and 20-F filings, companies are obliged to specify financial risks in some detail. For example, risks from swings in feedstock commodity prices. Why is safety risk so different, and why does it require less transparency?

Is Catastrophe Prevention part of Sustainable Development?

Since catastrophe prevention, or PSM, does not form part of the Global Reporting Initiative standards, I had to question myself as to why this should be. At their core, the standards encourage and specify disclosures on facts about an organisation's operations and their impacts on stakeholders.

The GRI Foundation Standard defines stakeholders as individuals or groups that have interests that are affected or **could be** affected by an organisation's activities. I emphasise the words 'could be'. Thus, stakeholders include employees, governments, local communities, shareholders and other investors, suppliers, trade unions.

In the context of Catastrophe Prevention, the employees in a hazardous facility and the community surrounding that hazardous facility are stakeholders. For transportation, the passengers onboard aeroplanes and trains and in airports are all stakeholders who could be impacted by a catastrophe.

The GRI foundation standard also defines impacts as 'referring to the effect an organisation has or could have on the economy, environment, and people, including effects on their human rights, as a result of the organisation's activities or business relationships. The impacts can the actual or potential, negative or positive, short-term or long-term, intended or unintended, and reversible or irreversible. These impacts indicate the organisation's contribution to sustainable development'. (Gates W.; How to Avoid a Climate Disaster – The Solutions We Have and the Breakthroughs We Need.)

The foundation standard also comments that 'if it is not feasible to address all identified impacts on the economy, environment, and people at once, the organisations should prioritise the order in which to address potential negative impacts based on their severity and likelihood. Thus, the probability/consequence concept discussed extensively in Section 3 of this book is also fundamental to the GRI standards.

The GRI is developing Industry Sector-specific standards and already has a standard GRI 11 for the Oil and Gas Sector. This standard applies to all sectors of the oil and gas industry. Our hypothetical fuels terminal is covered, and we might decide to provide a sustainability report for the terminal in line with GRI standards. As we might expect, topics covered include Greenhouse Gas Emissions, Water and Effluents, Closure and Rehabilitation, Occupational Health and Safety etc. However, there is a very relevant section 11.8 on Asset Integrity and Critical incident management (GRI Standard 11 – Oil and Gas Sector). There is no corresponding topic standard as there is for Emissions, and Occupational Health and Safety. Perhaps the suggested standard I developed in Appendix E could be used accordingly.

Sustainable development has been defined as 'development which meets the needs of the present without compromising the ability of future generations to meet their own needs'.

Conclusion – Catastrophe Prevention is a part of Sustainable Development – this is not yet recognised in the GRI standards.

It is also possible that, as pointed out above, the GRI standards have some links to Catastrophe Prevention which have not so far been well developed in actual sustainability reports (see Current Company Reporting Process Safety versus Sustainable Development below).

Sustainable Development Reporting-Necessity and Validity

An organisation is not obliged to do sustainable development reporting, but many organisations, especially large ones, are choosing to do so. They are challenged by stakeholders to state their impacts. This is a very promising sign and a drive for organisations to improve their impacts long term. If catastrophe prevention becomes integrated with sustainable development reporting, this will provide a valuable motivator for improvement.

Can we trust sustainable development reporting and catastrophe prevention reporting in like manner? Of course, organisations will do what they can to present their sustainable development reports positively. This is as it should be. However, I find the rigours of the GRI reporting framework quite robust. Stakeholders can and do ask questions based on an organisation's sustainable development disclosures. Deliberate attempts to conceal may be quickly uncovered by the curious.

Furthermore, the presence of many risks should be clear, especially, I hope to those who have read section 2 of this book. Risks associated with an operation are mostly obvious just by knowing a little about the organisation's operations. Then we look to the companies' disclosures to see what they say about those risks and how they are mitigated.

It is a requirement of the GRI standards that disclosures are comparable over time. For example, so that the amount of emissions can be compared year on year. Estimates are OK so long as the assumptions and methodology are referenced. Furthermore, the disclosures must be verifiable. This includes ensuring that the information systems can be examined as part of an external assurance process. The original sources of the reported information should be indicated and provide reliable evidence to support assumptions or calculations.

External assurance is encouraged!

Developing a Reporting Standard for Catastrophe Prevention

Using the GRI Standards as a guide, here are some key considerations in developing a standard for Catastrophe Prevention.

- Disclosures must be comparable over time.
- Capable of being verified by external assessors.
- Determination of the likelihood of potential negative and positive impacts either qualitatively or quantitatively. (GRI Standard 3 – Material Topic).
- Setting thresholds to define a cut-off point which determines which impacts it will report.

Based on this guidance, we can start to see what a Reporting Standard might require for Catastrophe Prevention. I will use process industry examples, but these can be relatively easily adapted to other industries.

Contents might include:

- Process Safety Performance Indicators – See Chapter 31 on the Governance of Safety.
- Hazards prior to mitigation which could result in a catastrophe.
- Residual risk given current mitigation measures i.e. outcome of a HAZOP.
- An indication of any plans, e.g. additional technology or controls, to reduce the residual risk.
- Expected future residual risks are mitigation.

In each case, we would use the risk matrix to define the threshold for reporting and perhaps confine this to high, very high and critical risks.

As mentioned above the GRI has developed a sector-specific standard for oil and gas, which includes a topic 11.8 'Asset Integrity and Critical Incident Management'. However, it does not point to a corresponding topic standard detailing Asset Integrity disclosures. It does cross-reference GRI topic standards on Occupational Safety and Health, but this does not cover the catastrophic risk. The oil and gas sector standard also cross-references Topic Standard 413 on Local Communities. The latter is more revealing on coverage of Catastrophic Risk and requires disclosure of Operations with significant actual and potential negative impacts on local communities. Amongst other requirements, disclosure is required on for each of the significant actual and potential negative economic, social, cultural, and/or environmental impacts on local communities and their rights, describe:

- the intensity or severity of the impact
- the likely duration of the impact
- the reversibility of the impact
- the scale of the impact.

Catastrophe prevention is buried here, and I don't see this practised in many of the sustainable development reports studied below, but this gives us also a link to the standard I suggest in Appendix E.

CURRENT COMPANY REPORTING PROCESS SAFETY VERSUS SUSTAINABLE DEVELOPMENT

During my research for this book, I examined a selection of sustainability reports from a number of well-known oil and gas and mining companies. I looked for parts of the report which would reflect useful data on the risks of that company with respect to catastrophic risk.

Materiality

An important concept in the context of company reporting is the topic of 'materiality'. There is an understandable tendency for companies to 'cherry pick' the good news for their reports. 'Materiality' originally had an accounting definition 'Information is material if omitting it is misstating it could influence decisions that users make Bellucci et al. (2019) 'Stakeholder Engagement and Sustainability Reporting'.

In looking to find public disclosure of catastrophic risks, there are a number of potential sources of information:

- Annual report and accounts
- Sustainability Report
- Form 10-K/20-F, a U.S. Securities and Exchange Commission requirement, the latter is for non-US and non -Canadian companies that have securities trading in the U.S.A., whilst the former is for US and Canadian companies.

Spills

The number of spills was also reported in over 75% of cases, and this might be expected since spills reporting is a disclosure required under GRI standard 306-3. However, this standard also requires disclosure of:

- Location of spill
- Volume of spill
- Material involved

The latter is rarely done, except for major spills, which have received media attention.

Process Safety

First a little terminology. There is a recognised standard on process safety event reporting known as API RP 754. This recommended practice is now widely used, at least in Europe, Australia and North America. This practice defines a process safety event as:

'An event that is potentially catastrophic, i.e. an event involving the release/loss of containment of hazardous materials that can result in large-scale health and environmental consequences.'

The guideline defines four tiers of events, with Tier 1 having the largest consequence. The details need not concern us here. The following descriptions are made with the aid of the CCPS Process Safety Metrics: Guide for Selecting Leading and Lagging Metrics.

We can loosely define a Tier 1 event as an incident that resulted in consequences which could include:

- A serious injury to an employee.
- Effect on community requiring them to evacuate or 'shelter-in-place'.
- Fire or explosion causing more the USD 100,000 cost to the company.
- A release of flammable, combustible, or toxic chemicals greater than predefined threshold quantities

A Tier 2 Event is an incident that resulted in consequences which could include:

- An employee recordable injury.
- Fire or explosion resulting in between USD 25,000 and USD 100,000 of cost to the company.
- A release of flammable, combustible or toxic chemicals greater than predefined threshold quantities (which are less than those defined for a Tier 1).

Examples for Threshold Quantities are: (Table 46.2)

A Tier 3 Event is a 'near miss' event. No damage occurred but could have in somewhat different circumstances. It represents the failure of one of the barriers we showed in the bow-tie diagrams shown in Section 2 Examples might include:

Table 46.2 Threshold Quantities for Tier 1 & 2 PSM Event Recording

Material	Tier 1 Threshold (kgs) outdoor	Tier 2 Threshold (kgs) outdoor
LPG, Propane, Butane	500	50
Gasoline	500	50
Chlorine	5	0.5

- Excursions to safe operating limits
- Inspection results outside of acceptable limits
- Demands on safety systems such as critical safety interlocks and pressure relief valves
- Loss of Process Containment below the volumes indicated for Tier 2.

A tier 4 event indicates failures in barriers that comprise Management System and Operating Discipline issues. An example might be the discovery that an operator was working on a process for which he/she had not been trained.

What I found was that, indeed, 75% of the reports I reviewed had a section on Process Safety. 75% also had information about trends in Process Safety Tier 1 and Tier 2 incidents.

Risk Matrix

None of the reports I examined included the risk matrix that they use. In Chapter 27, we considered the risk matrix as a statement of a company's risk appetite. Whilst many reports referred to either a risk matrix or risk appetite, they did not disclose any detail of their risk matrix in their reports.

Generalised Statements concerning Risk Management

Most reports include some generalised statements of risk. For example, the following from BP's Sustainability Report and Annual Report 2020:

'Technical integrity failure, natural disasters, extreme weather or a change in its frequency or severity, human error and other adverse events or conditions, including breach of digital security, could lead to loss of containment of hydrocarbons or other hazardous materials. This could also lead to constrained availability of resources used in our operating activities, as well as fires, explosions or other personal and process safety incidents, including when drilling wells, operating facilities and those associated with transportation by road, sea or pipeline. There can be no certainty that our operating management system or other policies and procedures will adequately identify all process safety, personal safety and environmental risks or that all our operating activities, including acquired businesses, will be conducted in conformance with these systems'.

I found that companies who had recent experiences of catastrophes (by recently, I mean 1–2 years) are more explicit with respect to particular risks related to that particular catastrophe. For example, from the Vale Sustainability Report and Annual Report 2019:

Incidental Case Study – Vale Reporting

This report gives links that can produce considerable detail with each geotechnical structure (primarily tailings dams) and the state of inspection listed. Following the Brumadinho disaster, the qualification for an acceptable inspection was upgraded in Brazilian law, a number of the dams do not meet current expectations. Priorities are set. Without the Brumadinho event, one has to speculate that this report would have been very different.

The 10-K report is more revealing:

'Decharacterization of upstream dams. Our key initiative is the decharacterization of all of our 30 upstream structures in Brazil, including dams, dikes and drained piles. An upstream structure is a structure raised using the upstream raising method, in which the body of the structure is built using the thick tailings deposited in the reservoir, by successively layering them up and in the direction opposite to the water flow (upstream). This is the same construction method as the Brumadinho dam. The term "decharacterization" means functionally reintegrating the structure and its contents into the environment so that the structure no longer serves its primary purpose of acting as a tailings containment. Recently approved laws and regulations require us to decharacterize all of our upstream structures on a specified timetable, based on projects agreed with authorities'.

The Vale disclosures continue 'The decharacterization of an upstream structure is a complex process and may take a long time to be concluded with the necessary safety precautions. The works related to the decharacterization process may influence the geotechnical stability conditions and increasing the risk of such structures. We are conducting detailed engineering studies for each structure to be decharacterized and may need to improve the construction or build additional containment structures (back-up dams) to safely proceed with the decharacterization. For instance, for certain dams, we have to build downstream back-up dams to ensure the retention of tailings in case of a dam rupture. All these projects are subject to further review and eventual approval by the relevant authorities.

Our decharacterization plan was updated in 2020 and currently comprises 25 geotechnical structures (including 13 dams, 10 dikes and 2 drained piles) built by the upstream raising method. Our plan also includes the construction of seven containment structures (back-up dams for certain dams, to retain tailings in case of dam rupture, protecting the area downstream from these dams during the decharacterization works.

Further useful details follow, although I did not find any reference to the particular risk of toxicity of sediments behind the dams.

Vale then proceeds to declare:

'We are promoting a cultural transformation, seeking to bring safety, people and reparation to the centre of our decisions. As part of this effort, we believe that five key behaviours must be present throughout the organisation:

- Obsession with safety and risk management.
- Open and transparent dialogue.
- Empowerment with responsibility
- Ownership for the whole.
- Active listening and engaging with society.

These latter key behaviours are particularly relevant to the next chapter of this book concerning High Reliability Organisations.

Perhaps the best reflection of the quality of sustainability reports comes from the Responsible Mining Foundation (RMI). The Responsible Mining Foundation is an independent research organisation based in Switzerland. They publish a Responsible Mining Index (RMI) Report, which is a biennial assessment of large globally-dispersed mining companies' policies and practices on economic development, business conduct, lifecycle management, community wellbeing, working conditions and environmental responsibility – with gender and human rights issues integrated throughout the report. What is relevant here are the parts of the RMI report relevant to catastrophic risk. Their report from 2020 makes the following key findings, (responsible Mining foundation (RMF), RMI Report 2020), with some simplification of the wording for the readership of this book.

1. Significant gaps remain with society's expectations

The performances of even the best-scoring companies fall considerably short of society's expectations in all six main areas of the RMI report research. Stronger efforts are required by all companies to ensure their practices are managed effectively in light of social expectations and Sustainable Development Goals (SDGs).

2. Some signs of progress, but mostly commitments

Since the RMI Report 2018, more companies have made and disclosed formal commitments on some economic, environmental, social and governance (EESG) issues. A few companies have developed new or stronger management standards. Yet many companies show little sign of movement, and much needs to be done to translate corporate commitments and standards into successful business practices.

3. Effectiveness requires persistence

Most companies are still not able to demonstrate that they track and publicly report on how effectively they are managing ESG issues. Even fewer companies show evidence of reviewing their performance and taking responsive actions where necessary. Once commitments are in place, it takes persistence to follow these through to execution.

4. Risk of SDG-washing

It is good to see that companies are increasingly aligning their sustainability reporting with the SDGs. However, this reporting is selective and risks the perception of SDG-washing as companies generally omit any mention of negative impacts potentially impeding the achievement of these internationally agreed objectives. It is essential that an honest picture emerges of the true challenges the mining sector faces in its support of the SDGs.

5. Mine-site data still missing

Many mine-sites do not disclose site-level data on issues of strong public interest for communities, workers, governments, and investors. And very rarely do mine sites evidence engagement with local stakeholders on EESG issues. To build trust with all stakeholders and reduce risks, companies will benefit from adopting responsible mine-site behaviour across all their operations and transparently sharing information.

6. External requirements drive performance

Stronger-performing and more transparent companies tend to be subject to specific requirements set by investors or producing country or home country governments. For example, the investor-led request for disclosure of information on tailings storage facilities has generated much more publicly available data of critical interest to shareholders, debt issuers, insurers and governments.

7. Severe adverse impacts must be addressed urgently

Events such as the Vale tailings disaster in Brumadinho are harsh reminders of the unacceptable risks faced by many communities, workers and environments in mining areas. Such tragedies, and other adverse impacts and fatalities, reflect very poorly on the industry as a whole and put into stark perspective any claims of responsible EESG management. The mining industry needs to prove that it prioritises EESG risk management over short-term considerations.

8. It can be done

The RMI reports that, although individual company results are generally very low, collectively, the companies prove that social expectations are achievable. If one company were to attain all the highest scores seen for every indicator, it would reach over 70% of the maximum achievable score. Similarly, for a mine site in the mine-site assessment, achieving all the best scores recorded would enable it to reach over 80%. Each company and mining operation is encouraged to adopt the responsible practices already being demonstrated across the sector.

Based on my reading of other non-mining sustainability reports, I suspect that such detailed research as is carried out by the Responsible Mining Foundation, if it were available, would show similar features, notably:

- 'Yet many companies show little sign of movement and much needs to be done to translate corporate commitments and standards into successful business practices'.
- Sustainability reporting is selective and risks the perception of SDG-washing as companies generally omit any mention of negative impacts potentially impeding the achievement of these internationally agreed objectives.
- Many companies do not disclose site-level data on issues of strong public interest for communities.

Chapter 47

In Pursuit of Genuinely High Reliability Organisations

COMPETING VIEWS ON CATASTROPHE PREVENTION

From my research, I propose three main competing views with regards to catastrophe prevention:

1. **Incidents will inevitably occur due to the increasing complexity and tight coupling of industrial activities.** We need to accept the consequences, or if the consequences of an accident with the facility are too dire, not carry out the activity at all. These are the 'Normal Accidents', a concept from Perrow (1984), which we have discussed before. With complex, tightly coupled systems, Perrow argues, incidents will inevitably occur. The only solution is to reduce or eliminate complex, tightly coupled industries. Nuclear power plants are one of the examples of complex, tightly coupled systems he targets in this regard.
2. **Incidents will inevitably occur due to the short-term thinking and short term needs of leadership both in industrial organisations and in regional and national governments.** Perrow (2007) 'The Next Catastrophe' termed this 'Executive Malfeasance'; however, the implication of deliberate wrongdoing may be misrepresented. It is not unusual for circumstances to arise where there are temptations to take short-term wins to the detriment of long-term goals.
3. **An organisational culture can be created, a 'High-Reliability Organisation (HRO)' where risks, even dire ones, can be managed and incidents eliminated.** This is the concept of HRO originally put forward by Weick and Sutcliffe. (Weick and Sutcliffe (2007). Managing the unexpected: Resilient performance in an age of uncertainty). This concept will now be described more clearly.

DOI: 10.1201/9781003360759-52

CHARACTERISTICS OF HIGH RELIABILITY ORGANISATIONS

Up until now, this book has not introduced the concept of such 'High Reliability Organisations. Such organisations are considered to have the following characteristics:

> *Preoccupation with failure*: HROs treat anomalies as symptoms of a problem with the system. The latent organisational weaknesses that contribute to small errors can also contribute to larger problems, so errors are reported promptly. Problems can be found and fixed.
>
> *Reluctance to simplify interpretations*: HROs take deliberate steps to comprehensively understand the work environment as well as a specific situation. They are conscious that the operating environment is very complex, so they look across system boundaries to determine the path of problems (where they started, where they may end up) and value a diversity of experience and opinions.
>
> *Sensitivity to operations*: HROs are continuously sensitive to unexpected changed conditions. They monitor the systems' safety and security barriers and controls to ensure they remain in place and operate as intended. Situational awareness is extremely important to HROs.
>
> *Commitment to resilience*: HROs develop the capability to detect, contain, and recover from errors. Errors will happen, but HROs are not paralysed by them.
>
> *Deference to expertise*: HROs follow a typical communication hierarchy during routine operations but defer to the person with the expertise to solve the problem during upset conditions. During a crisis, decisions are made at the front line, and authority migrates to the person who can solve the problem, regardless of their hierarchical rank. (based on HSE Research Report R899)

Readers may see, in this description, approaches which will considerably correct the failings indicated in Chapter 31 on the Governance of Safety. The approach supports some of the issues put forward as potential solutions in Chapter 43 on Leadership Mindset.

Whilst I find the concept of High Reliability Organisations (HROs) convincing, this book has not been a book about HROs. Whilst in developing HRO thinking, researchers evaluated a number of different organisations (HSE Report RR899 on High Reliability Organisations and Hopkins (2021)), I am not convinced, with the test of time, that any one of the example organisations could be considered High Reliability. Rather, for the examples of organisations approaching HRO, the consequences of a catastrophic incident are more severe. Take, for example, the nuclear industry, yet we still had Fukushima Daiichi (see Chapter 10). Take, for example, air traffic control, yet we still had Shoreham and the Boeing 757Max disasters (see Chapter 9).

Nevertheless, I value the five characteristics of HROs that Weick and Sutcliffe identified as responsible for the "mindfulness" that keeps them working well when facing unexpected situations.

After Anthony Hopkins had completed his excellent books on disasters such as Deepwater Horizon and others, he kindly put together his thoughts on achieving a High Reliability Organisation (Hopkins, 2021).

He describes HROs as organisations alert to "bad news" or warning signs that things may be about to go seriously wrong. The striking thing about HROs is that they are preoccupied with the possibility of failure. To use a now well-known expression, they exhibit "chronic unease" about how well they have their major hazards under control. They recognise that before every major accident, there were warning signs of what was to come, which, had they been attended to, would have prevented the accident from occurring. This is true for every major accident that has been studied systematically.

Consider this description of the US nuclear submarine organisation – probably the most celebrated of all HROs.

"One of the most amazing elements of the Nuclear Submarine culture is its self-enforced refusal to sweep problems under the rug. For decades the submarine culture has recognised the criticality of squeezing out every ounce of lessons learned from imperfect performance".

Leaders of HROs are suspicious of a steady stream of good news and are forever probing for the bad news that they know lies beneath the surface.

HROs have a policy of "challenging the green and embracing the red".

It is time to "Embrace the Red".

THE DEFEAT OF SHORT-TERMISM

Readers will be pessimistic about defeating short-termism. But society globally has made some remarkable achievements, maybe a little late, but achieved, nevertheless. For examples:

- Reduction of ozone-depleting agents (Montreal Protocol)
- Global concern about climate change
- National plans concerning sustainable development
- Progress in inequality on the basis of sex and ethnicity

I will take one example in just a little detail. The International Maritime Organization has worked to set a new limit on the sulphur content in the fuel oil used onboard ships. This has reduced the maximum sulphur content from 3.5% to 0.5%. This change has had a significant economic impact to the industry. Ships must either buy considerably more expensive fuel, or they have the option to install scrubbers on the engine exhaust to scrub out the sulphur oxide (Sox) emissions from entering the atmosphere. Oil refineries depended on selling the residual from crude oil distillation as heavy fuel oil for ships. In the absence of this route, they have had to reconfigure and, in some cases, install additional process plants (e.g. a Coker) to break down the heavy oil into other lighter products such as gasoline.

Yes, short-termism can be defeated. It takes all stakeholders, shareholders, employees, communities.

It takes you.

Chapter 48

Conclusions

So, this is the end of my book. I did not tell you before – I failed English at my first attempt at GCSE 'O' Level (these are examinations taken at the age of around 16 at the end of secondary school education, I think the equivalent school level in the USA I 'Junior High'). I am an engineer who marvelled at the inspired writing of Anthony Hopkins and Charles Perrow whose books I referenced many times during the last 47 Chapters. They had one advantage and one disadvantage. The advantage: as sociologists, reasoned writing is their core skill.

The disadvantage: they were neither engineers nor people with direct hands-on experience of the real world of managing high hazard operations.

This book was written to appeal to both technical and non-technical readers. It is sound advice that the writer should decide early on his/her audience and keep them firmly in mind throughout. In truth I stubbornly resisted and maintained two audiences. To those of you that reached this point, your determination is much appreciated.

I am an American Football fan. Following the racist issues in the USA which came to achieve much prominence in 2020, many players helmets were inscribed with various themes relating to racial injustice. My favourite, decidedly neutral, are the helmets inscribed "INSPIRE CHANGE". I do hope this book will do something to inspire change in the world of preventing catastrophes from man-made activities.

At the time of writing this conclusion, February 2022, we are threatened by the potential of nuclear war resulting from war in Ukraine. Nuclear war is not covered in this book. It would indeed be the catastrophe to beat all catastrophes. I leave it to the sociologists and political commentators to apply their expertise to this problem.

Greta Thunberg boldly declared that "No One is Too Small to Make a Difference" and goes on to say, "We must change almost everything in our current societies". She was not restricting her comments to sustainability and climate change alone. This book has been about catastrophe prevention as part of what needs to change.

So, what can you do? Where is your place in this change?

DOI: 10.1201/9781003360759-53

Leaders of all types are very sensitive to the mood of the population. Such leaders might be politicians, board members of large companies, civic councillors, trade union heads, town planners and so on. We can let them know our concern.

How to do that?

By asking informed well thought out questions and being persistent about getting an answer.

The first draft of this book was entitled "Is It Safe? The Questions You Should Ask; The Answers You Should Expect".

It is not about asking if something is safe. We can have real influence if our questions are well thought through and based on facts. They should be difficult questions which will make those responding think. I have been in the position of having to respond to such questions. I can tell you that those questions really influenced me. On reflection people were far too willing to accept my first reply.

We do not need a society of 'nosy parkers and 'busybodies'. Such people get some egotistical satisfaction from their often-superfluous enquiries.

We do need a society of enquiring minds. What gives you the right to ask these questions?

YOU ARE THE PERSON AT RISK.

Hypothetical Case Exercises – Maps and Diagrams

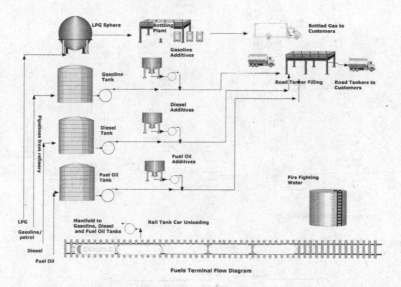

Figure A1 Fuels Terminal Flow Diagram.

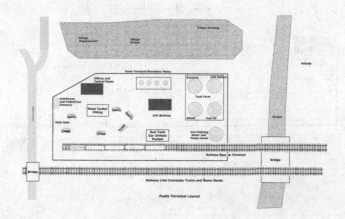

Figure A2 Fuels Terminal Plot Plan.

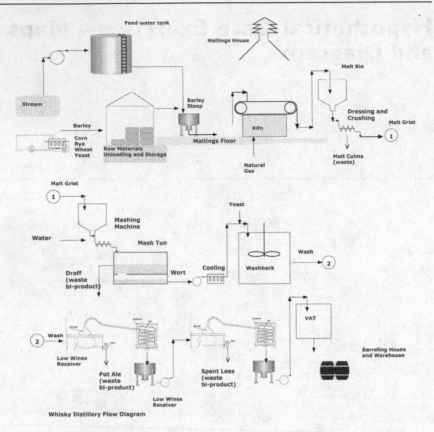

Figure A3 Whisky Distillery Flow Diagram.

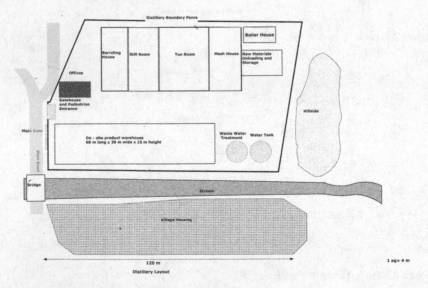

Figure A4 Whisky Distillery Plot Plan.

Appendix B

Seveso-Regulated Site Densities – Examples from Europe

SEVESO SITES IN EUROPE

Information on sites registered under the Seveso regulations (COMAH in the United Kingdom) can be useful in identifying the location and densities of hazardous facilities. This information locates only hazardous processes and hazardous materials storage above certain thresholds. Other hazardous facilities such as airports, nuclear power, and defence establishments will not necessarily show.

Information on the location of Seveso Sites is not always available. This is valuable information; people should know where these Seveso regulated sites are located.

In the following pages maps are shown giving the approximate locations of Seveso Tier 1 and Tier 2 sites in:

- England and Wales
- Scotland
- Belgium
- Netherlands

As examples.

Using available data, it is possible to produce similar maps of Germany, France, and many other EU countries. I have not so far found an equivalent source for the U.S.A.

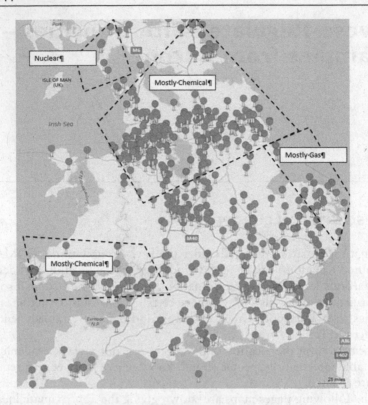

Figure B1 Seveso-Regulated Sites in England and Wales.

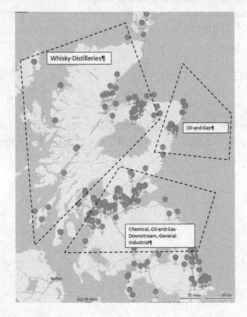

Figure B2 Seveso Sites in Scotland.

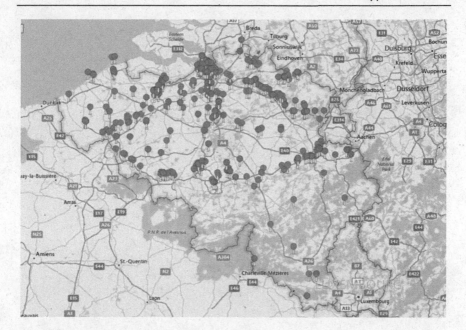

Figure B3 Seveso-Regulated Sites in Belgium.

Figure B4 Seveso-Regulated Sites in the Netherlands.

Appendix C

Mine Tailings Dam Failures

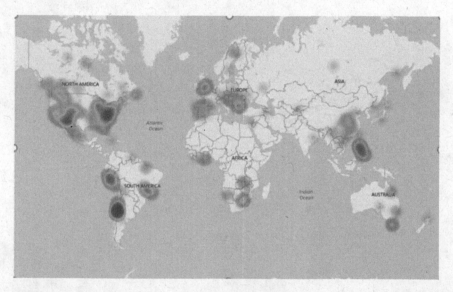

Figure C1 Mining Tailings Dam Failures – Heat Map (World).

Questionnaire used: Likelihood of Future Catastrophes – Your Opinion?

LIKELIHOOD OF FUTURE CATASTROPHES – YOUR OPINION?

What do you think is the probability of a future catastrophe, similar to the past catastrophes listed below? If you have not heard of this catastrophe, please move on to the next question:

1. **Catastrophe: Grenfell Tower Fire** (2017). U.K. 72 Fatalities.
   What is your opinion of the likelihood of a similar catastrophe to occur within the next 50 years?

Not possible	Unlikely	Possible (50/50)	Likely, no more than one	Very likely/ maybe several	OR Not heard of this incident
☐	☐	☐	☐	☐	☐

2. **Catastrophe: Bhopal Toxic Gas Release** (1984), India – over 2000 immediate fatalities, many long-term illnesses, and deaths.
   What is your opinion of the likelihood of a similar catastrophe to occur within the next 50 years?

Not possible	Unlikely	Possible (50/50)	Likely, no more than one	Very likely/ maybe several	OR Not heard of this incident
☐	☐	☐	☐	☐	☐

3. **Catastrophe: Gulf of Mexico Oil Spill** (otherwise known as BP **Deepwater Horizon** (2010), – 11 fatalities, massive oil spill.
   What is your opinion of the likelihood of a similar catastrophe to occur within the next 50 years?

Not possible	Unlikely	Possible (50/50)	Likely, no more than one	Very likely/ maybe several	OR	Not heard of this incident
☐	☐	☐	☐	☐		☐

4. **Catastrophe: Chernobyl Nuclear Power Plant Radiation Release** (1986), Ukraine – 56 immediate Fatalities, widespread radiation leakage.
What is your opinion of the likelihood of a similar catastrophe to occur within the next 50 years?

Not possible	Unlikely	Possible (50/50)	Likely, no more than one	Very likely/ maybe several	OR	Not heard of this incident
☐	☐	☐	☐	☐		☐

5. **Catastrophe: Sinking of the Costa Concordia** (2012), Italy – 32 Fatalities.
What is your opinion of the likelihood of a similar catastrophe to occur within the next 50 years?

Not possible	Unlikely	Possible (50/50)	Likely, no more than one	Very likely/ maybe several	OR	Not heard of this incident
☐	☐	☐	☐	☐		☐

6. **Catastrophe: Paddington Rail Crash** (otherwise known as Ladbroke Grove) (1999), U.K. – 31 Fatalities.
What is your opinion of the likelihood of a similar catastrophe to occur within the next 50 years?

Not possible	Unlikely	Possible (50/50)	Likely, no more than one	Very likely/ maybe several	OR	Not heard of this incident
☐	☐	☐	☐	☐		☐

7. **Catastrophe: Beirut Warehouse Explosion** (2020), Lebanon – Over 200 Fatalities, widespread damage.
What is your opinion of the likelihood of a similar catastrophe to occur within the next 50 years?

Not possible	Unlikely	Possible (50/50)	Likely, no more than one	Very likely/ maybe several	OR	Not heard of this incident
☐	☐	☐	☐	☐		☐

Suggested Reporting Standard on Catastrophe Prevention

Following the structure of the Global Reporting Initiative standards I suggest the following framework for a standard on Catastrophe Prevention. For the process industries what follows is a standard for process safety management reporting. I have extended the same ideas with suggestions for non-process industries such as transportation.

Some of the topics are included within the GRI standards. For example, the Oil and Gas Sector Standard 11–1 includes Asset Integrity and the Effluents and Waste Standard 306 includes Spills. However, in both cases I consider that these disclosures either need expanding or made more precise. I have also included some sector specific comments as a guide.

Terms are defined in the GRI Glossary.

To be meaningful it is suggested that all disclosures except those marked * are done on a site-by-site basis. These site topics are not amenable to aggregated disclosures in the same way as greenhouse gas emissions.

The author appreciates that work needs to be done to develop these suggestions further. In particular guidance on the thresholds below which reporting is not required need to be developed so that reporting is not unduly burdensome.

CONTENTS OF THE SUGGESTED STANDARD ON CATASTROPHE PREVENTION

Disclosure CP 1 – Organisations Risk Matrix*
Disclosure CP 2 – Hazards
Disclosure CP 3 – Current Residual Risks, including any effect on local communities
Disclosure CP 4 – Planned Risk Reduction Activities
Disclosure CP 5 – Leading and Lagging Indicators of Catastrophic Risk
Disclosure CP 6 – Significant Spills or unplanned Atmospheric Emissions
Disclosure CP 7 – Fires and Explosions
Disclosure CP 8 – Catastrophic Risk Assessment Status

Disclosure CP 9 – Equipment Inspection Status
Disclosure CP 10 – Major Emergency Response Plans
Disclosure CP 11 – Cyber Security Protection
Disclosure CP 12 – Identified Legacy Issues relating to Catastrophic Risk
Disclosure CP 13 – Protection against Natural Catastrophes, Sabotage and Terrorism
Disclosure CP 14 – External and Internal Audit Status

DISCLOSURE CP 1 – ORGANISATION'S RISK MATRIX *

The organisation shall declare their current Risk Matrix which will define the levels of High, Very High and Critical Risk in terms of Probability and Consequence. In the subsequent sections of this disclosure only Hazards and Risks with High, Very High or Critical Potential. Probability and Consequence levels may be defined either quantitatively or qualitatively.

DISCLOSURE CP 2 – HAZARDS

The reporting organisation shall report the following information on significant hazards which have the potential to create a risk with a High, Very High or Critical Potential.

These hazards may be related to materials which are one of the following:

- Toxic
- Flammable
- Potentially explosive
- Sources of radiation

Comment: The disclosure threshold of the quantities qualifying for the Seveso Tier 1 and Tier 2 classifications may be used as a guide.

These risks may also be related to particular processes or facilities such as:

- Exothermic or potentially exothermic chemical reactions
- Combustion, of fuels or waste products
- Open cast and underground mining
- Tailings dams
- Aircraft operations involving landing or taking off
- Transportation by rail or road
- Nuclear reactions either for power generation, research or military purposes.
- Dams for hydroelectricity and / or water storage
- Furnaces with a potential for explosions.

Disclosure of these hazards enables the further disclosures described below.

DISCLOSURE CP 3 – CURRENT RESIDUAL RISKS, INCLUDING ANY EFFECT ON LOCAL COMMUNITIES

The reporting organisation shall then report on High, Very High and Critical Residual Risks, given current methods of risk mitigation. In the process industries these would be the risks as output from risk analyses processes such as HAZOP. For operations which are not Process, similar risk assessment process will have been applied.

If the site has not taken a risk assessment in a form which generates residual risks, this should be explained.

Generic disclosures for similar facilities e.g. gasoline retail stations can be used with caution given that their location with regards to neighbouring communities are different.

DISCLOSURE CP 4 – PLANNED RISK REDUCTION ACTIVITIES

For each of the residual risks disclosed in CP 3, the organisation should disclose planned measures to reduce the risk to a level less than High. The nature of the measure and the planned timing of completion should be indicated.

DISCLOSURE CP 5 – LEADING AND LAGGING INDICATORS OF CATASTROPHIC RISK *

For Process Industries the organisation should report Tier 1 and Tier 2 indicators (e.g. Chapter 31). These are lagging indicators.

Tier 3 leading indicators should also be disclosed in a manner which allows year on year comparability.

For non-process industries a similar set of indicators can be developed and may well be currently in place as described in Chapter 31.

DISCLOSURE CP 6 – SIGNIFICANT SPILLS OR UNPLANNED ATMOSPHERIC EMISSIONS

The organisation shall report the following information:

• Number and Amount of any unplanned discharges of hazardous materials (as disclosed in CP 1).

- Whether these discharges entered the groundwater, river water or atmosphere, or whether they were retained and remediated on-site.

Comment: Note that a discharge which was retained by secondary containment, such as a tank leak which was retained by a bund wall would be a disclosed event, but one which was retained and remediated on site, unless the area within the bund was not impermeable (such as concrete) in which case there is a risk of groundwater contamination.

These unplanned discharges should appear in the numbers shown in CP 5. CP 6 requires more detail.

DISCLOSURE CP 7 – FIRES AND EXPLOSIONS

- Number of fires or explosions at the site, including the amount of flammable and/or toxic material involved.
- Extent of damage to surrounding operations and impact on local communities.
- Fire/explosion contained using on-site emergency response teams, or required external assistance from civic emergency response teams.

These fire/explosions should appear in the numbers shown in CP 5. CP 7 requires more detail.

DISCLOSURE CP 8 – CATASTROPHIC RISK ASSESSMENT STATUS

The reporting organisation should disclose their programme for on-going reassessment of major risks. For process industry organisations this would typically be their programme for process unit HazOps, for non-process industries similar programmes should be in place.

The disclosure would inform:

- Planned risk assessment review frequency
- Any overdue risk assessments with reference to the unit.
- Any units or parts of operations not covered by risk assessment reviews.

DISCLOSURE CP 9 – EQUIPMENT INSPECTION STATUS

The reporting organisation should disclose their inspection regime for equipment.

This disclosure would inform:

- Inspection frequencies for vessels and pipework containing high hazard materials disclosed in CP 2. For air transportation this would include airworthiness certification for aircraft, and IATA recertification for airports.
- Number of overdue inspections and the longest duration of overdue.
- Critical risks relating to equipment nearing end-of-life and planned replacement.

DISCLOSURE CP 10 – MAJOR EMERGENCY RESPONSE PLANS

The reporting organisation should disclose information relating emergency response plans relating to any risks which give a consequence Very Serious of Catastrophic Consequence according to the risk matrix disclosed in CP 1, irrespective of the probability assigned.

- Do major emergency response plans exist for each scenario? If not provide an explanation.
- Have the emergency response plans been shared with all public authorities which are involved in assisting with the response or may be affected by it?
- Has the emergency response plans been tested either by practical drill or table-top exercise, and were public authorities involved?

DISCLOSURE CP 11 – CYBER SECURITY PROTECTION

The reporting organisation should disclose how computer systems which control hazardous operations are protected from Cyber Security threats. Information disclosed would include the existence of good practices such as:

- Firewalls between business and process control computing/control software.
- Prevention of remote access from other organisations to process control computing/control software.
- Control of access to places where the process control computing can be programmed.
- Banning of flash cards/ disabling of USB ports on equipment relating to process control computing.

Comment: Only systems which control hazardous processes or operations need to be disclosed here, not business information computers.

DISCLOSURE CP 12 – IDENTIFIED LEGACY ISSUES RELATING TO CATASTROPHIC RISK

The organisation should disclose any catastrophic risks relating to past operations. Any legacy issues relating to current operations should already be covered in the above disclosures.

Past operations may include:

- Tailings dams containing tailings waste which are no longer in active use.
- Hazardous chemical waste storage areas on site and off site.
- Identified or potential chronic (long term) risks to human health as a result of past emissions from operations now owned by the organisation.

DISCLOSURE CP 13 – PROTECTION AGAINST NATURAL CATASTROPHES, SABOTAGE AND TERRORISM

The reporting organisation should disclose Very Serious of Catastrophic Consequence according to the risk matrix disclosed in CP 1, which could be caused by a natural catastrophe, sabotage or terrorism. The disclosure would include

- The nature of the threat and the major consequences which could result.
- The existence of plans to prevent such a consequence.
- Whether training has been carried out in the conduct of these plans.
- Whether drills have been taken to test the plan and ensure familiarity.

DISCLOSURE CP 14 – EXTERNAL AND INTERNAL AUDIT STATUS

The reporting organisation should disclose their auditing regime for major accident hazards.

This disclosure would state:

- Internal audit frequencies with respect to aspect involved in catastrophe prevention.
- Number of any overdue internal audits and the longest duration of overdue internal audit.
- Number of critical recommendations identified and the timing for correction.

- External audit frequencies (e.g. Seveso Audits) with respect to aspect involved in catastrophe prevention.
- Number of any overdue external audits and the longest duration of overdue internal audit.
- Number of critical recommendations identified from these external audits and the timing for correction.

GENERIC SUGGESTED CONTENT

Based on my experience the following are typical items which might form the content of a disclosure for CP3 CP4, CP12, and CP13

Disclosure Number	Disclosure Heading	Suggested Content
CP 3	Current Residual Risks	Safety critical operations which require human intervention. Operations which have been LOPA/SIL evaluated (see Chapter 27) and require additional automation/upgrade of automation equipment.
		Process equipment with pressure relief valves either relieving to atmosphere or to flare header which is undersized for maximum vapour scenario, especially where a number of vessels are relieving simultaneously.
CP 4	Planned Risk Reduction activities	Might include installation of additional SIL rated instrumentation and automation. Re-evaluation of flare and flare header loading, with upgrades.
CP 10	Major Emergency Response Plans	Potentially could involve additional emergency response equipment such as fixed fire suppression, foam makers, deluge equipment.
		Enhanced training, training materials, changes to emergency response team manning.
CP 13	Protection against natural catastrophes, sabotage and terrorism.	Criteria for plant/ facility shutdown ahead of storm. Security improvements, especially up manning when terrorist threat alert level is raised.

SECTOR SPECIFIC SUGGESTIONS

The following additional suggestions are made with regards to content for each of the above disclosures for particular industry sectors.

OIL AND GAS UPSTREAM

Disclosure Number	Disclosure Heading	Suggested Content
CP 2	Hazards	Generic list of toxic, flammable and potentially explosive materials including hazard of crude oil pollution. Drilling muds etc. Hazards of exploration and production drilling operations and of blowout / loss of well control. Oil sands tailings residues. Fracking and seismicity. Fracking chemicals. Radioactive products of drilling.
CP 3	Current Residual Risks	Dependant on operation. However, past history suggests that there are still significant risks of blowout/loss of well control, and of crude oil spillage. Vapour cloud engulfment with potential explosion on loss/overload of flare. Hydrogen sulphide exposure potential, including consideration of local communities.
CP 4	Planned Risk Reduction activities	Might include double Blow Out Preventers. Extra barriers in oil and gas well operations. Capping device (see OSPRAG Chapter 41)
CP 10	Major Emergency Response Plans	Abandon platform. Blow-out. Oil pollution event.
CP 12	Legacy issues relating to Catastrophic Risk	Previous spills not fully remediated, waste deposits, oil sands tailings ponds. Older extraction facilities in the later stages of life not being renewed/maintained. Oil wells which have been plugged and abandoned.
CP 13	Protection against natural catastrophes, sabotage and terrorism.	Statement with regards to platform/onshore wel operations demanning and shutdown in advance of hurricanes. Filling of atmospheric storage tanks. Skeleton crews to remain in place. Security around terrorism and sabotage vulnerable locations.

OIL AND GAS DOWNSTREAM

Disclosure Number	Disclosure Heading	Suggested Content
CP 2	Hazards	Generic list of toxic, flammable and potentially explosive materials crude oil, LPGs, naphtha, gasoline, diesel and fuel oil. Catalysts including cobalt molybdenum.

(Continued)

Disclosure Number	Disclosure Heading	Suggested Content
		Distillation operations of hydrocarbons at atmospheric, vacuum and elevated pressures. Potential for vapour cloud explosions from process equipment and boiling liquid expanding vapour explosion from LPG storage spheres. Materials above their auto ignition temperatures.
CP 3	Current Residual Risks	See Generic
CP 4	Planned Risk Reduction activities	See Generic
CP 10	Major Emergency Response Plans	See Generic
CP 12	Legacy issues relating to Catastrophic Risk	Plant which has been shutdown and abandoned in place – potential for structural collapse.
CP 13	Protection against natural catastrophes, sabotage and terrorism.	Filling of atmospheric storage tanks ahead of storm. Other means to stabilise empty atmospheric storage.

CHEMICAL SECTOR

Disclosure Number	Disclosure Heading	Suggested Content
CP 2	Hazards	Plant specific – chemical sector is too varied to have generic lists.
CP 3	Current Residual Risks	
CP 4	Planned Risk Reduction activities	
CP 10	Major Emergency Response Plans	
CP 12	Legacy issues relating to Catastrophic Risk	Waste and waste dumps from previous operations. Potential health effects from previous operations both from emissions and products. Groundwater contamination.
CP 13	Protection against natural catastrophes, sabotage and terrorism.	Procedure for assigning a 'ride-out' crew. Emergency cooling/heating which is sufficiently robust to maintain continuity during the storm.

MINING SECTOR

Disclosure Number	Disclosure Heading	Suggested Content
CP 2	Hazards	Include gases potentially released from the rock and ore as it is mined. For coal mines this will include methane and other gases (see Chapter 2.4)
CP 3	Current Residual Risks	Rock fall Underground fire and/or explosion Tailing dams failure
CP 4	Planned Risk Reduction activities	Extension of surveying process Increases in ventilation provision including standby.
CP 10	Major Emergency Response Plans	For rescuing from underground mine in the event of major rock fall, fire or explosion. To alert and rescue inhabitants of villages downstream of a tailings dam.
CP 12	Legacy issues relating to Catastrophic Risk	Abandoned tailings storage and tailings dams. Method for surveying old facilities should be specified, together with identification if emergency response plans are relevant to these abandoned facilities.
CP 13	Protection against natural catastrophes, sabotage and terrorism.	Criteria to de-man the man in the event of high water levels/pump failures. Procedures if a tailings dam is in danger of overtopping due to abnormal rainfall.

TRANSPORTATION SECTOR – AIR

Disclosure Number	Disclosure Heading	Suggested Content
CP 2	Hazards	Fast moving aircraft. Kerosene and other hydrocarbon fuels. Airport fuel storages.
CP 3	Current Residual Risks	Aircraft collisions whilst on the ground Aircraft failure in the air.
CP 4	Planned Risk Reduction activities	
CP 10	Major Emergency Response Plans	For aircraft emergency landing For airport evacuation/fire
CP 12	Legacy issues relating to Catastrophic Risk	–

(Continued)

Disclosure Number	Disclosure Heading	Suggested Content
CP 13	Protection against natural catastrophes, sabotage and terrorism.	Pilot psychological monitoring. Two in cockpit policy. Airport security personnel and detection devices.

TRANSPORTATION SECTOR – RAIL

Disclosure Number	Disclosure Heading	Suggested Content
CP 2	Hazards	Fast moving locomotives and rolling stock Hazardous goods in transit including materials often carried such as LPG, crude oil, ammonia
CP 3	Current Residual Risks	Derailment Train Collision
CP 4	Planned Risk Reduction activities	Installation of TPWS/ATP (see Chapter 9) Derailment risk reduction such as installation of check rails.
CP 10	Major Emergency Response Plans	Passenger train collision/derailment at any point including tunnels Freight train collision/derailment at any point including tunnels with emergency provision for hazardous goods
CP 12	Legacy issues relating to Catastrophic Risk	Disused railway lines – potential for collapse of embankments.
CP 13	Protection against natural catastrophes, sabotage and terrorism.	Security measures for railway lines, sidings, terminals.

NUCLEAR POWER SECTOR

Disclosure Number	Disclosure Heading	Suggested Content
CP 2	Hazards	Radioactive materials
CP 3	Current Residual Risks	Reactor meltdown or partial meltdown Release of radioactive materials from fuel rod handling. Release from spend fuel rod storage

(Continued)

Disclosure Number	Disclosure Heading	Suggested Content
CP 4	Planned Risk Reduction activities	Decoupling of potential common mode sources of failure. Additional sources of emergency power.
CP 10	Major Emergency Response Plans	Local community evacuation/shelter in place plans. Communications with potentially affected communities. Alignment with local authorities. Training and drills.
CP 12	Legacy issues relating to Catastrophic Risk	Spent fuel. Abandoned facilities still containing radiations sources.
CP 13	Protection against natural catastrophes, sabotage and terrorism.	Tsunami and flooding – prevention of common mode failures e.g. emergency electricity. Consideration of advanced terrorism threats such as aircraft impact.

PROCESS INDUSTRY INFRASTRUCTURE SECTOR (TERMINALS, PIPELINES, WAREHOUSES)

Disclosure Number	Disclosure Heading	Suggested Content
CP 2	Hazards	For oil and gas generic hazards may be used. For warehouses materials may be diverse and need to be warehouse specific. Threshold guidance for disclosure may be Seveso Tier 2 quantities.
CP 3	Current Residual Risks	Overfill of storage tanks. Pipeline leakage especially from underground corrosion. Pipeline rupture due to collision or excavation work. Warehouse fire/explosion.
CP 4	Planned Risk Reduction activities	Improved instrumentation of storage tank levels with independent hi hi level to meet SIL assessment. Additional pipeline protection against collision. Enhanced communication with owners of properties along route of pipeline with regards to excavation.

(Continued)

Disclosure Number	Disclosure Heading	Suggested Content
		Improved cathodic protection of underground pipelines.
		Further fire/smoke detection and automatic deluge for warehouse facilities.
CP 10	Major Emergency Response Plans	Response plans for pipeline rupture applicable along pipeline length.
		Response to warehouse fire given variable contents of warehouse. Consideration of toxicity of smoke from burning contents – liaison with hospitals and medical profession.
CP 12	Legacy issues relating to Catastrophic Risk	Abandoned stock in warehouse, especially where ownership is in doubt.
CP 13	Protection against natural catastrophes, sabotage and terrorism.	Protection and security of remote, unmanned facilities.

EXPLOSIVES MANUFACTURE, MILITARY AND DEFENCE

Disclosure Number	Disclosure Heading	Suggested Content
CP 2	Hazards	Explosives. Nuclear materials. Seveso Tier 2 thresholds might be useful guidance.
CP 3	Current Residual Risks	Premature explosions. Release of nuclear materials.
CP 4	Planned Risk Reduction activities	–
CP 10	Major Emergency Response Plans	Local community evacuation/shelter in place plans. Communications with potentially affected communities. Alignment with local authorities. Training and drills.
CP 12	Legacy issues relating to Catastrophic Risk	Abandoned facilities.
CP 13	Protection against natural catastrophes, sabotage and terrorism.	Security against terrorist attack.

Example Responses to Suggested Reporting Standard on Catastrophe Prevention

Using the suggested Reporting standard on Catastrophe Prevention, I have suggested below some answers which might be given by the management of our hypothetical fuels terminal.

HYPOTHETICAL FUELS TERMINAL

Disclosure CP 1 – Organisation's Risk Matrix *

'The organisation shall declare their current Risk Matrix which will define the levels of High, Very High and Critical Risk in terms of Probability and Consequence. In the subsequent sections of this disclosure only Hazards and Risks with High, Very High or Critical Potential. Probability and Consequence levels may be defined either quantitatively or qualitatively.'

The fuels terminal has adopted the same risk matrix as used in Chapter 31. (Table F1 and Table F2)

Probability	Consequence				
	Very Minor	Minor	Serious	Very Serious	Catastrophic
Frequent	3	2	2	1	1
Occasional	3	3	2	2	1
Seldom	4	3	3	2	2
Unlikely	4	4	3	3	2
Very Unlikely	4	4	4	3	3

Figure F1 Fuels Terminal – Risk Matrix.

Table F1 Probability Classifications in Risk Matrices

Probability Classification	Frequency
Frequently	Once per annum
Occasional	Once every ten years
Seldom	Once every 100 years
Unlikely	Once every 1000 years
Very Unlikely	Once every 10000 years

Table F2 Consequences

Consequence Classification	Personal Safety	Environmental	Financial Impact
Catastrophic	Multiple deaths	Serious ecological with long term pollution and/or health risks to local community	Eur 100 million
Very Serious	One fatality or permanently disabling injuries	Serious excursion to environmental permit. Corrective measures outside of the site necessary	Eur 20–100 million
Serious	Serious injury, work disability, lost workday case	Serious excursion to environmental permit. Corrective measures outside limited to actions within the site boundary	Eur 5–20 million
Minor	Temporary limited work including injuries due to irritation	Slight permit exceedance. No environmental damage outside of site boundaries	Eur 100,000 to 500,000
Very Minor	Medical Treatment, first aid, short illness	Limited damage or emission within stie boundaries	Eur 10,000 to 100,000

For this disclosure only those risks rated high, very high, or critical are relevance.

Disclosure CP 2 – Hazards

'The reporting organisation shall report the following information on significant hazards which have the potential to create a risk with a High, Very High or Critical Potential'.

The terminal stores flammable and highly flammable materials as shown in Table F3.

Table F3 Hazardous Materials

Material	Quantity kgs
Gasoline	10,000
Diesel	10,000
Heavy Fuel Oil	10,000
LPG	10,000

There are hazardous operations on site involving:

- Locomotives and rolling stock movement
- Filling of storage tanks
- Unloading of rail tank cars with the above materials
- Loading of road tankers with the above materials
- Filling of bottles of LPG

Disclosure CP 3 – Current Residual Risks, including any effect on local communities

'*The reporting organisation shall then report on High, Very High and Critical Residual Risks, given current methods of risk mitigation*'.

1. There is a high risk of overfilling any of the storage tanks since there is only one level instrument on each tank.
2. There is a high risk of fire since vapours from storage tanks being filled, and road tankers being filled are vented to atmosphere.
3. There is a very high risk of explosion in the bottling plant since there are no means of detecting LPG discharge into the bottling plant from faulty bottling plant valves.

These risks could impact the neighbouring village school and supermarket.

Disclosure CP 4 – Planned Risk Reduction Activities

'For each of the residual risks disclosed in CP 3 the organisation should disclose planned measures to reduce the risk to a level less than High. The nature of the measure and the planned timing of completion should be indicated'.

1. Independent high-high level instruments for the storage tanks are planned to be installed in the next three years.
2. A Vapour Recovery Unit is planned for recovering vapours emitted during tank car filling and storage tank filling. This is recognised as

urgent for legal compliance emissions control and for risk reduction. A unit is planned for commissioning next year.

3. The bottling plant is to be fitted with flammable gas detectors, and a remote isolation valve which can be actuated from the control room and from a switch immediately outside the bottling plant. Installation is planned within 4 years.

Disclosure CP 5 – Leading and Lagging Indicators of Catastrophic Risk *

'For Process Industries the organisation should report Tier 1 and Tier 2 indicators (e.g. Chapter 31). These are lagging indicators'.

Tier 3 leading indicators should also be disclosed in a manner which allows year on year comparability.' (Table F4)

Table F4 Leading and Lagging Indicators

Indicator	2020	2021	2022
Tier 1	0	0	0
Tier 2	1	0	2
Tier 3			
Overdue Pressel Vessel Inspections	1	0	0
Overdue Pipework Inspections	0	1	4
Overdue Instrument and Interlock Testing	2	1	1

Disclosure CP 6 – Significant Spills or unplanned Atmospheric Emissions

'The organisation shall report the following information:

- Number and Amount of any unplanned discharges of hazardous materials (as disclosed in CP 1).
- Whether these discharges entered the groundwater, river water or atmosphere, or whether they were retained and remediated on-site.'

The Tier 2 incidents involved

1. 2020 Unplanned release of over around 300 kgs of LPG from the bottling plant.
2. 2022 Leak of 400 kgs of gasoline following a leak at the outlet pipe of the gasoline tank due to a corroded pipe. All of the material was retained within the bund wall of the storage tank and remediated.

Disclosure CP 7 – Fires and Explosions

' *Number of fires or explosions at the site, including the amount of flammable and/or toxic material involved.*

Extent of damage to surrounding operations and impact on local communities.'

Fire/explosion contained using on-site emergency response teams, or required external assistance from civic emergency response teams'

2022 A fire at the rail tank car unloading pump causing USD 20,000 of damage. Fire occurred due to gasoline spillage from a faulty pump seal. The fire was extinguished by the on-site emergency response team. The local civil defence team was called but did not need to take any action as the fire was extinguished by the time they arrived.

Some smoke drifted in the direction of the local school who were advised to close all windows.

Disclosure CP 8 – Catastrophic Risk Assessment Status

'*The reporting organisation should disclose their programme for on-going reassessment of major risks. For process industry organisations this would typically be their programme for process unit HazOps, for non-process industries similar programmes should be in place.*

The disclosure would inform:

- *Planned risk assessment review frequency*
- *Any overdue risk assessments with reference to the unit.*
- *Any units or parts of operations not covered by risk assessment reviews.*

The site is divided into three units each with the HAZOP review frequency as shown in Table F5.

Table F5 HAZOP Review Frequency

Unit	Review Frequency	Year of Last HAZOP
Rail operations and rail tank car unloading	5 years	2020
Storage Tank operations	5 years	2015
Road tanker filling including additive operations	5 years	2021
Bottling plant	5 years	2022

Disclosure CP 9 – Equipment Inspection Status

'The reporting organisation should disclose their inspection regime for equipment.
   This disclosure would inform:

- Inspection frequencies for vessels and pipework containing high hazard materials disclosed in CP 2. For air transportation this would include airworthiness certification for aircraft, and IATA recertification for airports.
- Number of overdue inspections and the longest duration of overdue.
- Critical risks relating to equipment nearing end-of-life and planned replacement.'

Inspection Frequencies and status is as shown in Table (Table F6)

Table F6 Inspection Status

Unit	Inspection Frequency	Overdue (end 2022)
Pressure Vessels	10 years	OK
Storage Tanks	10 years	OK
Pipework Inspections	3 years	4 overdue
Interlock and Instrumentation	3 years	1 overdue

Disclosure CP 10 – Major Emergency Response Plans

'The reporting organisation should disclose information relating emergency response plans relating to any risks which give a consequence Very Serious of Catastrophic Consequence according to the risk matrix disclosed in CP 1, irrespective of the probability assigned. (Table F7)

- Do major emergency response plans exist for each scenario. If not provide an explanation?
- Have the emergency response plans been shared with all public authorities which are involved in assisting with the response or may be affected by it?
- Has the emergency response plans been tested either by practical drill or table-top exercise, and were public authorities involved?'

Table F7 Emergency Response Plan Status

Risk/Scenario	Emergency Response Plan in Place	Reviewed with Local Authority/Fire Brigade	Drill Completed?
LPG Tank BLEVE	No	No	No
Fire at Storage Tanks	Yes	Yes	2021 with local fire brigade
Fire at rail unloading sidings	Yes	Yes	2021 own fire brigade
Fire at Bottling Plant	Yes	Yes	2021 own fire brigade
Fire at road loading facilities	Yes	Yes	2020 own fire brigade

Note BLEVE – Boiling Liquid Expanding Vapour Explosion.

The likelihood of an LPG tank BLEVE is seen as Very Unlikely, and an ER has not been prepared.

Disclosure CP 11 – Cyber Security Protection

'*The reporting organisation should disclose how computer systems which control hazardous operations are protected from Cyber Security threats*'.

1. There are firewalls between business and process control computing/control software.
2. No remote accessory to the fuel terminals process control software is necessary and it is not permitted.
3. There is a separate room where the terminals process control computers are programmed. Access is only permitted under authority of the terminal manager.

Disclosure CP 12 – Identified Legacy Issues relating to Catastrophic Risk

'*The organisation should disclose any catastrophic risks relating to past operations. Any legacy issues relating to current operations should already be covered in the above disclosures.*'

- In the area to the south of the storage tanks, a former tank was dismantled and removed in 2000. It is understood that there may be some groundwater contamination in this location.

Disclosure CP 13 – Protection against Natural Catastrophes, Sabotage and Terrorism

'The reporting organisation should disclose Very Serious of Catastrophic Consequence according to the risk matrix disclosed in CP 1, which could be caused by a natural catastrophe, sabotage or terrorism'.

- The terminal is not in a location subject to hurricanes.
- The site is contained within a continuous 2 metre fence.
- There are no security cameras.
- Some flooding has occurred in the past around the unloading pumps, but this has not cause damage.

Disclosure CP 14 – External and Internal Audit Status

' The reporting organisation should disclose their auditing regime for major accident hazards.
 This disclosure would state:

- *Internal audit frequencies with respect to aspect involved in catastrophe prevention.*
- *Number of any overdue internal audits and the longest duration of overdue internal audit.*
- *Number of critical recommendations identified and the timing for correction.*
- *External audit frequencies (e.g. Seveso Audits) with respect to aspect involved in catastrophe prevention.*
- *Number of any overdue external audits and the longest duration of overdue internal audit.*
- *Number of critical recommendations identified from these external audits and the timing for correction'.*

The terminal is not subject to any internal or external audits. It is a standalone fuels terminal.

It is understood that a Seveso audit is due, but none has so far taken place.

Question to readers. If you had a stakeholder interest in this terminal, e.g. a teacher in the school, what questions would this statement lead you to ask?

References

Abdushkour, H. et al. 2018. *Comparative Review of Collision Avoidance Systems in Maritime and Aviation.* University of Strathclyde.

Abdussamie, N. et al. 2018. "Risk Assessment of LNG and FLNG Vessels during Manoeuvring in Open Sea." *Journal of Ocean Engineering and Science* 2018 (2018): 56–66.

Aircraft Accidents Investigation Branch. 2011. "Aircraft Accident Report 2/2011 – Aerospatiale (Eurocopter) AS332 L2 Super Puma, G-REDL, 1 April 2009."

Aircraft Accidents Investigation Branch. 2012. "Report on the Accidents to Eurocopter EC225 LP Super Puma G-REDW 34 nm East of Aberdeen, Scotland on 10 May 2012 and G-CHCN 32 nm Southwest of Sumburgh, Shetland Islands on 22 October 2012."

Aircraft Accident Investigation Branch. 2017. "Aircraft Accident Report 1/2017 Report on the Accident the Hawker Hunter T7, G-BXFI near Shoreham Airport on 22 August 2015."

Aircraft Accidents Investigation Branch. 2020. "Report on the Accident to Piper PA-46-310P Malibu, N264DB AAR 1/2020," 20 January 2019.

Airline Ratings. 2022. "Airline Ratings 2022." Accessed 2 August 2022. https://www.airlineratings.com/

Allianz Global Corporate and Specialty. 2020. "Safety and Shipping Review 2020." Accessed 1 August 2022. https://www.iims.org.uk/wp-content/uploads/2020/07/AGCS-Safety-Shipping-Review-2020.pdf

Aviation Safety Management Software. 2022. "List of Key Performance Indicators for Airports/Aerodromes." Accessed 1 August 2022. https://asmpro.com/Modules/SafetyAssurance/ListofAirportKeyPerformanceIndicators.aspx

Atomic Weapons Establishment. "AWE." Accessed 5 August 2022. https://www.awe.co.uk.

Atomic Weapons Establishment. 2019. "AWE Aldermaston 'Consequences Report." British Crown Copyright. 2019.

Atomic Weapons Establishment. 2021. "Minutes of the 101st AWE local Liaison Committee Meeting, Wednesday 17 March 2021." Accessed February 2022 https://democracy.reading.gov.uk/documents/s17607/AWE%20Local%20Liaison%20Committee%20minutes.pdf

Atomic Energy Society of Japan 2014. *The Fukushima Daiichi Nuclear Accident – Final Report of the AESJ Investigation Committee.* Springer, 2015.

American Petroleum Institute. 2016. "Steels for Hydrogen Service at Elevated Temperatures and Pressures in Petroleum Refineries and Petrochemical Plants." API Recommended Practice 941. 2016.

Aviation Safety Network. 2022. "Airline Safety." Accessed 5 August 2022. https://aviation-safety.net

Azam, Shahid and Li, Qiren. 2010. "Tailings Dam Failures: A Review of the Last One Hundred Years." *Geotechnical News*, December 2010.

Bahr, N.J. 2015. System *Safety Engineering and Risk Assessment – A Practical Approach*. 2nd Edition: Taylor and Francis, 2017.

Basingstoke Gazette. 4 February 2012. "Fires, False Alarms and Chemical Leaks Reported at AWE Aldermaston." Accessed 5 August 2022. https://www.basingstokegazette.co.uk/news/9513089.fires-false-alarms-and-chemical-leaks-reported-at-awe-aldermaston/

Basingstoke Observer. 18 July 2016. "Safety Fears under the Spotlight again at AWE." http://www.basingstokeobserver.co.uk/safety-fears-under-the-spotlight-again-at-awe

BBC News. 24 February 2014. "AWE Aldermaston Nuclear Waste Deadline Expires." Accessed 5 August 2022. https://www.bbc.co.uk/news/uk-england-berkshire-26324595

BBC News. 18 June 2021. "Bosley Mill: Fine for Owner over Explosion which Killed Four People." Accessed March 2022. https://www.bbc.co.uk/news/uk-england-stoke-staffordshire-57528427

Bellucci, M. and Manetti, G. 2019. *Stakeholder Engagement and Sustainability Reporting*. Routledge.

BP. 2021. "BP Annual Report and Form 20-F."

BP. 2020. "BP Sustainability Report 2020."

Bretherick. 2006. *Bretherick's Handbook of Reactive Chemical Hazards*. Elsevier.

British Geological Survey. 2020. *Directory of Mines and Quarries*. UK research and Innovation.

Browning, Jackson. 1993. "Union Carbide: Disaster at Bhopal." Reprinted from Crisis Response: Inside Stories on Managing under Siege. Gottschalk J, Visible Ink Press.

Bureau of Safety and Environmental Enforcement. 2013. "Summary of Blowout Preventer (BOP) Failure Mode Effect Criticality Analyses."

Carnegie Institute for International Peace. 2012. "Why Fukushima Was Preventable." Accessed July 2022. https://carnegieendowment.org/2012/03/06/why-fukushima-was-preventable-pub-47361.

Casey, R. 2019. "Limitations and Misuse of LOPA." *Loss Prevention Bulletin* 265 (February 2019): 13–16. Institution of Chemical Engineers.

Centre for Chemical Process Safety. 2018. *Process Safety Metrics: Guide for Selecting Leading and Lagging Metrics*. AIChemE.

Centre for Chemical Process Safety and the Energy Institute. 2018. *Bow Ties in Risk Management – A Concept Book for Process Safety*. Wiley.

CGE Risk. "Bow Tie Instruction Videos." Accessed March 2022. https://www.youtube.com/watch?v=XvGALc1nSRs&t=76s

Civil Aviation Authority. 2016. "UK Civil Air Display Review: Actions that Impact on UK Civil Air Displays in 2016." CAP 1371. CAA.

Congressional Research Service. October 2019. "Attacks on Saudi Oil Facilities:

Effects and Responses." Accessed July 2022. https://crsreports.congress.gov/product/pdf/IN/IN11173

Davies, A. and Adams, C. 2015. "To Err Is Human: Human Error and Workplace Safety." Safety and Health Practitioner, 13 July 2015.

Det Norske Veritas. 2011. "Assessment of Freight Train Derailment Risk Reduction Measures: Part A Final Report – Report for the European Railway Agency." Report No: BA000777/01.

Distillery Trail. 2016. "What Is the Angel's Share." Accessed June 2021. https://www.distillerytrail.com/blog/what-is-the-angels-share/

DSS+ "Elements of the DSS Process Safety Management System." Accessed 5 July 2022 https://www.dsslearning.com/process-safety-management-system/

DSS+. "The DSS Bradley Curve." Accessed 5 August 2022. https://www.consultdss.com/bradley-curve/

DSS+. "The DSS Approach to Contractor Management." Accessed 5 March 2022. https://www.consultdss.com/contractor-management/

DSS+. "Incident Investigations: Getting Started." Accessed January 2022. https://www.dsslearning.com/incident-investigation-getting-started/INV001/

DSS+. "dss⁺ STOP® Improving Behavior through Observation and Communication." Accessed 5 August 2022. https://www.dsslearning.com/dss-stop.

Dust Safety Science. "A Dark Day in Cheshire: Bosley Mill Explosion." https://dustsafetyscience.com/a-dark-day-in-cheshire-bosley-mill-explosion/

Dustwatch. "Dust Particles and Lower Explosive Limits (LEL)." Accessed March 2022 https://www.dustwatch.com/dust-particles-lower-explosive-limits/

Energy Storage News. 29 July 2020. "Arizona Battery Fire's Lessons Can Be Learned by Industry to Prevent Further Incidents, DNV GL Says." Accessed 5 August 2022. https://www.energy-storage.news/arizona-battery-fires-lessons-can-be-learned-by-industry-to-prevent-further-incidents-dnv-gl-says/

Engineering Equipment and Materials Users' Association. 2013. "Alarm Systems: A Guide to Design, Management and Procurement." EEMUA Publication 191.

Engineering Toolbox. "Flanges – API vs ASME/ANSI." Accessed March 2022. https://www.engineeringtoolbox.com/api-flanges-d_746.htm

Entirely Safe. 4 March 2022. "Oil Rig Fire in the Caspian Sea." Accessed February 2022. http://entirelysafe.com/incident/oil-rig-fire-in-the-caspian-sea/

European Agency for Railways. 2020. "Report on Railway Safety and Interoperability in the EU 2020."

European Union Agency for Railways. 2020. "Report on Railway Safety and Interoperability in the EU, 2020." Publications Office, https://data.europa.eu/doi/10.2821/30980

European Commission Disaster Risk Management Knowledge Centre. "Technological Risk: Natech" Understanding Disaster Risk: Hazard Related Risk Issues – Chapter 3 Section 4: https://drmkc.jrc.ec.europa.eu/portals/0/Knowledge/ScienceforDRM/ch03_s04/ch03_s04_subch0314.pdf.

European Center for Constitutional and Human Rights. October 2019. "The Safety Business: TUV SUDS Role in the Brumadinho Dam Failure in Brazil." Accessed 5 August 2022. https://www.ecchr.eu/fileadmin/Fallbeschreibungen/Case_Report_Brumadinho_ECCHR_MISEREOR_20191014_EN.pdf.

Federal Aviation Authority. 2022. "Events with Smoke, Fire, Extreme Heat or Explosion Involving Lithium Batteries." Accessed 5 August 2022. https://www.faa.gov/hazmat/resources/lithium_batteries/media/Battery_incident_chart.pdf

Fire Safety Journal. 2013. "An Integrated Approach for Fire and Explosion Consequence Modelling – Dadashzadeh et al." *Fire Safety Journal* 61 (2013): 324–336.

Gardner, D. 2008. Risk: *Why We Fear the Things We Shouldn't – and Put Ourselves in Greater Danger*. Toronto: Emblem.

Gates, W. 2021. *How to Avoid a Climate Disaster – The Solutions We Have and the Breakthroughs We Need*. Allen Lane.

Gerstein, M. and Ellsberg, M. 2008. *Flirting with Disaster – Why Accidents Are Rarely Accidental*. Sterling Publishing Co., Inc.

Getreading. 30 July 2014. "AWE Aldermaston Remains on Nuclear Regulator 'Special Measures' List" – Accessed July 2022. https://www.getreading.co.uk/news/awe-aldermaston-remains-nuclear-regulator-7535041

Global Reporting Initiative. 2021. *GRI 1 – Foundation*. GRI.

Global Reporting Initiative. 2021. *GRI 3 – Material Topics*. GRI.

Global Reporting Initiative. 2021. *GRI 11 – Oil and Gas Sector*. GRI.

Global Reporting Initiative. 2018. *GRI 303 – Water and Effluents*. GRI.

Global Reporting Initiative. 2016. *GRI 305 – Emissions*. GRI.

Global Reporting Initiative. 2018. *GRI 403 – Occupational Health and Safety*. GRI.

Global Reporting Initiative. 2016. *GRI 413 – Local Communities*. GRI.

GlobalSecurity.Org. n.d. "AWE Aldermaston – United Kingdom Nuclear Forces." Accessed 5 August 2022. https://www.globalsecurity.org/wmd/world/uk/awe_aldermaston.htm

Goose, M. 2010. "Representing Societal Risk as an FN Curve and Calculating the Expectation Value." Accessed January 2022. http://www.alarp.plus.com/SREV.pdf

Gov.uk. 2022. "Derailment of a Passenger Train at Carmont on 12 August 2020." Accessed 5 August 2022. https://www.gov.uk/government/news/report-022022-derailment-of-a-passenger-train-at-carmont

Grenfell Tower Inquiry 2019. "Phase 1 Report Overview." Accessed August 2022. https://www.grenfelltowerinquiry.org.uk/phase-1-report

Grenfell Tower Public Inquiry Update. "Meeting 18 December 2019 – Item 19/59."

Haddon-Cave, C. 2009. "The Nimrod Review." Crown Copyright.

Hagley Museum. "Hagley Museum and Library." Accessed January 2022. https://www.hagley.org

Hazardex. 3 December 2021. "Methane Data Falsified at Coal Mine where Deadly Blast Killed 51, Putin says." Accessed February 2022. https://www.hazardexonthenet.net/article/188432/Methane-data-falsified-at-coal-mine-where-deadly-blast-killed-51--Putin-says.aspx

Health and Safety Commission. 2000. "The Ladbroke Grove Rail Inquiry – Part 1 Report."

Health and Safety Commission. 2001. "The Ladbroke Grove Rail Inquiry – Part 2 Report."

Health and Safety Executive. 2001. "Best Practice for Risk-based Inspection as a Part of Plant Integrity Management." CRR 363/2001.

Health and Safety Executive. 2011. *Buncefield: Why Did It Happen?*. Competent Authority Strategic Management Group.

Health and Safety Executive. 2011. "A Guide to the Control of Major Accident Hazards Regulations (COMAH) 2015." L111.

Health and Safety Executive. 2002. "DSEAR Regulations." Accessed January 2022. https://www.hse.gov.uk/fireandexplosion/dsear-regulations.htm

Health and Safety Executive. 2009. "A Review of Layers of Protection Analysis (LOPA) Analyses of Overfill of Fuel Storage Tanks." Research Report RR716.

Health and Safety Executive. 2021. "Non-fatal Injuries at Work in Great Britain." Accessed July 2022. https://www.hse.gov.uk/statistics/causinj/index.htm

Health and Safety Executive. 2001. "Reducing Risks, Protecting People."

Health and Safety Executive. 2022. "Fatal Injuries in Agriculture, Forestry and Fishing in Great Britain 2020/21." Accessed July 2022. https://www.hse.gov.uk/agriculture/resources/fatal.htm

Health and Safety Executive. 2011. "High Reliability Organisations – A Review of the Literature." Research Report RR899.

Health and Safety Executive. 2003. "'Transport Fatal Accidents and FN-Curves: 1967–2001." Research Report 073.

Health and Safety Executive. 1988. "The Tolerability of Risk from Nuclear Power Stations."

Health and Safety Executive. 1999. "Reducing Error and Influencing Behaviour." HSG 48.

Health and Safety Executive. 2013. "Reporting Accidents and Incidents at Work." INDG453.

Health and Safety Executive. 1999. "The Management of Health and Safety at Work Regulations 1999." Accessed 5 August 2022. https://www.legislation.gov.uk/uksi/1999/3242/contents/made

Health and Safety Executive. 1999. "The Cost to Britain of Workplace Accident and Work-related Ill Health in 1995/96." HSG101.

Health and Safety Laboratory. 2003. "Potential Explosion Hazards due to Evaporating Ethanol in Whisky Distilleries." HSL/04/04.

Health and Safety Laboratory. 2004. "Human Vulnerability to Thermal Radiation Offshore." HSL/2003/08.

Hofstede, Geert. 1980. *Culture's Consequences: International Differences in Work-Related Values.* Beverly Hills, CA: Sage, 2001.

Hofstede Insights. "Country Comparison." Accessed 5 August 2022. https://www.hofstede-insights.com/country-comparison/

Hopkins, Andrew. 2000. *Lessons from Longford. The Esso Gas Plant Explosion.* CCH Australia.

Hopkins, Andrew. 2009. *Failure to Learn: The BP Texas City Refinery Disaster.* CCH Australia.

Hopkins, Andrew. 2012. *Disastrous Decisions: The Human and Organisational Causes of the Gulf of Mexico Blowout.* CCH Australia.

Hopkins, Andrew. 2021. *A Practical Guide to Becoming a "High Reliability" Organisation.* Australian Institute of Health and Safety.

Hughes, T. 2007. *Workplace Risk Assessment: A Pilot Study on Attitudes to Risk Assessment and Its Effectiveness in Countries with Differing Cultural and Legislative Backgrounds.* MSc Research Paper, University of Portsmouth Occupational Health and Safety Management.

Hu, Y. et al. 2013. "A Case Study of Electrostatic Accidents in the Process of Oil-gas Storage and Transportation." Journal of Physics: Conference Series,

Volume 418, Issue 1, article id. 012037 (2013). https://ui.adsabs.harvard.edu/link_gateway/2013JPhCS.418a2037H/doi:10.1088/1742-6596/418/1/012037

HR Wallingford. 2019. "A Review of the Risks Posed by the Failure of Tailings Dams." Accessed January 2022. https://damsat.com/wp-content/uploads/2019/01/BE-090-Tailings-dams-R1-Secured.pdf

IMCA Safety Flash 19/19 concerning incinerator safety. https://imcaweb.blob.core.windows.net/wp-uploads/2020/06/IMCASF-19-19.pdf. Downloaded 2022.

International Association of Oil and Gas Producers. 2008. "Asset Integrity – the Key to Managing Major Incident Risks." Report 415.

International Association of Oil and Gas Producers. 2016. "Standardisation of Barrier Definitions." Report 544.

International Atomic Energy Agency. 1992. "INSAG-7 The Chernobyl Accident: Updating of INSAG-1 A Report by the International Nuclear Safety Advisory Group." Safety Series No. 75-INSAG-7.

International Electrotechnical Commission. 2014. "Management of Alarms for the Process Industries." IEC-62682.

International Society for Automation. "Management of Alarm Systems for Process Industries." ANSI/ISA-18.2-2009.

IOSH Magazine. 7 October 2021. "Potato Factory Fined £300,000 Following a Catalogue of Errors." Accessed February 2022. https://www.ioshmagazine.com/2021/10/07/potato-factory-fined-ps300000-following-catalogue-errors

IOSH Magazine. 3 December 2021. "Car Salvage Firm Boss Jailed for Ignoring HSE Notices." Accessed 5 August 2022. https://www.ioshmagazine.com/2021/12/03/car-salvage-firm-boss-jailed-ignoring-hse-notices

IOSH Magazine. 2021 "First Line of Defence." Nov/Dec 2021.

IOSH Magazine. 2021. "Strengthen your Safety Culture." Sep/Oct 2021.

IOSH Magazine. "Never Again? – Grenfell Tower." May/June 2022.

John W. Kennedy Company Blog. 2018 "Fire at Gas Stations: Some Facts, Some Statistics and Some Prevention Tips!" Access February 2022. https://jwkblog.com/wordpress/fire-at-gas-stations-some-facts-some-statistics-and-some-prevention-tips/

Karthikeyan, B. 2019, March. "Risk Based Process Safety – What Happens When Organizations Don't Follow Risk-based Process Safety Guidelines." *Journal of Chemical Engineering Progress*. AIChE.

Kinnane, Adrian. 2002. *Du Pont: From the Banks of the Brandywine to the Miracles of Science*. E.I. Du Pont de Nemours and Company.

Kernick, Gill. 2022. Catastrophe *and Systemic Change – Learnings for the Grenfell Tower Fire and Other Disasters*. London Publishing Partnership.

Kirwan, Barry. 1994. *A Guide to Practical Human Reliability Assessment*. CRC Press.

Kletz, Trevor. 2009. *What Went Wrong?: Case Histories of Process Plant Disasters and How They Could Have Been Avoided*. Elsevier.

Landucci et al. 2011. "The Viareggio LPG Accident: Lessons Learnt." *Journal of Loss Prevention in the Process Industries* 24, no. 4 (July 2011): 466–476. Elsevier.

Leduc M. 2004. "Sudden Drydock Flooding. A Tragic Event in Dubai, March 2002." Accessed 5 August 2022. http://www.dieselduck.info/images/drydock/index.htm

Le, Shuttle, 2021. "Eurotunnel's Dangerous Goods Guide 2021." Accessed January 2022. https://www.eurotunnelfreight.com/uploadedfiles/xnt/adr_2021_uk.pdf

Lees, F.P. 2014. *Process Safety Essentials – Hazard Identification, Assessment and Control*. Butterworth Heinemann.

Lloyd's Markets Association. 2016. "Major Causes of Losses in the Onshore Oil, Gas and Petrochemical Industries."

Marine Accident Investigation Branch Report. 2015. "Major Leak of Liquified Propane from the Liquid Gas Carrier Ennerdale whilst alongside Fawley Marine Terminal 17 October 2006." MAIB Report No 10/2007.

Marine Accident Investigation Branch Report. 2015. "Collision between Aggregates Dredger Bowbelle and Passenger Vessel Marchioness Resulting in Marchioness Sinking with Loss of 51 lives." MAIB Report June 1990.

Marsh. 2019. "100 Largest Losses in the Hydrocarbon Industry." 26th Edition 1974–2019.

Marsh. 2015. "Fire Pre-Plans." Risk Engineering Position Paper No 2.

Marsh. 2016. "Marsh BLAST Powered by MaxLoss^{TM}." Accessed February 2022. https://www.marsh.com/au/industries/energy-and-power/products/marsh-blast-powered-by-maxloss.html

Marsh. 2015. "Management of Change." Risk Engineering Position Paper No 5.

Methley Archive. n.d. "St Aiden's Flood." Accessed 5 August 2021. https://www.methleyarchive.org/minnet-collection/st-aidens-flood-methley-mppt037/.

Metalpedia. n.d. "Environmental Effects of Molybdenum." Accessed January 2022. http://metalpedia.asianmetal.com/metal/molybdenum/health.shtml

National Fire Chiefs Council. 2013. "Waste Management and Recycling Fires." Accessed July 2022. http://www.cfoa.org.uk/17512

National Ocean Industries Association, 2018. "Hurricanes and the Offshore Oil and Natural Gas Industry." Accessed June 2022. http://www.noia.org/wp-content/uploads/2018/05/2018-Hurricane-Fact-Sheet-Web.pdf

NBC News. 16 March 2011. "What Are the Odds? US Nuke Plants Ranked by Quake Risk." Accessed February 2022. https://www.nbcnews.com/id/wbna42103936

National Fire Protection Association. 2020. "Fires in or at Service or Gas Stations: Supporting Tables." Accessed May 2022. https://www.nfpa.org/News-and-Research/Data-research-and-tools/Building-and-Life-Safety/Service-or-Gas-Station-Fires

Norilsk Nickel. 2021. "Nornickel 2021 Sustainability Report." Accessed February 2022. https://www.nornickel.com/upload/iblock/b30/z64sq9ea81xktkh2je8uvir8kikumfia/NN_CSO2021_ENG_2807.pdf

Nuclear Information Service. November 2009. "NIS Briefing Note – Project Pegasus – AWE Aldermaston's Proposed Enriched Uranium Facility." Accessed December 2021. https://www.nuclearinfo.org/wp-content/uploads/2020/09/AWE-EUF-briefing-November-2009.pdf

Nuclear Information Service. March 2015. "Warhead Factory's Flagship Construction Project Placed 'on Hold'" – Nuclear Information Service. Accessed December 2021 https://www.nuclearinfo.org/article/awe-aldermaston/warhead-factorys-flagship-construction-project-placed-hold

Nuclear Information Service. October 2013. "Environment Agency Warns Atomic Weapons Establishment Following Radioactive Effluent Investigation."

Accessed December 2021. https://www.nuclearinfo.org/article/awe-aldermaston/environment-agency-warns-atomic-weapons-establishment-following-radioactive

Nuclear Weapons Archive. n.d. "Britain's Nuclear Weapons – British Nuclear Facilities." Accessed December 2021. https://nuclearweaponarchive.org/Uk/UKFacility.html

Office for Nuclear Regulation. July 2019. "Improvements Required at AWE." Accessed December 2021. https://news.onr.org.uk/2019/07/regulator-requires-improvements-at-awe/

Oil and Gas Journal. About Fracking: https://oilandgasinfo.ca/all-about-fracking/induced-seismicity/. Downloaded 2022.

OilandGas.org.uk. "The OSPRAG Well Capping Device." Accessed February 2022. https://www.youtube.com/watch?v=fM-aUf1X_0o

Occupational Safety and Health Administration. 2013. "Process Safety Management of Highly Hazardous Chemicals" 1910.119. Accessed November 2021. https://www.osha.gov/laws-regs/regulations/standardnumber/1910/1910.119

Occupational Safety and Health Consultants Register. n.d. "What Does a Health and Safety Manager Do?." Accessed 5 August 2022. https://www.oshcr.org/what-does-a-health-and-safety-manager-do/

Perrow, Charles. 1984. *Normal Accidents: Living in High-Risk Technologies.* New York: Basic Books.

Perrow, Charles. 2007. *The Next Catastrophe: Reducing Our Vulnerabilities to Natural, Industrial and Terrorist Disasters.* Princeton University Press.

Publicatiereeks Gevaarlijke Stoffen (Dutch Government Publication). 2005. "Guidelines for Quantitative Risk Assessment." Publication Series on Dangerous Substances PGS 3/CPR18. Accessed March 2022. https://content.publicatiereeksgevaarlijkestoffen.nl/documents/PGS3/PGS3-1999-v0.1-quantitative-risk-assessment.pdf

Radakovic, Nenad. 2017. *Towards the Pedagogy of Risk: Teaching and Learning Risk in the Context of Secondary Mathematics.* Doctor of Philosophy Thesis. Ontario Institute for Studies in Education, University of Toronto.

Rausand, Marvin. 2011. *Risk Assessment: Theory, Methods, and Applications.* Wiley.

Redmill, Felix et al. 1999. *System Safety: HAZOP and Software HAZOP.* Wiley.

Responsible Mining Foundation. 2019. "Research Insight on Tailings Dam Failures 2018." Accessed November 2021. https://www.responsibleminingfoundation.org/app/uploads/2019/02/ENG-RMF-Research-Insight-on-Tailings-Dam-Failures.pdf

Responsible Mining Foundation. 2022. "Responsible Mining Index Report 2022." Accessed 5 August 2022. https://2022.responsibleminingindex.org/en

Rolls Royce https://www.rolls-royce.com/~/media/Files/R/Rolls-Royce/documents/customers/nuclear/smr-brochure-july-2017.pdf

Rockwell Automation. 2013. *Process Safebook 1 – Functional Safety in the Process Industry – Principles, Standards and Implementation.* Rockwell Automation Inc.

Ropeik, David and Gray, George. 2002. *Risk – A Practical Guide for Deciding What's Really Safe and What's Really Dangerous in the World Around You.* Houghton Mifflin Company.

Shell Plc. 2022. "Health, Security, Safety, the Environment, and Social Performance Commitment and Policy." Accessed 5 August 2022. https://www.shell.com/sustainability/our-approach/commitments-policies-and-standards.

Stanton, Neville, and Baber, Christopher. "Modelling of Human Alarm Handling Response Times: A Case Study of the Ladbroke Grove Rail Accident in the UK." *Ergonomics* 51 no. 4 (2008): 330–423 https://www.researchgate.net/deref/http%3A%2F%2Fdx.doi.org%2F10.1080%2F00140130701695419.

Stern, Gerald. 2008. *The Buffalo Creak Disaster – How the Survivors of One of the Worst Disasters in Coal Mining History Brought Suit against the Coal Company – And Won.* Vintage.

The Chemical Engineer. 2018. "Piper Alpha: The Disaster in Detail." Accessed November 2021. https://www.thechemicalengineer.com/features/piper-alpha-the-disaster-in-detail/

The Chemical Engineer Magazine. "Bill Gates' Nuclear Company Picks Coal Site for New Plant." January 2022.

The Chemical Engineer Magazine. "Fire on Industrial Estate." October 2021.

The Chemical Engineer. "Piper Alpha – What Have We Learned?" Loss Prevention Bulletin 261, June 2018, p. 3.

The Spirits Business. 2014. "The World's 10 Worst Distillery Disasters." Accessed January 2022. https://www.thespiritsbusiness.com/2014/08/the-worlds-10-worst-distillery-disasters/

Tweedale, Geoffrey. 2001. *Magic Mineral to Killer Dust: Turner and Newall and the Asbestos Hazard.* Oxford University Press.

UK Essays. 2019. "Soma Mine Disaster Health and Safety in Coal Mining." Accessed November 2021. https://www.ukessays.com/essays/environmental-studies/soma-mine-disaster-health-and-safety-in-coal-mining.php

United States Environmental Protection Agency and National Oceanographic and Atmospheric Administration. "Aloha® Software." Version 5.4.7. Accessed July 2022. https://www.epa.gov/cameo/aloha-software.

United States Chemical and Hazard Investigation Board n.d. "Bio Lab Chemical Release." Accessed 5 August 2022. https://www.csb.gov/bio-lab-chemical-release-/

United States Chemical and Hazard Investigation Board, 2015. "Final Investigation Report – Caribbean Petroleum Tank Terminal Explosion and Multiple Tank Fires – 2010." Report No. 2010.02.I.PR.

United States Chemical and Hazard Investigation Board. 2014. "Investigation Report – Catastrophic Rupture of Heat Exchanger (Seven Fatalities) – Tesoro Anacortes 2010." Report No. 2010-08-I-WA.

United States Chemical and Hazard Investigation Board. 2015. "Final Investigation Report." Chevron Richmond Refinery Pipe Rupture and Fire – 2012." Report No. 2012-03-I-CA.

United States Chemical and Hazard Investigation Board. 2019. "Chemical Safety Board Releases Factual Update and New Animation Detailing the Events of the Massive Explosion and Fire at the PES Refinery in Philadelphia."

United States Chemical and Hazard Investigation Board. 2010. "Case Study – Explosion and Fire in West Carrollton, Ohio – 2009." Report No 2009-10-I-OH.

United States Chemical and Hazard Investigation Board 2004. "Chlorine Release (16 Medically Evaluated, Community Evacuated) – DPC 2003." Report No 2004-02-I-AZ.

United States Chemical and Hazard Investigation Board. 2016. "Investigation Report – Drilling Rig – Explosion and Fire at the Macondo Well." Report No 2010-10-I-OS.

United States Chemical and Hazard Investigation Board. 2017. "Investigation Report – Chemical Spill Contaminates Public Water Supply in Charleston, West Virginia – Freedom Industries – 2014." Report No 2014-01-I-WV.

United States Chemical and Hazard Investigation Board. 2009. "Imperial Sugar Company Dust Explosion and Fire – Imperial Sugar – 2008."

United States Chemical and Hazard Investigation Board. 2017. "Key Lessons for Preventing Inadvertent Mixing during Chemical Unloading Operations – MGPI Processing 2016." Report No 2017-01-I-KS.

United States Chemical and Hazard Investigation Board. 2017. "Investigation Report – Tank Explosions at Midland Resource Recovery – 2017." Report No 2017-06-I-WV.

United States Chemical and Hazard Investigation Board. 2017. "Investigation Report – Organic Peroxide Decomposition, Release, and Fire at Arkema Crosby Following Hurricane Harvey Flooding." Report No 2017-08-I-TX.

United States Chemical and Hazard Investigation Board. 2021. "Update on Poultry Plant Incident – Foundation Food."

United States Chemical and Hazard Investigation Board. 2011. "Investigative Study – Public Safety at Oil and Gas Storage Facilities." Report No 2011-H-1.

United States Chemical and Hazard Investigation Board. 2011. "Investigation Report – Refinery Explosion and Fire – BP Texas City, 2005." Report No 2005-04-I-TX.

United States Chemical and Hazard Investigation Board. 2014. "Case Study – Tesoro Martinez Refinery – Process Safety Culture." Report No 2014-02-I-CA.

United States Chemical and Hazard Investigation Board. 2019. "Investigation Report – Toxic Chemical Release at the DuPont La Porte Chemical Facility – 2015." Report No 2015-01-I-TX.

United States Chemical and Hazard Investigation Board. 2013. "Investigation Report – West Fertilizer Company Fire and Explosion – 2013." Report No 2013-02-I-TX.

UK Government. "Flood Risk Maps 2019." Accessed 5 August 2022. https://www.gov.uk/government/publications/flood-risk-maps-2019

Vale. 2020. "Vale Annual Report", "Vale Sustainability Report", "Vale 10-K." Accessed 5 August 2019. http://www.vale.com/brasil/EN/investors/information-market/annual-reports/Pages/default.aspx

Vanem, Erik, Antão, Pedro, Castillo Comas, Francisco Del and Skjong, Rolf. 2007. "Formal Safety Assessment of LNG Tankers." SAFEDOR Project IP-516278 HAZID for LNG.

Vice.com/ 2020 "One of the Country's Only Hydrogen Fuel Cell Plants Suffers Huge Explosion." Accessed 5 August 2022. https://www.vice.com/en/article/y3m9ab/one-of-the-countrys-only-hydrogen-fuel-cell-plants-suffers-huge-explosion

Weick, Karl, and Sutcliffe, Kathleen. 2007. *Managing the Unexpected: Resilient Performance in an Age of Uncertainty.* 2nd edition. San Francisco: Jossey-Bass.

West Berkshire Council. 2020. *What You Should Do if There Is a Radiation Emergency at the AWE Aldermaston or Burghfield Sites.* Radiation Emergency Preparedness and Public Information Regulations.

Whatdotheyknow.com. 2017. "Top and Lower Tier Comah Sites in the UK" Reply to a Letter to the HSE under the UK Freedom of Information Act Requesting Data on COMAH Sites." Accessed December 2019: https://www.whatdotheyknow.com/cy/request/top_and_lower_tier_comah_sites_i

Willey, Ronald. 2014. *Consider the Role of Safety Layers in the Bhopal Disaster.* AIChemE.

Wise Global Training. 2015. *Introduction to Oil and Gas Operational Safety.* Routledge/Taylor and Francis.

Wise. "Chronology of Major Tailings Dam Failures (from 1960)." Accessed 5 August 2022. https://www.wise-uranium.org/mdaf.html Downloaded 2022.

World Health Organisation. 2016. "Dioxins and Their Effects on Human Health." Accessed 5 August 2022. https://www.who.int/news-room/fact-sheets/detail/dioxins-and-their-effects-on-human-health

WorldData.Info. n.d. "Tsunamis in Greece." Accessed January 2022. https://www.worlddata.info/europe/greece/tsunamis.php

World Nuclear Organisation. March 2022. "Safety of Nuclear Power Reactors." Accessed 7 August 2022. https://www.world-nuclear.org/information-library/safety-and-security/safety-of-plants/safety-of-nuclear-power-reactors.aspx

World Nuclear Organisation. April 2022. "Chernobyl Accident 1986." Accessed 7 August 2022. https://www.world-nuclear.org/information-library/safety-and-security/safety-of-plants/chernobyl-accident.aspx

World Nuclear Organisation. 2016. "Boiling Water Reactor." Accessed 7 August 2022. https://www.world-nuclear.org/gallery/reactor-diagrams/boiling-water-reactor.aspx

World Nuclear News. 2008. "Seimaszko Found Guilty in Davis-Besse Case." Accessed 7 August 2022. https://www.world-nuclear-news.org/RS_Siemaszko_found_guily_in_Davis-Besse_case_2708081.html

Wikipedia. 2022. "Accident Triangle." Last modified 20 June 2022. https://en.wikipedia.org/wiki/Accident_triangle

Wikipedia. 2022. "Asbestos." Last modified 5 August 2022. https://en.wikipedia.org/wiki/Asbestos

Wikipedia. 2022. "Atomic Weapons Establishment." Last modified 10 May 2022. https://en.wikipedia.org/wiki/Atomic_Weapons_Establishment

Wikipedia. 2022. "Aviation Accidents and Incidents." Last modified 4 August 2022. https://en.wikipedia.org/wiki/Aviation_accidents_and_incidents

Wikipedia. 2022. "2020 Beirut Explosion." Last modified 5 August 2022. https://en.wikipedia.org/wiki/2020_Beirut_explosion

Wikipedia. 2022. "Biosafety Levels." Last modified 29 July 2022. https://en.wikipedia.org/wiki/Biosafety_level

Wikipedia. 2022. "Catastrophe de Ghislenghien." Last modified 19 July 2022. https://fr.wikipedia.org/wiki/Catastrophe_de_Ghislenghien

Wikipedia. 2022. "Cecil Kelley Criticality Accident." Last modified 10 June 2022. https://en.wikipedia.org/wiki/Cecil_Kelley_criticality_accident

Wikipedia. 2022. "Channel Tunnel Fire 1996." Last modified 18 March 2022. https://en.wikipedia.org/wiki/1996_Channel_Tunnel_fire

Wikipedia. 2022. "Chernobyl Disaster." Last modified 25 July 2022. https://en.wikipedia.org/wiki/Chernobyl_disaster

Wikipedia. 2022. "Costa Concordia." Last modified 2 July 2022. https://en.wikipedia.org/wiki/Costa_Concordia

Wikipedia. 2022. "Derrybrien Landslide 2003." Last modified 5 April 2022. https://en.wikipedia.org/wiki/2003_Derrybrien_landslide

Wikipedia. 2022. "Germanwings Flight 9525." Last modified 5 April 2022. https://en.wikipedia.org/wiki/Germanwings_Flight_9525

Wikipedia. 2022. "Global Reporting Initiative." Last modified 21 June 2022. https://en.wikipedia.org/wiki/Global_Reporting_Initiative

Wikipedia. 2022. "Guadalajara Explosions." Last modified 3 August 2022. https://.en.wikipedia.org/wiki/1992_Guadalajara_explosions

Wikipedia. 2022. "ISO 31000." Last modified 16 June 2022. https://en.wikipedia.org/wiki/ISO_31000

Wikipedia. 2022. "List of Food Contamination Incidents." Last modified 29 July 2022. https://en.wikipedia.org/wiki/List_of_food_contamination_incidents

Wikipedia. 2022. "Jilin Baoyuanfeng Poultry Plant Fire." Last modified 9 June 2022. https://en.wikipedia.org/wiki/Jilin_Baoyuanfeng_poultry_plant_fire

Wikipedia. 2022. "List of Industrial Disasters." Last modified 1 August 2022. https://en.wikipedia.org/wiki/List_of_industrial_disasters

Wikipedia. 2022. "List of Rail Accidents in the United Kingdom." Last modified 30 July 2022. https://en.wikipedia.org/wiki/List_of_rail_accidents_in_the_United_Kingdom

Wikipedia. 2022. "2004 Madrid Train Bombings." Last modified 19 July 2022. https://en.wikipedia.org/wiki/2004_Madrid_train_bombings

Wikipedia. 2022. "Marburg Virus Disease." Last modified 29 July 2022. https://en.wikipedia.org/wiki/Marburg_virus_disease

Wikipedia. 2022. "Marchioness Disaster." Last modified 29 June 2022. https://en.wikipedia.org/wiki/Marchioness_disaster

Wikipedia. 2022. "Minot Train Derailment." Last modified 19 April 2022. https://en.wikipedia.org/wiki/Minot_train_derailment

Wikipedia. 2022. "Monju Nuclear Power Plant." Last modified 13 July 2022. https://en.wikipedia.org/wiki/Monju_Nuclear_Power_Plant

Wikipedia. 2022. "Mont Blanc Tunnel Fire." Last modified 1 August 2022. https://en.wikipedia.org/wiki/Mont_Blanc_tunnel_fire

Wikipedia. 2022. "Non-Destructive testing." Last modified 16 June 2022. https://en.wikipedia.org/wiki/Nondestructive_testing

Wikipedia. 2022. "Norilsk Oil Spill." Last modified 6 May 2022. https://en.wikipedia.org/wiki/Norilsk_oil_spill

Wikipedia. 2022. "Oil and Gas Fields of the North Sea." Last modified 30 July 2022. https://en.wikipedia.org/wiki/List_of_oil_and_gas_fields_of_the_North_Sea

Wikipedia. 2022. "Perfluorooctanoic Acid." Last modified 4 August 2022. https://en.wikipedia.org/wiki/Perfluorooctanoic_acid

Wikipedia. 2021. "Prudhoe Bay Oil Spill." Last modified 5 July 2021. https://en.wikipedia.org/wiki/Prudhoe_Bay_oil_spill

Wikipedia. 2021. "Qingdao Oil Pipeline Explosion." Last modified 5 July 2021. https://en.wikipedia.org/wiki/2013_Qingdao_oil_pipeline_explosion

Wikipedia. 2021. "Return Period." Last modified 18 September 2021. https://en.wikipedia.org/wiki/Return_period

Wikipedia. 2021. "Richter Magnitude Scale." Last modified 27 July 2021. https://en.wikipedia.org/wiki/Richter_magnitude_scale

Wikipedia. 2021. "Sayano-Shushenskaya Power Station." Last modified 16 July 2021. https://en.wikipedia.org/wiki/Sayano-Shushenskaya_power_station_accident

Wikipedia. 2021. "Soma Mine Disaster." Last modified 29 July 2021. https://en.wikipedia.org/wiki/Soma_mine_disaster

Wikipedia. 2022. "Stuxnet." Last modified 29 July 2022. https://en.wikipedia.org/wiki/Stuxnet

Wikipedia. 2021. "Transport Accidents." Last modified 21 October 2021. https://en.wikipedia.org/wiki/Transport_accidents

Wikipedia. 2016. "Two Barrier Rule." Last modified 11 December 2016. https://en.wikipedia.org/wiki/Two-barrier_rule

Wikipedia. 2022. "UPS Airlines Flight 6." Last modified 3 August 2022. https://en.wikipedia.org/wiki/UPS_Airlines_Flight_6

Wikipedia. 2022. "Viareggio Train Derailment." Last modified 10 July 2022. https://en.wikipedia.org/wiki/Viareggio_train_derailment

Wikipedia. 2022. "Wuhan Laboratory for Virology." Last modified 11 July 2022. https://en.wikipedia.org/wiki/Wuhan_Institute_of_Virology

Index

Index: Notes from Trevor's Files

Index to Hypothetical Case Exercises (Whisky Distillery and Fuels Terminal)